MATHEMATICS ACTIVITIES

for Elementary School Teachers

A Problem-Solving Approach

THIRD EDITION

MATHEMATICS ACTIVITIES

for Elementary School Teachers

A Problem-Solving Approach

THIRD EDITION

Dan Dolan
Wesleyan University

Jim Williamson
University of Montana

Mari Muri
Connecticut Department
of Education

An imprint of Addison Wesley Longman, Inc.

Reading, Massachusetts • Menlo Park, California • New York • Harlow, England
Don Mills, Ontario • Sydney • Mexico City • Madrid • Amsterdam

Library of Congress Cataloging-in Publication Data

Dolan, Dan.
Mathematics activities for elementary school teachers: a problem-solving approach.—3rd ed. / Dan Dolan, Jim Williamson, Mari Muri.
p. cm.
ISBN 0-201-44096-2 (pbk.)
1. Mathematics—study and teaching (Elementary) I. Williamson, Jim. II. Muri, Mari. III. Title.
QA135.5.D64 1997
372.7—dc20 96-26206
CIP

1 2 3 4 5 6 7 8 9 10 - CRS 00 99 98 97

Preface

ABOUT THIS BOOK

Mathematics Activities for Elementary School Teachers provides a hands-on, manipulative-based, problem-solving approach to learning and teaching elementary mathematics. The activities in the book were developed to correspond to the chapters in *A Problem Solving Approach to Mathematics for Elementary School Teachers*, Sixth Edition, by Rick Billstein, Shlomo Libeskind, and Johnny Lott (Addison Wesley Longman, 1997). Although this manual was designed to supplement the textbook, it can be used to develop students' understanding of mathematical concepts in a variety of settings:

- mathematics content courses for preservice elementary teachers, grades K–8,
- mathematics methods courses for preservice elementary teachers, grades K–8,
- inservice courses and staff development workshops for elementary and middle-school teachers.

The activities in this book demonstrate an alternative approach to the traditional teaching and learning of mathematics. They are based on a constructivist philosophy and sequenced in a developmentally appropriate manner. The activities may be used to

- develop a mathematical concept,
- reinforce a concept that has been previously taught,
- illustrate applications of mathematical concepts in contextual situations, and
- promote the construction of knowledge of mathematical concepts.

Although the book has been designed for preservice and inservice classes, it can also serve as a resource for use with elementary students at various grade levels, K–8.

A NEW APPROACH

Everybody Counts: A Report to the Nation on the Future of Mathematics Education (Mathematical Sciences Education Board, 1989) states: "Those who would teach mathematics need to learn contemporary mathematics appropriate to the grades they will teach, in a style consistent with the way in which they will be expected to teach. . . . The content of the special mathematics courses for prospective elementary and middle school teachers must be infused with examples of mathematics in the world that the child sees (sports, architecture, house, and home), examples that illustrate change, quantity, shape, chance, and dimension."

"The teaching we envision differs significantly from what many teachers themselves have experienced as students."

Professional Teaching Standards NCTM, 1991

In *Everybody Counts, The Curriculum and Evaluation Standards for School Mathematics*, the *Professional Teaching Standards,* and the *Assessment Standards for School Mathematics* (National Council of Teachers of Mathematics, (1989, 1991, and 1995), knowing mathematics is described as having the ability to use mathematics in meaningful ways. In the process of learning mathematics, teachers must be involved in doing mathematics—investigating, conjecturing, discussing, and validating—in order to develop confidence in their own mathematical ability and to be able to instill an appreciation of its value in their students.

A Call for Change: Recommendation for the Preparation of Teachers of Mathematics (Mathematical Association of America, 1991) asserts that: "Collegiate mathematics classrooms must become a place where students actively do mathematics rather than simply learn about it." It further states: "Teaching collegiate mathematics must change to enable learners to grapple with the development of their own mathematical knowledge. . . . the collegiate curriculum in mathematics must be open to new ways of presenting mathematics."

"WHAT students learn is fundamentally connected with HOW they learn it."

Professional Teaching Standards NCTM, 1991

These mathematics reform documents call for a basic restructuring of the curriculum, instructional practices, and assessment systems for mathematics in grades K–16 and of the program for the preparation of teachers of mathematics. As stated in the *Curriculum and Evaluation Standards*: "To equip students for productive lives in the Information Age, the definition of success in mathematics—the objective of mathematics—must be transformed."

This transformation necessitates dramatic changes in the ways teachers learn and teach mathematics. As noted in *A Call for Change,* alternative methods must be presented to preservice and inservice teachers so that they can learn and practice them while they are in the learning process themselves.

GOALS OF THE BOOK

This book was written to engage preservice and inservice teachers in *doing* mathematics rather than simply *reading about* mathematics. It is not intended to be a textbook for a mathematics content course at the collegiate level or to provide all of the mathematical content necessary for such a course. Rather, it is intended to be used as a companion to a textbook, such as *A Problem Solving Approach to Mathematics for Elementary School Teachers,* to provide hands-on, manipulative-based activities that involve elementary teachers in discovering mathematical concepts, doing real problem-solving, and exploring mathematical concepts in interesting, stimulating, real-world settings.

The content and instructional approach of this activities manual embodies the spirit and intent of the three NCTM standards documents. By engaging students in meaningful mathematical tasks, group work, hands-on activities, and classroom discourse, these materials have captured the essence of the first standard for the professional development of teachers in the *Teaching Standards.*

STANDARD 1

Mathematics and mathematics education instructors in preservice and continuing education programs should model good mathematics teaching by doing the following:

- posing worthwhile mathematical tasks,
- engaging teachers in mathematical discourse,
- enhancing mathematical discourse through the use of a variety of tools, including calculators, computers, and physical and pictorial models,
- creating learning environments that support and encourage mathematical reasoning and teachers' dispositions and abilities to do mathematics,
- expecting and encouraging teachers to take intellectual risks in doing mathematics and to work independently and collaboratively,
- representing mathematics as an ongoing human activity, and
- affirming and supporting full participation and continued study of mathematics by all students.

If students learn concepts through a problem-solving approach, develop ideas from the concrete level to the abstract level, and connect multiple representations of mathematical ideas, they will gain a broader and deeper understanding of mathematics and begin to construct their own meanings of mathematical concepts.

The activities pay careful attention to the recommendations regarding mathematics courses for elementary teachers contained in *A Call for Change.* In addition, they promote the major goals of the NCTM standards documents. These goals state that students should

- become mathematical problem solvers,
- learn to communicate mathematically,
- learn to reason mathematically,
- become confident in their ability to do mathematics, and
- learn to value mathematics.

"If students are to develop a disposition to do mathematics, it is essential that the teacher communicate a love of mathematics and a spirit of doing it."

Professional Teaching Standards NCTM, 1991

These goals are consciously reflected in every activity in this book. All of the activities have been tested extensively with preservice teachers, teachers in professional development programs, and elementary students. The activities have proven to be effective and stimulating with all groups. We hope that as you engage in the activities, you will enjoy exploring mathematics, become confident in your ability to do mathematics, and come to love it as we do.

Introduction

CONTENT

This book contains activities designed to provide preservice and inservice teachers with opportunities to explore mathematical ideas using a problem-solving approach and a variety of manipulative materials. For the most part, the activities do not require expensive or single-use materials. Full-color replicas of some commercial manipulatives that can be used at many levels and in a variety of settings are included in the Appendix.

"There is little value in telling students how exciting mathematics is if they are not actively engaged in doing mathematics themselves."

Professional Teaching Standards NCTM, 1991

A problem-solving approach to learning enhances the development, reinforcement, and application of mathematical concepts in a way that practicing rote skills cannot. It also engages teachers in learning mathematics in a way we trust they will apply when teaching their elementary students.

Each activity contains investigations that involve students in doing mathematics. These materials can be adapted for use with elementary students at a later time. Each activity is preceded by a brief instructional plan that provides some direction for the course instructor.

ORGANIZATION

The instructional plan that precedes each activity has the following elements:

- **Purpose** outlines the major mathematical concepts that are developed in the activity and describes which of the three major objectives—developing, reinforcing, or applying a concept—the activity is designed to meet.

- **Materials** describes any special equipment or supplies that are needed for the activity. Models of several manipulatives and other materials are in the Appendix.

- **Grouping** describes the classroom setting for the activity. Many activities can be completed on an individual basis. However, we have encouraged working in pairs or small groups whenever appropriate in order to model the recommendations in *A Call for Change* and the *Professional Standards for Teaching Mathematics*, and to model how mathematics should be taught in the elementary grades.

Once an activity has been completed, the instructor should review and extend it by facilitating whole-class discussion of the results, summarizing the activity, and formalizing the mathematical content.

- **Getting Started** provides an introduction to the activity and explains any preparation necessary before using it. Several activities include more than one section. In some cases, it may not be essential to do all of them. Since the purpose of these activities is not only to teach mathematical concepts, but also to illustrate how the concepts should be taught in the elementary grades, it is important that students keep the following questions in mind and formulate answers to them as they are completing each activity:

  A. How could this activity be used with elementary students?

  B. At which grade level would the activity be appropriate?

  C. Could the activity be used at more than one grade level?

  D. What adaptations would be necessary to make the activity suitable for use with elementary students at various grade levels?

  E. How would the content and instructional approach of this activity be important in the mathematical development of elementary students?

- **Extensions** present suggestions and ideas for extending the activity to other mathematical topics or making connections between the mathematical concepts in the activity and their application in the real world or other curricular areas. Included are some or all of the following:

  - additional questions and problems to explore,

  - questions to extend ideas in selected problems or the entire activity, and

  - other activities, problems, and information related to the concepts in the activity.

TIME REQUIREMENTS

The activities in this book can replace much of the lecture time that is usually devoted to teaching topics. Many concepts are developed in the activities by working from the concrete level to the abstract level. Through a carefully guided discussion of the results of an activity, students will develop their own mental constructs of the concepts being presented.

APPENDICES

- **Manipulatives.** Pages A-1 through A-15 are printed one-sided on heavy stock and in color so that you can cut them out for use with many different activities. We suggest that you store each individual set in sealable plastic bags or covered storage containers.

- **Activity Masters.** Pages A-17 through A-48 contain models, activity recording sheets, graph paper, nets for polyhedrons, activity cards, etc., that are to be used as part of selected activities. They should be photocopied for use with an activity and the master retained, as it may be used in a later activity.

About The Authors

DAN DOLAN Dan Dolan is the Director of the Project to Increase Mastery in Mathematics and Science (PIMMS) at Wesleyan University, Connecticut. From 1981 to 1991, he was the State Mathematics and Computer Education Specialist in the Montana State Office of Public Instruction. He has been actively involved in education for 35 years and has taught mathematics at all levels from grade six through university teacher education. He has authored numerous articles and co-authored three books. He presents inservice workshops and conference sessions throughout the country and serves as an editorial consultant for several publishers. He served on the Mathematical Sciences Education Board and the Board of Directors of the National Council of Teachers of Mathematics and was a member of the 5–8 writing group for the NCTM *Curriculum and Evaluation Standards for School Mathematics.* He recently completed terms on the national advisory panels for Montana's Mathematics and Science Systemic Initiative, the Six Through Eight Mathematics Curriculum Development Project, and the Annenberg K–4 Video Library Project.

MARI MURI Mari Muri is a Mathematics Consultant for the Connecticut State Department of Education and a co-principal investigator of the CT State Systemic Initiative, Project CONNSTRUCT. She plays a principal role in the design and implementation of the CT Mathematics Assessment Program at grades 4, 6, 8, and 10. The performance assessment components of these tests are a model of the type of assessment called for in the NCTM *Assessment Standards* and other mathematics reform documents. Prior to her current position, she was the Mathematics Instructional Consultant for the Killingly (CT) Public Schools. She has been involved in education for 17 years and has taught mathematics at all levels from elementary to university teacher education. She was a member of the writing team for the NCTM *Assessment Standards* and currently serves on the National Advisory Board for the Annenberg/WGBH Assessment Video Library Project. In 1994, she received the Outstanding Educational Leader of the Year Award from the CT chapter of ASCD. She presents workshops and conference sessions both regionally and nationally, serves as president of the Association of State Supervisors of Mathematics, and is a member of the editorial panel for *Mathematics Teaching in the Middle School.*

JIM WILLIAMSON Jim Williamson is an Assistant Professor of Mathematics at the University of Montana and Assistant Director and Lead Writer for the

Six Through Eight Mathematics (STEM) Project that is developing integrated mathematics curriculum and instructional materials for grades 6–8. Prior to joining STEM, he served as the interim Mathematics and Computer Education Specialist for the Montana Office of Public Instruction, as Visiting Instructor of Mathematics at Montana State University, and as the Mathematics Specialist for the Billings (MT) Public Schools from 1984 to 1989. He has been involved in mathematics education for thirty years and has taught mathematics at all levels from fourth grade through university. He was awarded a Presidential Award for Excellence in Mathematics Teaching in 1984. He has authored several articles and co-authored two books. He presents inservice workshops and conference sessions throughout the country. He served on the NCTM MATHCOUNTS committee that developed tests and coaching materials. He chaired the national competition committee and the MATHCOUNTS Calculator Pilot Project for one year each.

Contents

Chapter 1
Tools for Problem Solving

Solving a problem is *to search for some means* to attain some clearly conceived but not immediately attainable solution. If the solution by its simple presence does not instantaneously suggest the means, we have to search for the means, reflecting consciously how to find the solution.

To solve a problem is to find such a means.
George Polya

"A teacher of mathematics has a great opportunity. If the teacher fills the allotted time with drilling students with routine operations, the teacher kills their interest, hampers their intellectual development, and misuses the opportunity. But if the teacher challenges the curiosity of the students by setting them problems proportionate to their knowledge, and helps them solve their problems with stimulating questions, the teacher may give them a taste for, and some means of, independent thinking."

George Polya
How to Solve It, 1957

"Problem solving should be the central focus of the mathematics curriculum. As such, it is a primary goal of all mathematics instruction and an integral part of all mathematical activity. Problem solving is not a distinct topic but a process that should permeate the entire program and provide the context in which concepts and skills can be learned."
—*Curriculum and Evaluation Standards for School Mathematics*

The activities in this chapter are designed to help you improve your problem-solving skills. Good problem solvers need to know a variety of techniques for solving problems, so the primary focus of the activities is on developing and applying a variety of problem-solving strategies: Look for a Pattern, Make a Table, Use Logical Reasoning, Make a Model, and Simplify the Problem.

As you complete the activities, you will be learning to make conjectures based on observations and data, to verify and generalize the conjectures, and to communicate your results to others. This is the essence of the problem-solving process. The problem-solving strategies and process are the tools you will use throughout this book to explore and develop mathematical concepts.

Activity 1: Describing Patterns

PURPOSE Describe geometric patterns.

MATERIALS Pattern blocks

GROUPING Work in pairs.

GETTING STARTED Work together to list four to six different ways to describe each pattern. After completing the descriptions for the patterns below, each pair reads one description to the class, while others eliminate it from their list. Continue until all descriptions have been read. In how many ways was each pattern described?

Example:

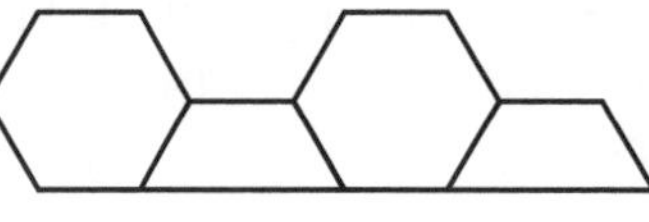

a. Yellow, red, yellow, red, . . .
b. Caution light, stop light, . . .
c. Hexagon, trapezoid, hexagon, trapezoid, . . .
d. Whole, half, whole, half, . . .
e. Six sides, four sides, six sides, four sides, . . .
f. Perimeter of 6, perimeter of 4, . . .

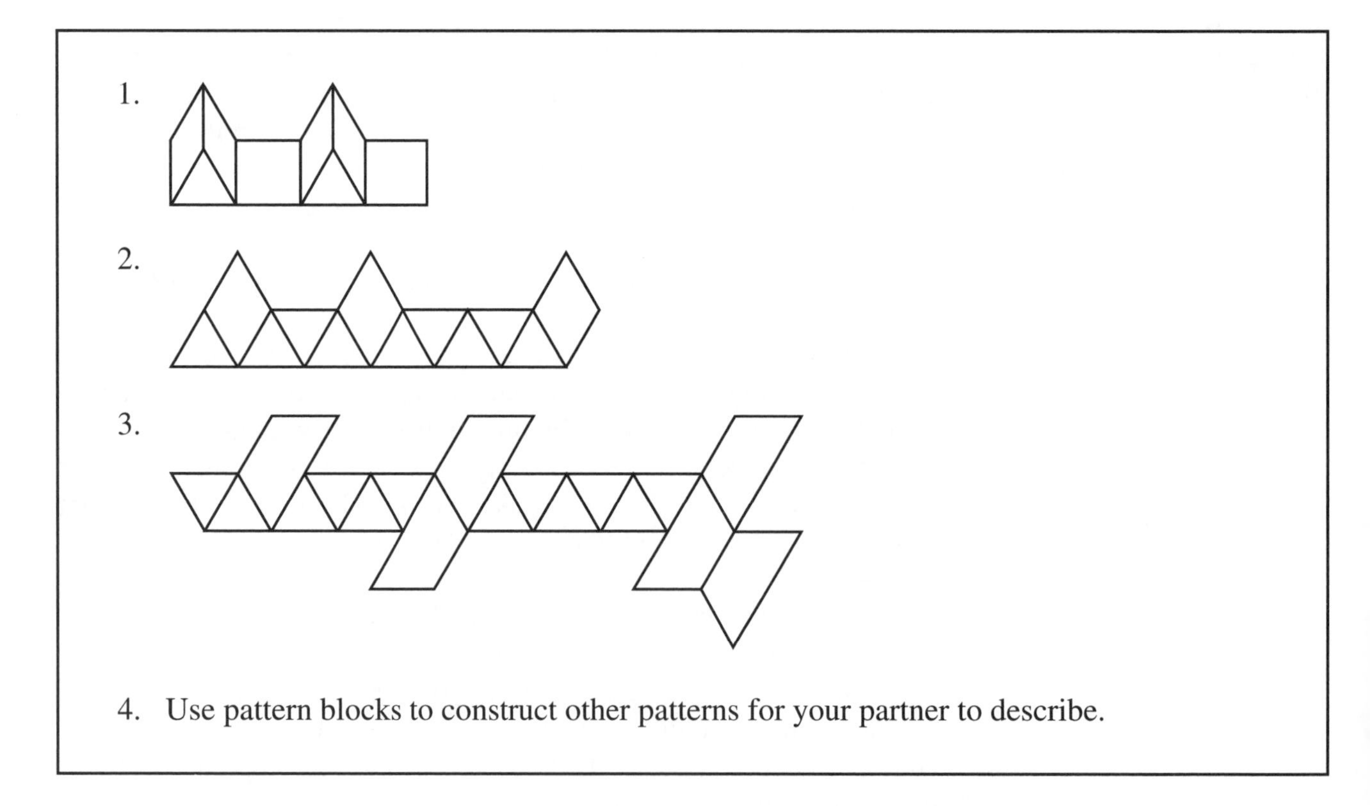

4. Use pattern blocks to construct other patterns for your partner to describe.

EXTENSIONS Verbally describe a pattern to your partner, as in the previous example. Your partner uses pattern blocks to construct the pattern.

Activity 2: Pictorial Patterns

PURPOSE Identify and extend pictorial patterns.

GROUPING Work individually or in groups of 2–3.

GETTING STARTED Look for a pattern in the successive pictures. Then draw pictures in the blanks to extend the pattern.

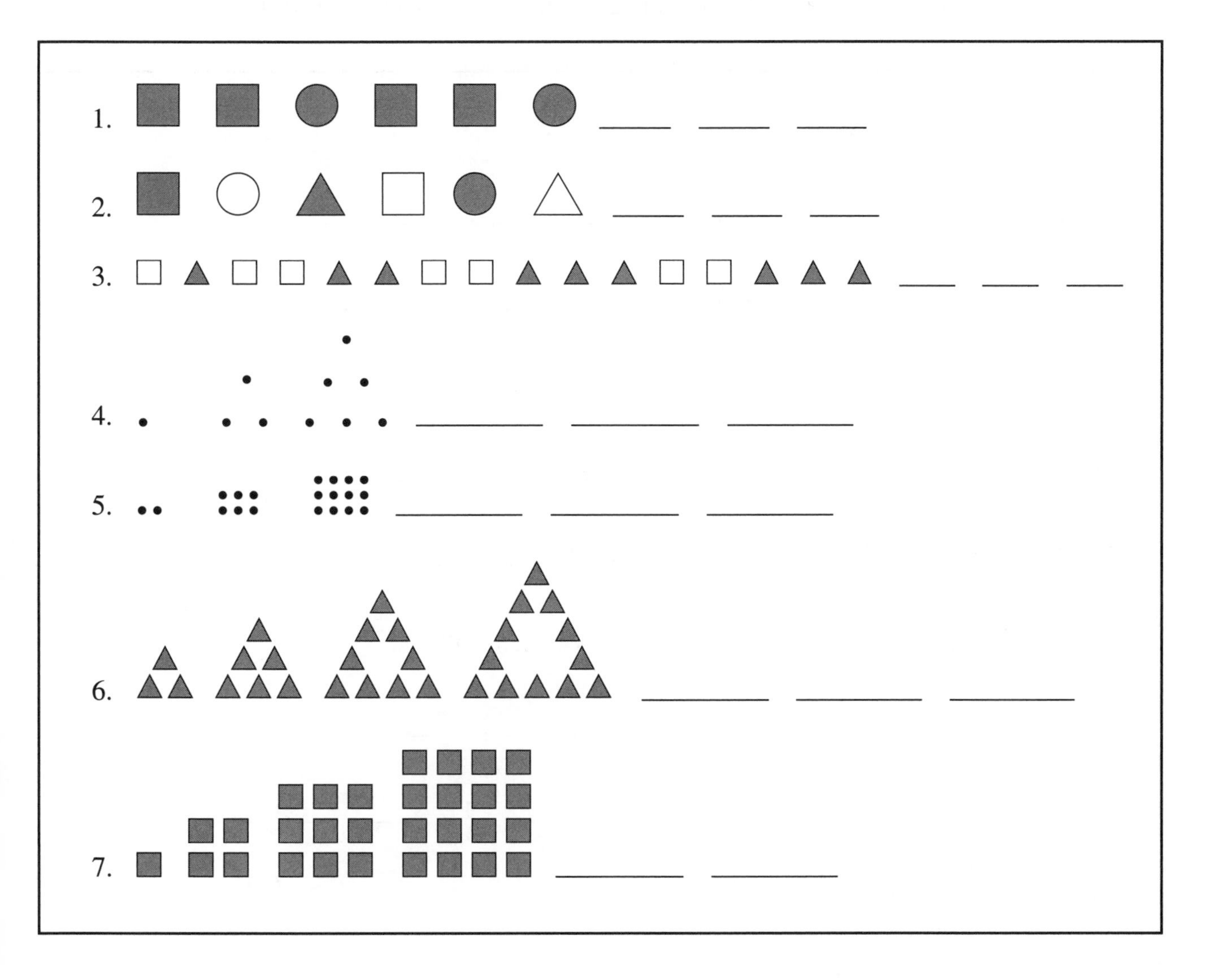

EXTENSIONS Make up several pictorial patterns of your own and give them to a classmate to extend.

Activity 3: Number Patterns

PURPOSE Identify and extend numerical patterns.

GROUPING Work individually or in groups of 2–3.

GETTING STARTED Fill in the blanks with the numbers that complete the sequence and briefly explain the rule you used. In some cases, an intermediate term or the last term is given so that you can check your work.

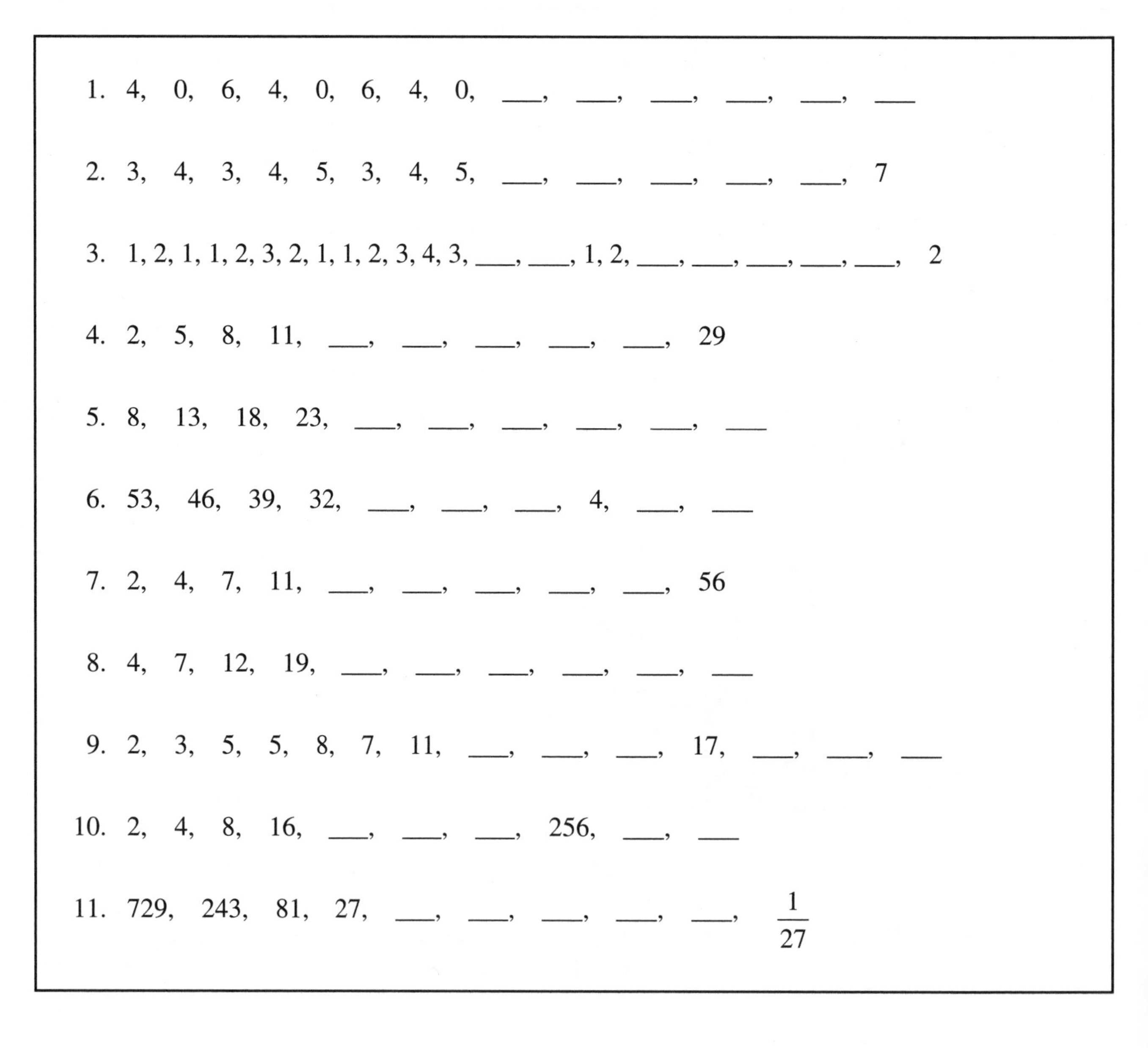

1. 4, 0, 6, 4, 0, 6, 4, 0, ___, ___, ___, ___, ___, ___
2. 3, 4, 3, 4, 5, 3, 4, 5, ___, ___, ___, ___, ___, 7
3. 1, 2, 1, 1, 2, 3, 2, 1, 1, 2, 3, 4, 3, ___, ___, 1, 2, ___, ___, ___, ___, ___, 2
4. 2, 5, 8, 11, ___, ___, ___, ___, ___, 29
5. 8, 13, 18, 23, ___, ___, ___, ___, ___, ___
6. 53, 46, 39, 32, ___, ___, ___, 4, ___, ___
7. 2, 4, 7, 11, ___, ___, ___, ___, ___, 56
8. 4, 7, 12, 19, ___, ___, ___, ___, ___, ___
9. 2, 3, 5, 5, 8, 7, 11, ___, ___, ___, 17, ___, ___, ___
10. 2, 4, 8, 16, ___, ___, ___, 256, ___, ___
11. 729, 243, 81, 27, ___, ___, ___, ___, ___, $\frac{1}{27}$

EXTENSIONS Make up several number patterns of your own and give them to a classmate to complete.

Activity 4: Making Sequences

PURPOSE Explore the relationship between the terms of a sequence and the term numbers.

GROUPING Work individually or in groups of 2–3.

GETTING STARTED Use the given rule to determine the first eight terms of each sequence.

Example: Each term is the term number times 4 plus 1.

First Term	Second Term	Third Term	Fourth Term
$1 \times 4 + 1$	$2 \times 4 + 1$	$3 \times 4 + 1$	$4 \times 4 + 1$
5	9	13	17

1. Each term is the term number times 6, minus 2.

1	2	3	4	5	6	7	8	9
____	10	____	____	____	____	____	____	52

2. Each term is the term number times $^{-}3$, plus 47.

1	2	3	4	5	6	7	8	9
____	____	38	____	____	____	____	____	20

3. Each term is the term number cubed plus 5.

1	2	3	4	5	6	7	8	9
____	____	____	____	____	____	____	____	734

4. Each term is the term number squared times the next term number.

1	2	3	4	5	6	7	8	9
____	____	____	____	____	____	____	____	810

5. Find the difference between the successive terms in Exercises 1 and 2. What do you observe? Does the same thing occur in Exercises 3 and 4?

EXTENSIONS Make up your own rules for generating a sequence. Be sure to include some rules for increasing and decreasing sequences in which the difference between successive terms is constant, as well as some rules in which the differences are not constant. Give them to a classmate to solve.

Activity 5: Constant Differences

PURPOSE Develop a procedure for determining a rule (function) that describes the general term of an arithmetic sequence.

GROUPING Work individually or in groups of 2–3.

GETTING STARTED Fill in each blank below to discover a method for determining the function that generates an arithmetic sequence.

Term Number	1	2	3					
Term	4	11	18					

Difference __ __ __ __ __ __ __

What is the constant difference? _______

Term Number		Constant Difference			What Was Done?		To Get	
1	×	7	→	7	________	=	4	First Term
2	×	7	→	_____	________	=	11	Second Term
3	×	_____	→	_____	________	=	_____	Third Term
10	×	_____	→	_____	________	=	_____	Tenth Term
50	×	_____	→	_____	________	=	_____	Fiftieth Term

Write a sentence, like those in Activity 4, that states the rule to generate the terms in the sequence.

Write the rule as an equation.

Activity 6: What's the Rule?

PURPOSE Determine the rule to find the general term in an arithmetic sequence.

GROUPING Work individually.

GETTING STARTED In each of the following sequences:

1. Fill in the missing numbers.
2. Find the rule that generates the terms in the sequence.
3. Determine the 25th and 100th term of the sequence.

	Sequence	Rule	25th Term	100th Term
1.	9, 13, 17, 21, ____, ____, ____, ____, . . .	☐	____	____
2.	3, 8, 13, 18, ____, ____, ____, ____, . . .	☐	____	____
3.	$^{-}3$, $^{-}1$, 1, 3, ____, ____, ____, ____, . . .	☐	____	____
4.	98, 96, 94, 92, ____, ____, ____, ____, . . .	☐	____	____
5.	77, 74, 71, 68, ____, ____, ____, ____, . . .	☐	____	____
6.	2, 9, 16, 23, ____, ____, ____, ____, . . .	☐	____	____
7.	10, 9.6, 9.2, 8.8, ____, ____, ____, ____, . . .	☐	____	____

EXTENSIONS Make up your own sequences in which the difference between successive terms is constant. Be sure to include both increasing and decreasing sequences. For each sequence, have a classmate fill in the missing terms, determine a rule for generating the terms, and find the 25th and 100th terms.

Activity 7: Polygons and Diagonals

PURPOSE Extend the method of differences to sequences in which the difference between successive terms is not constant.

GROUPING Work individually or in pairs.

GETTING STARTED The method of differences can often be used to analyze and extend sequences even if the difference between terms is not constant. Work the following exercises to learn how.

How many diagonals can be drawn in a dodecagon, a 12-sided polygon?

1. Draw the diagonals in each of the following polygons and record the results in the table. The first two have been done for you.

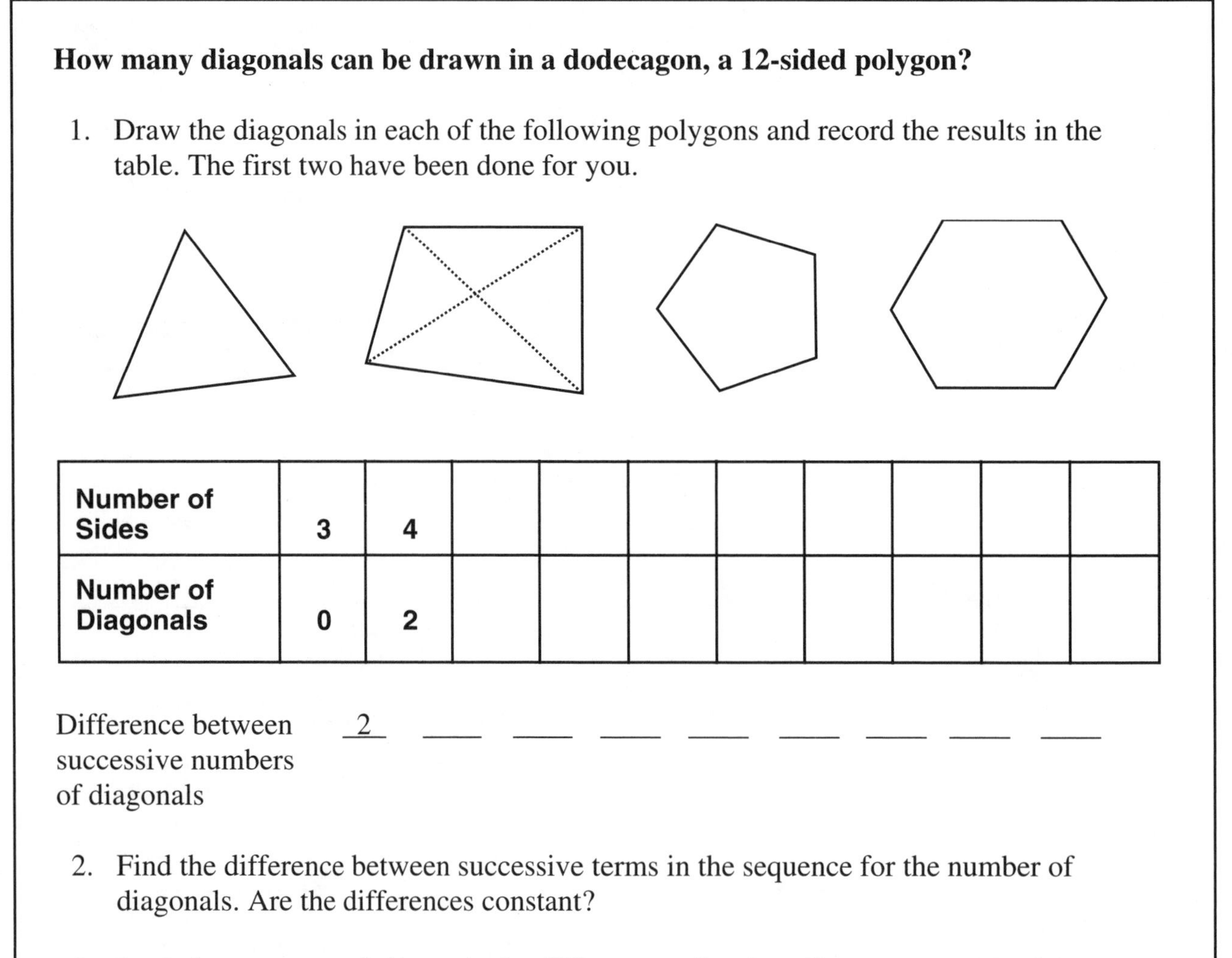

Number of Sides	3	4								
Number of Diagonals	0	2								

Difference between successive numbers of diagonals 2 ____ ____ ____ ____ ____ ____ ____ ____

2. Find the difference between successive terms in the sequence for the number of diagonals. Are the differences constant?

3. Look for a pattern of change in the differences. Continue that pattern to get the next terms in the sequence for the number of diagonals.

The problem of finding the number of diagonals in a dodecagon shows that in order to extend a sequence it is sometimes necessary to look for a pattern in the differences between successive terms. Complete the following exercises to see how this can be done.

1.

Term Number	1	2	3	4	5	6	7	8	9
Term	1	4	9	16	25	____	____	____	____
1st Difference	+3	+5	+7	+9	____	____	____	____	
2nd Difference	+2	+2	+2	____	____	____	____		

2.

Term Number	1	2	3	4	5	6	7	8	9
Term	3	8	15	24	35	____	____	____	____

3.

Term Number	1	2	3	4	5	6	7	8	9
Term	0	2	7	15	26	____	____	____	____

1.

Term Number	1	2	3	4	5	6	7	8	9
Term	2	10	30	68	130	____	____	____	____
1st Difference	+8	____	____	+62	____	____	____	____	
2nd Difference	+12	____	+24	____	____	____	____		
3rd Difference	+6	____	____	____	____	____			

2.

Term Number	1	2	3	4	5	6	7	8	9
Term	2	4	8	16	____	____	____	____	186

3. Compare the sequence in Exercise 2 to the sequence in Exercise 10 of Activity 3. What can you conclude from these two exercises?

Activity 8: Missing Person Reports

PURPOSE Use the process of elimination to identify which person is being described by a set of clues.

MATERIALS One set of attribute people pieces (page A-1) per person

GROUPING One clue reader; groups of 3–8 students

GETTING STARTED Each participant lays out a set of attribute people pieces. The "sergeant" reads the clue card. Each "detective" tries to locate the correct piece by moving aside those identified by the clues. "Detectives" hold up the identified person for evaluation by the "sergeant."

Missing Person Report 1:
This person is large.
He is wearing a dotted jacket.
The jacket is red and makes this clown look very funny.

Missing Person Report 2:
This person is a hobo.
He was last seen wearing a blue jacket.
The jacket has stripes.
This person is large.

Missing Person Report 3:
This person was last seen wearing red.
He is small.
He does not have stripes on his jacket.
He is a hobo.

Missing Person Report 4:
This person is not a hobo.
This person is not small.
This person is not wearing stripes.
This person is not wearing blue or red.

Missing Person Report 5:
This small person is not a clown.
He is wearing stripes of his favorite color.
He is not fond of green or red.

Missing Person Report 6:
This person is not a hobo.
This person is large.
The dots on his jacket are small and red.

EXTENSIONS Write additional missing person reports similar to the ones above and have a classmate solve them.

Activity 9: Who Dunit?

PURPOSE Introduce the elimination problem-solving strategy.

GROUPING Work individually or in groups of 3–4.

Inspector Bob E. Sleuth of the London Police Force is investigating a £1 million robbery at the Two Dot Diamond Exchange. The following suspects are in custody:

- "Green-faced" Larry. He gets so car sick, the police had to walk him to the station.
- "Gun Shy" Gordon. He has been afraid of guns since he shot off his big toe as a boy.
- "Loud Mouth" Louise. She is so shy, she leaves her Aunt Jane's house only at night to rent "Wild World of Wrestling" videos at the corner store.
- "Tombstone" Teri. She works the graveyard shift running a forklift at a warehouse.
- "Lefty" McCoy. He lost his left arm in a demolition derby accident.

Use these clues to help Inspector Sleuth solve the crime.

a. The salesclerk told police the robber had a large handgun.
b. A waiting taxi whisked the robber away.
c. The robber wore a large trench coat and a ski mask.
d. The robber clowned around in front of the security cameras.
e. The manager said the robber nervously twiddled his or her thumbs while the clerk stuffed diamonds into some sacks.

Who should be booked and held over for trial?

EXTENSIONS Explain how the clues helped you to eliminate the innocent suspects and to determine the likely suspect.

"How often have I said to you that when you have eliminated the impossible, whatever remains, however improbable, must be the truth?"

Sherlock Holmes, *The Sign of Four*

Activity 10: What's the Number?

PURPOSE Apply the elimination problem-solving strategy.

GROUPING Work individually or in groups of 3–4.

GETTING STARTED Use the process of elimination to solve the following number puzzles. Keep a record of the order in which you used the clues and the reasons you used them in that order.

Circle the number below that is described by the following clues.

a. The sum of the digits is 14.
b. The number is a multiple of 5.
c. The number is in the thousands.
d. The number is not odd.
e. The number is less than 2411.

2660 2570 905
1580 1058 1922
1355 1455 770
2290 2435 1770
1832 860 1680

Solve the following number riddle.

a. I am a positive integer.
b. All my digits are odd.
c. I am equal to the sum of the cubes of my digits.
d. I am less than 300.

Who am I? _______________

Rebecca has a collection of basketball cards. When she puts them in piles of two, she has one card left over. When she puts them in piles of three or four, there is also one card left over, but when she puts them in piles of five there are no cards left over.

If Rebecca has fewer than 100 basketball cards, what are the possible numbers of cards she could have?

Activity 11: Eliminate the Impossible

PURPOSE Introduce the method of indirect reasoning.

GROUPING Work individually or in groups of 2–3.

Andrea was visiting her Uncle Ralph, who has a large gumball collection. When she asked if she could have some, he said yes, if she could solve a problem for him. He told her that he has three jars, each covered so that no one could see the color of the gumballs. One jar is labeled red, the second green, and the third red-green. However, he said, no jar has the correct label on it. She could reach into one jar and take one gumball. Then she had to tell him the correct color of the gumballs in each jar. She reached into the jar labeled red-green and pulled out a red gumball.

1. Are there any green gumballs in that jar? Why?

2. What is the correct label for the jar labeled red-green? Explain your answer.

3. Can the jar labeled red contain red and green gumballs? Why?

4. What are the correct labels for each of the jars?

Jorge claims that he has a certain combination of U.S. coins and he cannot make change for a dollar, half dollar, quarter, dime, or nickel. Is this possible? If so, what is the greatest amount of money Jorge could have, and what coins would they be? He does not have any silver dollars.

Total amount: ________

Coins: ____________________________

Students in the fifth grade were playing a trivia game involving states, state birds, and state flowers. They knew that in Alaska, Alabama, Oklahoma, and Minnesota, the flowers are the camellia, forget-me-not, pink-and-white lady's slipper, and mistletoe. The state birds are the common loon, yellowhammer, willow ptarmigan, and scissor-tailed flycatcher. No one knew which bird or flower matched which state. They called the library and received the following clues. Use the clues to complete the table below.

a. The flycatcher loves to nest in the mistletoe.
b. The forget-me-not is from the northernmost state.
c. Loons and lady's slippers go together, but Minnesota and mistletoe do not.
d. The yellowhammer is from a southeastern state.
e. The willow ptarmigan is not from the camellia state.

State				
Flower				
Bird				

Which of the clues were the key(s) to solving the puzzle? Explain your reasoning.

Each year, the Calaveras County Frog Jumping Contest is held at Angel's Camp, California. In last year's contest, four large bullfrogs—Flying Freddie, Sailing Susie, Jumping Joe, and Leaping Liz—captured the first four places. Each frog was decorated with a brightly colored bow before the competition began. From the following clues, determine which frog won each place and the color of its bow.

a. Joe placed next to the frog with the purple bow.
b. The frog with the yellow bow won, and the frog with the purple bow was second.
c. The colors on Freddie's bow and Susie's bow mix to form orange.
d. The color of the remaining bow was green.

Construct a table similar to the one above to help organize your work.

Activity 12: Candies, Couples, and a Quiz

PURPOSE Use the elimination strategy in problem-solving.

GROUPING Work individually or in groups of 2–3.

GETTING STARTED In the following problems, construct a set of possible solutions based on certain clues, and then eliminate according to the other clues, or construct a table to help organize your work as illustrated on the previous pages.

Mike said to Linda, "Bet you can't guess the number of candies I have in this sack."

"Give me a clue," she said.

"I have more than 50 but fewer than 125. If you divide them into piles of 8, there are 2 left over. If you divide them into piles of 7, there is 1 left over," said Mike.

"Oh, that's easy, you have __________ candies!!" said Linda.

What is Linda's answer, and how did she get it so easily?

Four married couples are celebrating Thanksgiving together. The wives' names are Jolene, Jane, Marie, and Chris. Their husbands are Bob, Lyle, Lee, and Rick.

Examine the following clues and determine who is married to whom.

a. Rick is Jolene's brother.
b. Marie has two brothers, but her husband is an only child.
c. Lyle is married to Chris.
d. Jolene and Lee were once engaged but broke up when Lee met his present wife.

A social studies quiz consists of five true-false questions.

a. There are more true than false answers.
b. Questions 1 and 5 have opposite answers.
c. No three consecutive answers are the same.
d. Josh knows the correct answer to the second question.
e. From these clues, Josh can determine the correct answer to each question.

What are the correct answers to each of the five questions on the quiz?

Activity 13: Ten People in a Canoe

PURPOSE Introduce the simplify problem-solving strategy and apply the make a table, make a model, and patterns strategies.

MATERIALS Five each of two different-colored squares (page A-7) for each group

GROUPING Work individually or in groups of 2–3.

Ten people are fishing from a canoe. The seats in the canoe are just wide enough for one person to sit on, and the center seat is empty. The five people in the front of the canoe want to change seats and fish from the back of the canoe, and the five people in the back of the canoe want to fish from the front. Because the canoe is so narrow, only one person may move at a time. A person changing seats may move to the next empty seat, or step over one other person to reach an empty seat. Any other move will capsize the canoe.

What is the minimum number of moves needed to exchange the five people in the front with the five in the back?

HINT: Sometimes, the best approach to solving a problem is to simplify it by considering easier cases of the same problem. Use squares of two different colors to represent the people in the canoe and a model like the one below to represent the seats.

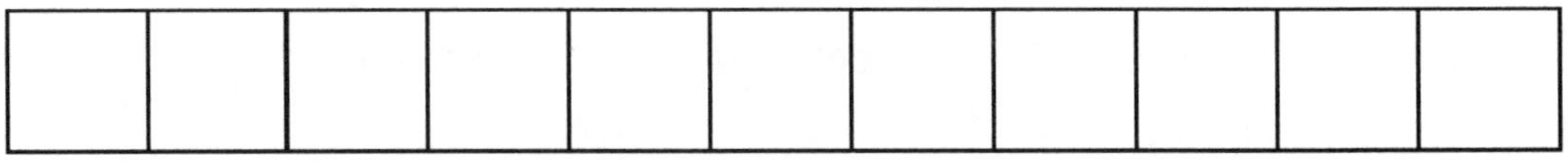

Simplify the problem by solving easier cases. The solution to the problem for two people in the canoe is shown below.

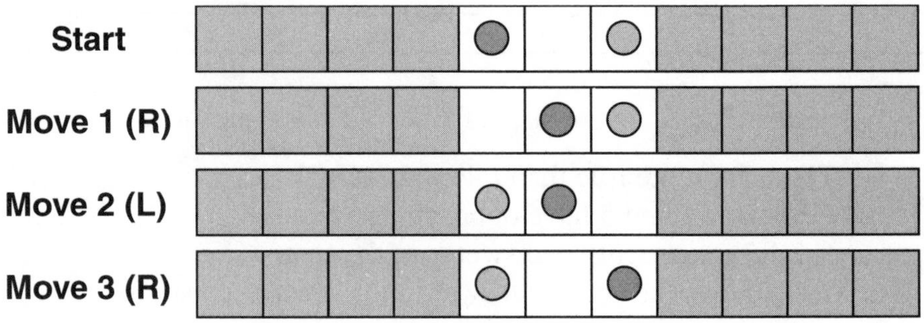

Start

Move 1 (R)

Move 2 (L)

Move 3 (R)

1. Solve the problem for four people. Record the results in the table below.

Number of People	2	4	6	8	10
Number of Pairs	1	2	3	4	5
Minimum Number of Moves	3				
Sequence of Moves	RLR				

2. Complete the table. Look for two patterns, one for how the moves should be made and one for the minimum number of moves.

3. Describe the patterns you found.

4. How many people in the canoe would produce the following sequence of moves?

   R LL RRR LLLL RRRRR LLLLLL RRRRRR LLLLLL RRRRR LLLL RRR LL R

5. If 30 people were in the canoe, how many moves would be needed for them to change places?

6. How would the results change if there was an odd number of people in the canoe?

EXTENSIONS

The Legend of the Tower of Brahma

It is said that in a temple at Benares, India, the priests work continuously moving golden disks from one diamond needle to another. It seems that when the world was created, the priests of Benares were given three diamond needles and 64 golden disks. The priests were told that they were to place the disks on one of the needles in increasing order of size and then move the whole pile to one of the other two needles, moving only one disk at a time and never placing a larger disk on top of a smaller one. According to the legend, God told the priests, "When you finish moving the pile, the world will end."

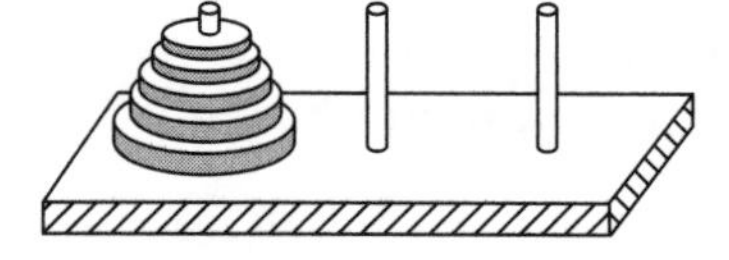

We can simulate the priests' problem by using coins, Cuisenaire rods, or different-sized squares cut from paper to represent the disks. Each peg can be represented by a square in a model like the one below.

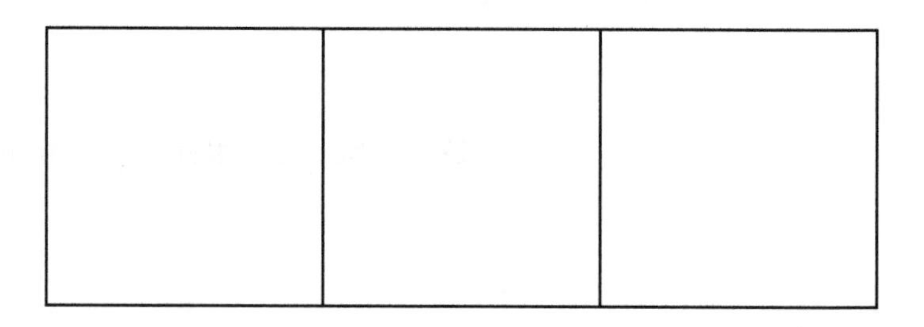

Stack the objects in one of the squares in ascending order of size. The goal is to move the stack of objects from one square to another in the fewest possible moves. There are two rules: (a) only one object may be moved at a time, and (b) a larger object may never be placed on top of a smaller one.

1. What is the minimum number of moves required to move 5 objects from one square to another? **HINT:** Look for two patterns, one for how the moves should be made and one for the minimum number of moves.

2. Just for fun, pretend the priests move one disk every second without stopping. How long will it take them to move:

   a. 10 disks?

   b. 30 disks?

   c. 50 disks?

   d. All 64 disks?

Chapter Summary

The study of patterns and functions is a central theme in mathematics. In this chapter, you learned various ways to analyze patterns. First you learned to recognize different types of patterns:

a. Patterns, such as those in Exercise 1 of Activity 1 and Exercise 1 of Activity 3, that have a repeating core:

ABAC ABAC

ABC ABC

b. Patterns, like those in Exercise 3 of Activity 2 and Exercise 2 of Activity 3, that have a growing core:

ABA ABBA ABBBA ...

AB ABC ABCD ...

c. Patterns that grow:

•, • •, • • •, ...

1, 3, 5, 7, ...

2, 4, 8, 16, ...

d. Nested patterns, like those in Exercise 3 of Activity 1 and Exercise 9 of Activity 3, that combine two or more patterns into one:

Pattern A	1,		3,		5,		7,		...
Pattern B		2,		4,		8,		16,	...
Nested pattern	1,	2,	3,	4,	5,	8,	7,	16,	...

While studying patterns, you learned some new terminology: *term, term number,* and *sequence*. You also learned that for many sequences, there is a rule (function) that relates the term of the sequence to the term number.

One method you learned for analyzing and extending numeric sequences was to examine the differences between the successive terms of the sequence. You found that in the cases where the differences are constant, you can use the difference to generate a rule (function) for the general term of the sequence.

The methods you used to analyze and extend patterns based on your observations are examples of *inductive reasoning*. You discovered the limitations of the inductive reasoning process in Activity 7. No matter how many initial terms of a sequence you may know, there is generally more than one way to extend it.

Activities 8–12 introduced the *logical reasoning* problem-solving strategy. Most of the activities began with a set of clues, or premises, that were accepted as true. By reasoning logically from these premises, you were able to conclude something about a number or situation. This process of deriving a conclusion by reasoning logically from a set of known premises is called *deductive reasoning*.

Usually, you were able to reason *directly* from the premises to the conclusion. However, in Activity 11 you had to test possible solutions by assuming they were true. If the assumption led to a contradiction of a known fact, then you knew that the proposed solution was not correct. This method, which was introduced through elimination, is known as *indirect reasoning*—it is used extensively in mathematics.

Logical reasoning is the cornerstone on which mathematics is built. New mathematics is often discovered via inductive reasoning. But before a conjecture arrived at inductively is accepted as a fact, it must first be verified using deductive reasoning. It is this standard of proof that distinguishes mathematics from the other sciences.

The activities in this chapter were intended to provide only an informal introduction to inductive and deductive reasoning. You will learn more about these techniques later in this book and use them throughout the rest of it.

Activity 13 integrated many problem-solving techniques. You learned to simulate problems that could not be experienced firsthand, to apply the patterns strategy, and to use tables as an organizer. You also learned a new problem-solving strategy—*simplify the problem*—whereby you begin with a simple case of the problem and work through successively more complex cases until a general method of solution is discovered. This is a technique closely related to inductive reasoning. It will be used extensively for investigating new mathematical concepts.

Chapter 2
Sets, Functions, and Logic

Logic and reasoning are critical to developing your *mathematical power*, that is, the ability to know and do mathematics. Using the concepts associated with sets, identifying properties of the elements of a set, and distinguishing those properties that sort elements into different sets provide an underlying structure for logical reasoning.

The activities in this chapter will engage you in sorting and classifying objects, describing their properties (attributes), and explaining their similarities and differences. All of these activities promote communication in mathematics. Explaining your reasoning, defending your conjectures, and evaluating input from others will enhance your understanding of mathematical ideas.

While you are involved in these sorting, classifying, and discriminating problem-solving situations, reflect on the connections between these processes in science and mathematics. You will find that the direct and indirect reasoning skills you develop in this chapter will be an important asset in other problem settings throughout the book.

The concept of function is one of the central unifying themes in mathematics. The study of arithmetic, algebra, geometry, probability, and statistics relies on generalizing patterns and developing mathematical models (functions) that can be used to describe real-world situations. Two activities in this chapter extend the idea of describing a pattern with an informal "rule" to the mathematical concept of function.

Activity 1: Attribute Matrix

PURPOSE Identify common attributes and complete an established pattern on a matrix.

MATERIALS One set of attribute pieces (page A-5) and matrix cards (pages A-17–A-19)

GROUPING Work individually or in groups of 2–3.

GETTING STARTED Some attribute pieces have been placed in the following matrix according to a pattern. Choose appropriate pieces from the remaining attribute pieces and place them in the blank cells of the matrix to continue the established pattern.

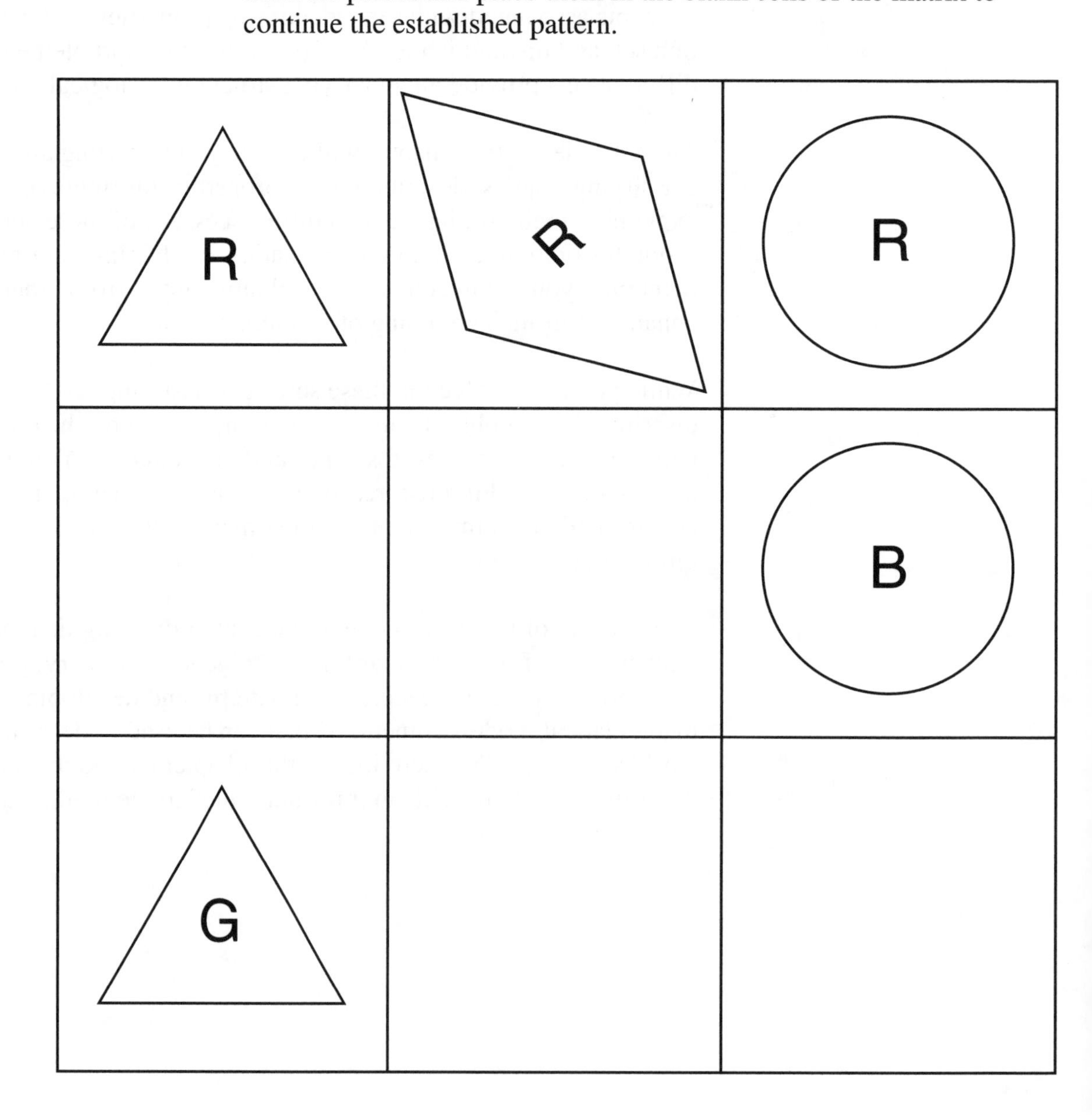

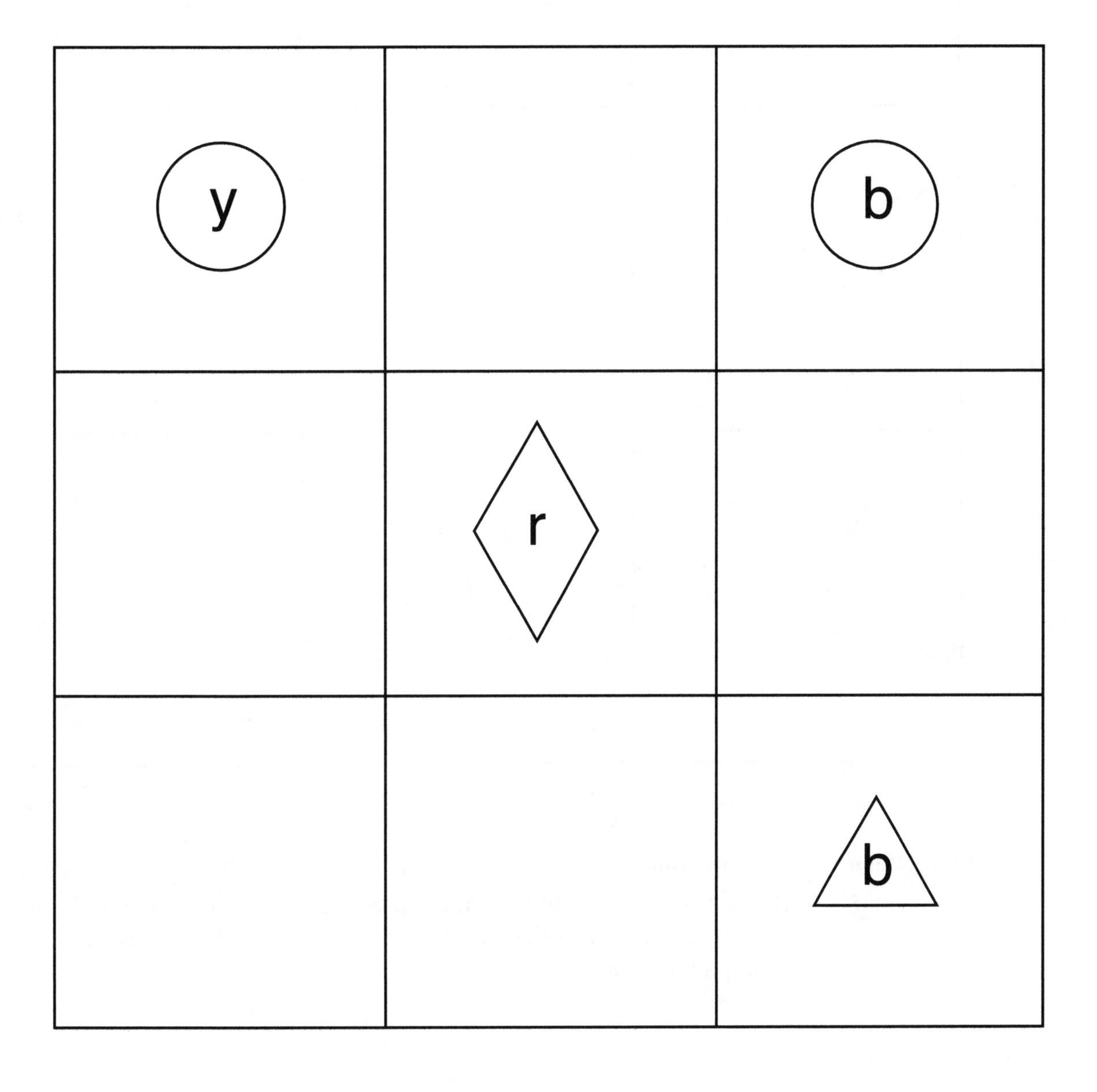
y
b
r
b

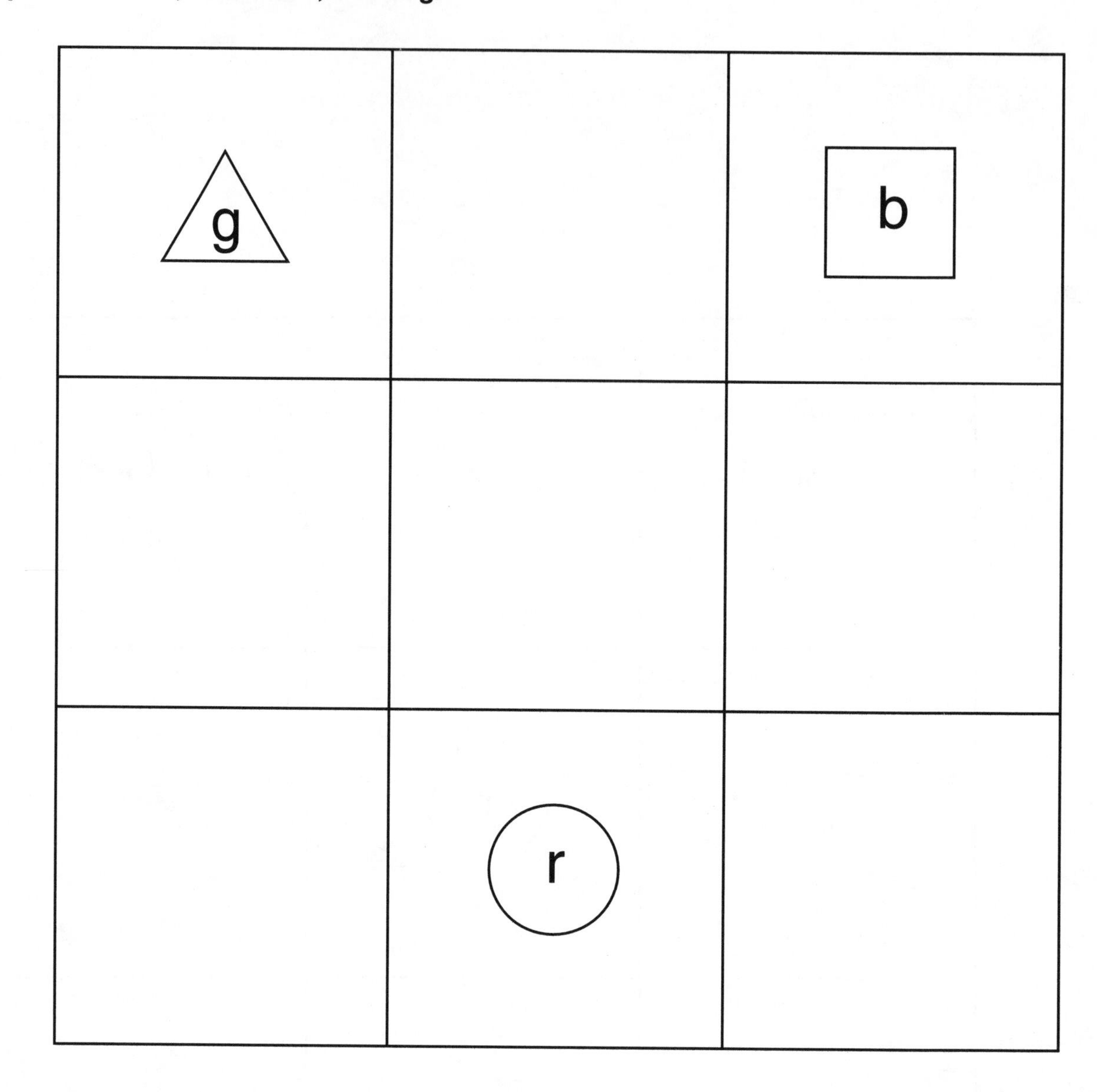

EXTENSIONS Use the blank matrix cards on pages A-17–A-19. Begin a pattern on a matrix by using as few attribute pieces as possible. Make additional attribute matrix puzzles on the 2×2, 3×3, and 4×4 cards to be completed by a classmate.

Activity 2: Sort and Classify

PURPOSE Introduce the concept of an attribute through sorting and classifying attribute pieces.

MATERIALS Set of attribute pieces (page A-5) or attribute people pieces (page A-1)

GROUPING Work individually or in pairs.

1. Sort the attribute pieces according to size, shape, and color and complete the following table.

Sort the Pieces by	Number of Piles	Number in Each Pile
Size		
Shape		
Color		

2. How many attribute pieces are there? Explain how you can answer this question without counting the pieces one by one.

In Exercises 1–3, a set of four attribute pieces is given. In each set, one of the pieces is different from the others. For each problem, decide which piece does not belong in the given set. Explain your decision.

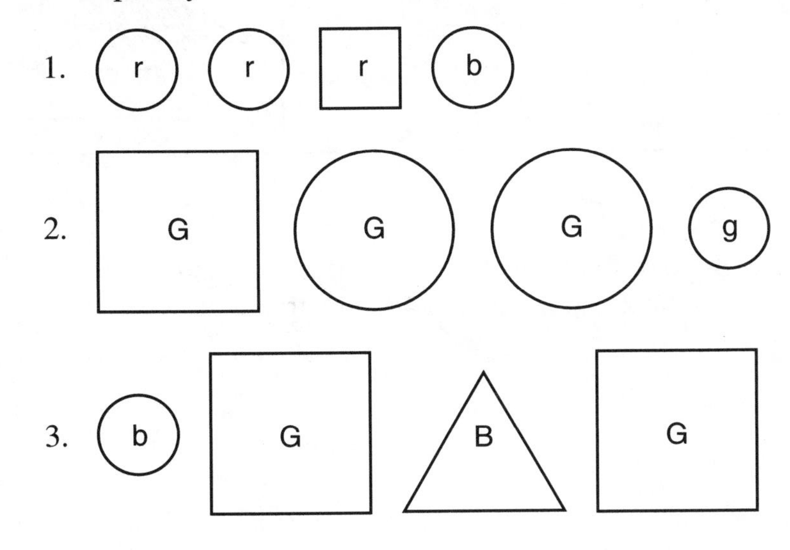

Activity 3: Matrix Arrangements

PURPOSE Distinguish attribute pieces that differ from each other by one attribute, two attributes, and three attributes.

MATERIALS Set of attribute pieces (page A-5) or attribute people pieces (page A-1) and matrix cards (pages A-17–A-19)

GROUPING Work individually or in pairs.

GETTING STARTED Choose a matrix card and place attribute pieces in the cells so that each piece is different from the adjoining piece, horizontally and vertically (but not diagonally) in exactly one way. Begin with one difference on the 2×2 matrix and progress to the 4×4 matrix.

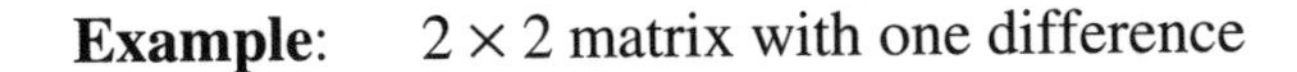

Example: 2×2 matrix with one difference

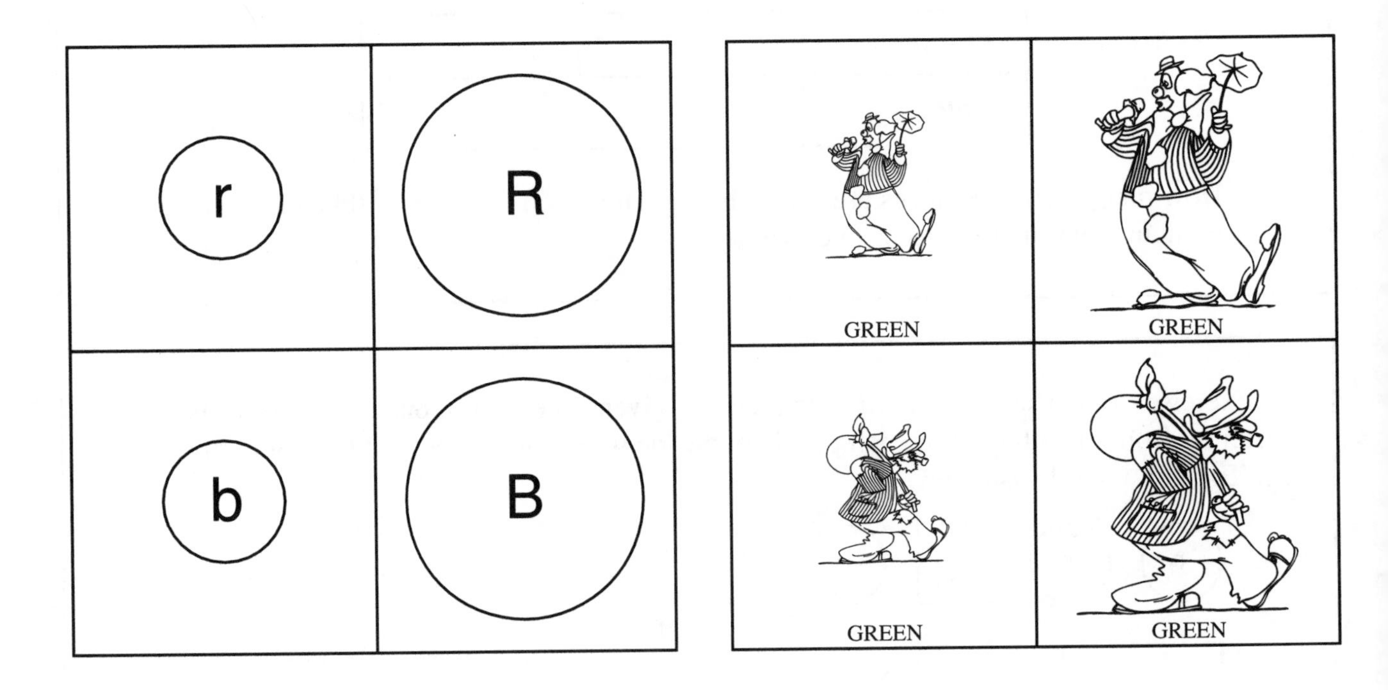

EXTENSIONS Proceed to more difficult levels of two differences and three differences on the 2×2, 3×3, and 4×4 matrix cards.

Activity 4: Difference Puzzles

PURPOSE Explore attribute differences by completing a puzzle.

MATERIALS One set of attribute pieces or attribute people pieces and the difference puzzles on pages A-20–A-22

GROUPING Work individually or in pairs.

GETTING STARTED Complete the puzzle by placing the attribute pieces in the cells of the difference puzzle so that adjacent cells differ by the number of attributes shown in the diamonds between the cells. Use the blank matrix page A-22 to make up other difference puzzles for other students.

Activity 5: What's in the Loop?

PURPOSE Use the elimination problem-solving strategy and logical reasoning to determine the identity of an attribute label card.

MATERIALS A piece of yarn or string to make a large loop and a set of attribute label cards (pages A-23–A-24)
Do not use any of the cards that include the word **NOT**.

GROUPING Work in pairs or in teams of two students each.

GETTING STARTED Arrange the yarn or string into a loop between the players. Shuffle the label cards. One student picks a label card from the pile, looks at it, and places it face down next to the loop without showing it to the other player(s). The second student chooses an attribute piece or people piece and places it in the loop. The first student states, "Yes, it may stay in the loop," or "No, it may not," based on the label card. Play continues until the second student can correctly identify the exact attribute on the label card.

Example: The face-down card is **GREEN**.

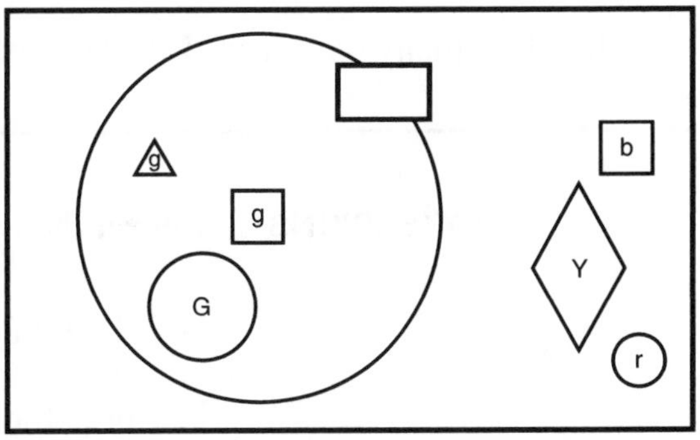

g
g
G
b
Y
r

EXTENSIONS Repeat the activity using all of the attribute label cards, including the **NOT** cards.

Activity 6: Intersecting Loops

PURPOSE Explore the set concepts of union, intersection, complement, and the universal set.

MATERIALS One set of attribute pieces or people pieces, two loops of yarn or string, and a set of label cards
Do not use any of the cards that include the word **NOT**.

GROUPING Work in pairs or teams of two students each.

GETTING STARTED Form two loops of yarn or string between the players and place one label card face up on each loop as shown. Students take turns placing pieces in the appropriate loop or outside the loops.

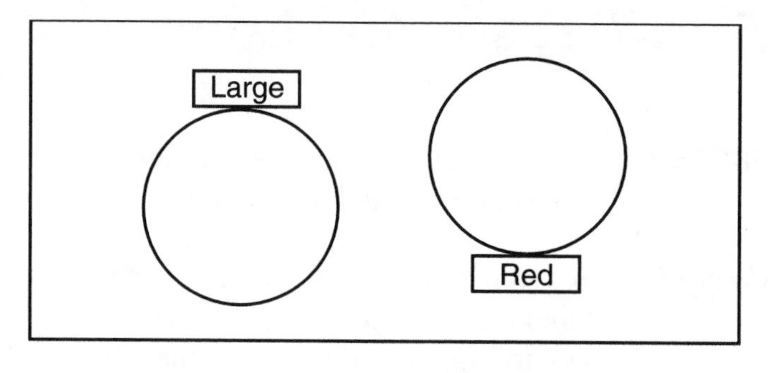

1. Given the labels **Large** and **Red**, determine how the loops must be arranged to allow correct placement of all the attribute pieces. Make a drawing of the correct placement of the loops.

2. How many pieces are inside both loops, that is, how many pieces are either RED or LARGE?

3. How many pieces are both RED and LARGE?

4. How many pieces are LARGE but not RED?

EXTENSIONS Repeat the activity with

a. two loops and different pairs of label cards, including the NOT cards.
b. three loops and three label cards.

Activity 7: What's in the Loops?

PURPOSE Use indirect reasoning to determine the attributes described on the label cards.

MATERIALS One set of attribute pieces or attribute people pieces, two loops of yarn or string, and one set of label cards

GROUPING Work individually or in teams of two students each.

GETTING STARTED Shuffle the label cards and place the deck between the players. Overlap the loops as shown below. One student picks two label cards and, without showing them to the other player(s), places them face down, one on each loop as shown. The other student(s) chooses an attribute piece and places it in one of the four regions. The first student then indicates, "Yes, that is a correct placement," or "No, it is not a correct placement," according to the labels on the cards that have been placed on the loops. Play continues until the second student can correctly identify the label cards. Students then switch roles.

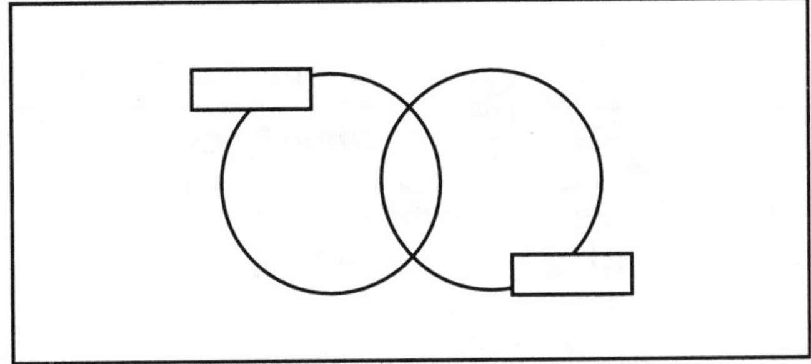

1. What labels could be used for the loops so that the intersection of the two sets is empty?

2. What labels could be used for the two loops so that one loop would be contained in the other? (One set is a subset of the other.)

EXTENSIONS Repeat the activity using three loops and three label cards.

Activity 8: Relating Patterns and Functions

PURPOSE Introduce the concept of functions through patterns and the make a table strategy.

MATERIALS Cuisenaire rods

GROUPING Work individually.

GETTING STARTED Construct each model with Cuisenaire rods and extend the pattern to include the next two figures.

Red Rods	1	2	3			
White Rods	2	3				

1. How many white rods are there with 10 red rods? ______
2. How is the number of white rods in each figure related to the number of red rods?

Light Green Rods	1	2			
White Rods	2	4	6		10

1. How many green rods are there with 10 white rods? ____________
2. How is the number of white rods in each figure related to the number of green rods?
3. Write two equations for the pattern to describe the relationship between the number of white rods and the number of green rods in each figure.

Yellow Rods	1	2				
White Rods						

1. Determine a function to describe the relationship between the number of white rods and the number of yellow rods in each figure.

EXTENSIONS Create some patterns with Cuisenaire rods or other materials such as square tiles or cubes. Give the pattern to a classmate to extend to the next two figures. Your classmate should then determine the function you used to determine the pattern.

Activity 9: What's My Function?

PURPOSE Determine the function relating input and output values.

MATERIALS One set of function cards (page A-25)

GROUPING Work in pairs.

GETTING STARTED

- In this game, players alternate turns trying to guess the function on one of the cards.
- A player scores 5 points if the function is guessed on the first try, 3 points if it is guessed on the second try, and 1 point if it is guessed on the third try. Once a player has determined the function, 1 additional point is scored for each correct way in which he or she can express the function. The player who has scored the most points after all the cards have been used is the winner.
- To start the game, shuffle the cards and place them face down on the table.
- One player draws a card from the top of the deck without showing it to the other player. The other player tries to guess the function on the card by giving input values one at a time. The player holding the card applies the function on the card to the input value and responds with the result.
- The guesser can give at most 10 input values and can guess the rule at any time. However, only 3 guesses per card are allowed.

Example:

What's My Function?

Multiply input by 3 and add 1.

$y = 3x + 1$

$x \to 3x + 1$ — x is mapped to $3x + 1$; x is paired with $3x + 1$

$f(x) = 3x + 1$

Input	Output	Guess	Response
2	7	Add 5 to the input.	Incorrect.
0	1	(No guess)	
3	10	Multiply input by 3 and add 1. $y = 3x + 1$ $f(x) = 3x + 1$	Correct.

Score: Guessed function on second try: 3 points
Expressed function three ways: 3 points
Total **6 points**

1. What strategies for choosing input values might make it easier to guess the function?
2. Find a function that could produce each of the following input-output tables. Then fill in the blanks in the table and write the function rule.

a.

Input	**3**	**6**	**4**		**0**
Output	**12**	**27**	**17**	**2**	

b.

Input		**4**	**7**	**10**	**8**
Output	**2**	**17**	**50**	**101**	

EXTENSIONS Make up function cards of your own and use them to play the game.

Chapter Summary

While the study of patterns is a central theme in mathematics, the fields of mathematics, such as arithmetic, geometry, and algebra, are characterized by the study of particular sets of objects. This study entails an analysis of the elements of a set, their attributes, and what makes a set uniquely different from others.

In Activity 2, you sorted elements of a set according to certain attributes. Sorting and classifying were visited again with increasing complexity in subsequent activities. These activities develop an informal understanding of how precise definitions of terms—so important in mathematics—are generated. For example, in the classification of quadrilaterals, squares and rectangles have several common attributes. An attribute that distinguishes squares from rectangles is that squares have congruent adjacent sides or they have perpendicular diagonals. Either attribute uniquely distinguishes squares as a special subset of rectangles.

Problem 2 of the first set of exercises in Activity 2 introduced the Fundamental Counting Principle. If we know the number of each of the attributes—for example, two sizes, three colors, four shapes—then the total number of pieces is the product of the numbers of the attributes: $2 \times 3 \times 4 = 24$. The Fundamental Counting Principle is important in probability and will be explored in more depth in Chapter 8.

Activity 2, which involves sorting and classifying pieces, and Activities 3 and 4, which deal with identifying differences and similarities among pieces, introduced skills that are basic to the study of science as well as mathematics. While mathematicians classify geometric shapes according to attributes, scientists sort all living creatures into a hierarchy of classifications. Each class differs from the previous one by some attribute that makes it unique from all other living creatures. It is important to recognize this connection between the fundamental processes of mathematics and science and how activities such as these can help in developing basic understanding of both subjects in the elementary grades.

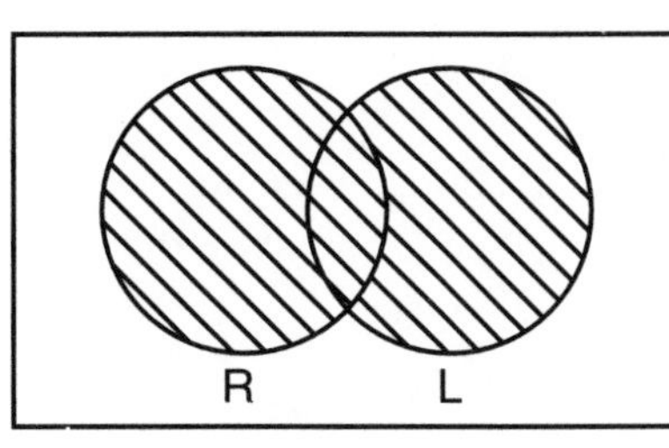

RED or LARGE
Union

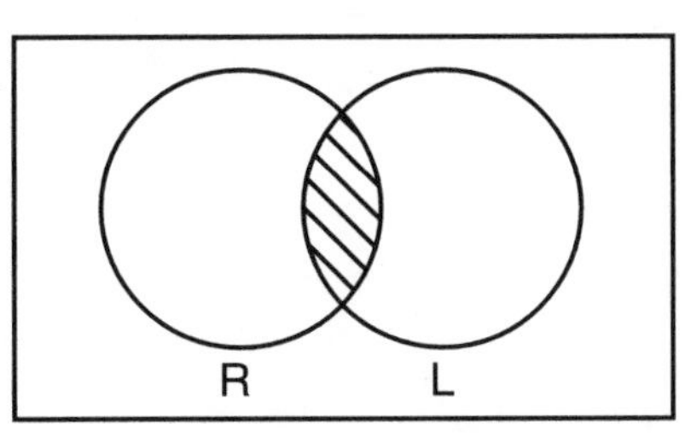

RED and LARGE
Intersection

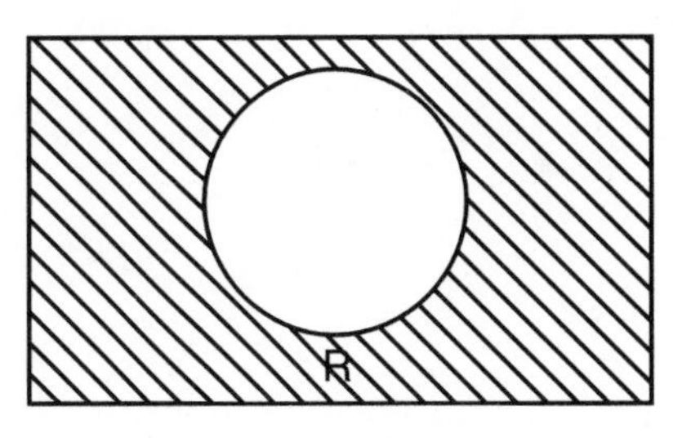

Not RED
Complement

In Activities 5–7, you encountered the set concepts of *union*, *intersection*, *complement*, and the *empty set*. The set of all pieces provided an example of a *universal set,* while the word OR described the *union* of sets; AND, the *intersection* of sets; and NOT, the *complement*. These concepts are illustrated by the following diagrams.

In Activity 5, when you chose to place a particular attribute piece in a loop, you guessed (made an assumption) that a certain label belonged to that loop. Given a *yes* answer, you gained information about the correct label. Given a *no* answer, you also gained information about which label was NOT correct. Thus you were able to eliminate some labels and reduce the number of pieces to try in order to complete the problem. Activity 7 extended this use of the elimination problem-solving strategy and the use of indirect reasoning to determine the proper label for the sets.

Activities 8 and 9 extended the ideas learned in Activities 4–6 in Chapter 1. The mathematical concept of function was introduced through an activity involving patterns, much like those in Chapter 1, and a game that applied the guess and check strategy to input and output values in order to identify various functions. The concept of function is revisited in many activities throughout the book.

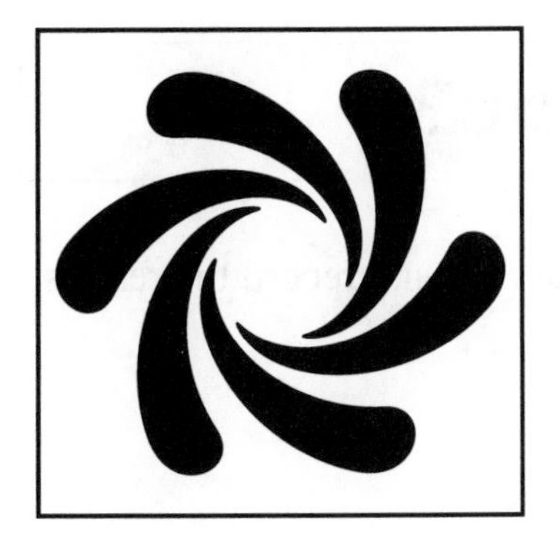

Chapter 3
Numeration and Computation

Give a man a fish, and he eats for a day. Teach him how to fish, and he eats for a lifetime.

"Programs that provide limited developmental work, that emphasize symbol manipulation and computational rules, and that rely heavily on paper-pencil worksheets do not fit the natural learning patterns of children and do not contribute to important aspects of children's mathematical development."

—*Curriculum and Evaluation Standards for School Mathematics*

The development of number sense, operation sense, place-value concepts, and an understanding of the algorithms for the basic operations are among the most important tasks in teaching elementary mathematics. The elementary teacher promotes greater understanding of number and operation sense through increased emphasis of these concepts and by making connections among concrete models, numbers, and algorithms for the basic operations.

This conceptual approach to teaching mathematics allows students to construct their own ideas of numbers and computation. The use of manipulatives provides a problem-solving environment for learning in which students are constantly discussing, conjecturing, and asking new questions as they review their work.

The activities in this chapter provide you with new and interesting situations to increase your understanding of number sense, place value, and algorithms for operations. In each, you will explore some mathematical concepts in a hands-on, problem-solving setting.

Activity 1: How Do the Beans Add Up?

PURPOSE Practice simple addition with concrete materials and record the results.

MATERIALS One set of beansticks (page A-26)

GROUPING Work individually or in pairs.

GETTING STARTED Choose *target numbers* and record them in the squares (see examples below). Choose any three beansticks, place them on the outline shown on the worksheet, and record the number of beans on the tag attached to the stick. Form sums equal to each *target number* using one, two, or three of the tagged beansticks. Record the number on each tag in a bean below the target box. Leave beans blank if all are not used or when there is no solution. Worksheet 1-B is similar to 1-A. Worksheet 1-C provides the opportunity to determine and record more than one solution for each target number.

Example:

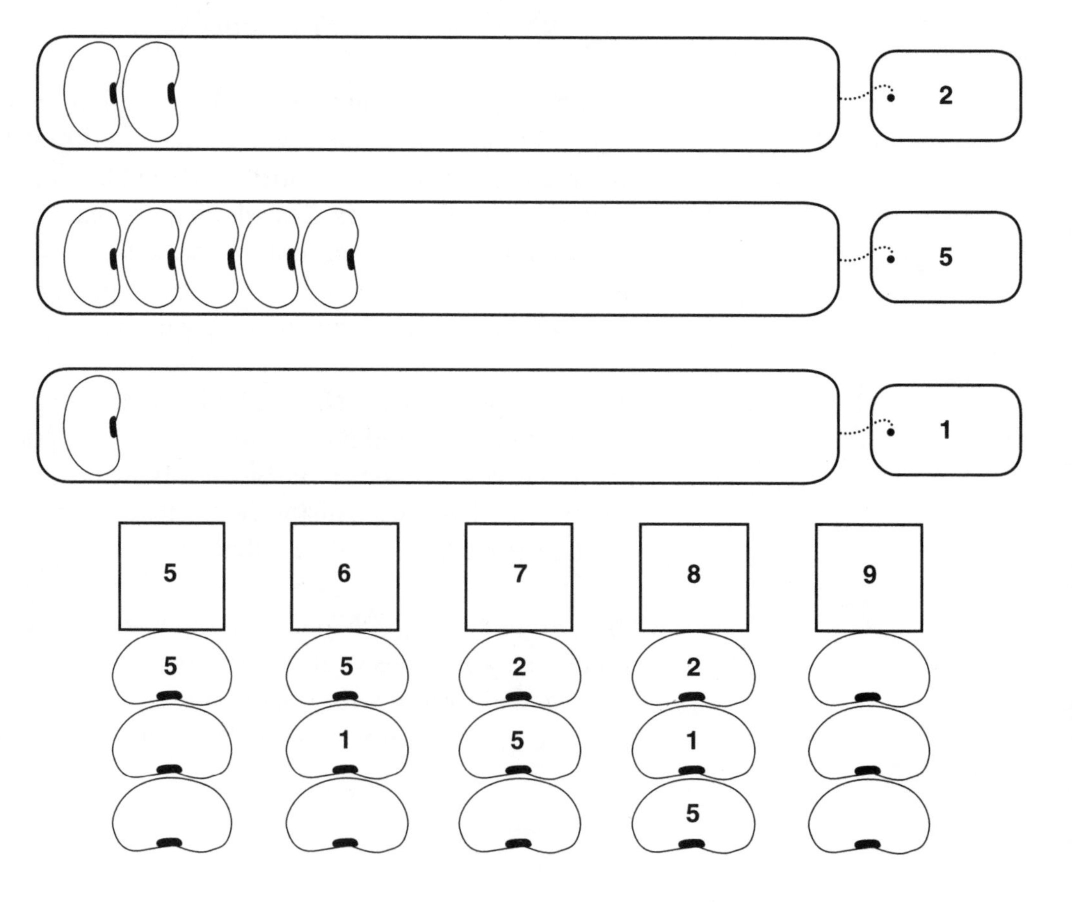

EXTENSIONS Use different beansticks. Use different target numbers. Make a page for your partner by labeling the beanstick tags and choosing different target numbers.

WORKSHEET 1-A

WORKSHEET 1-B

WORKSHEET 1-C

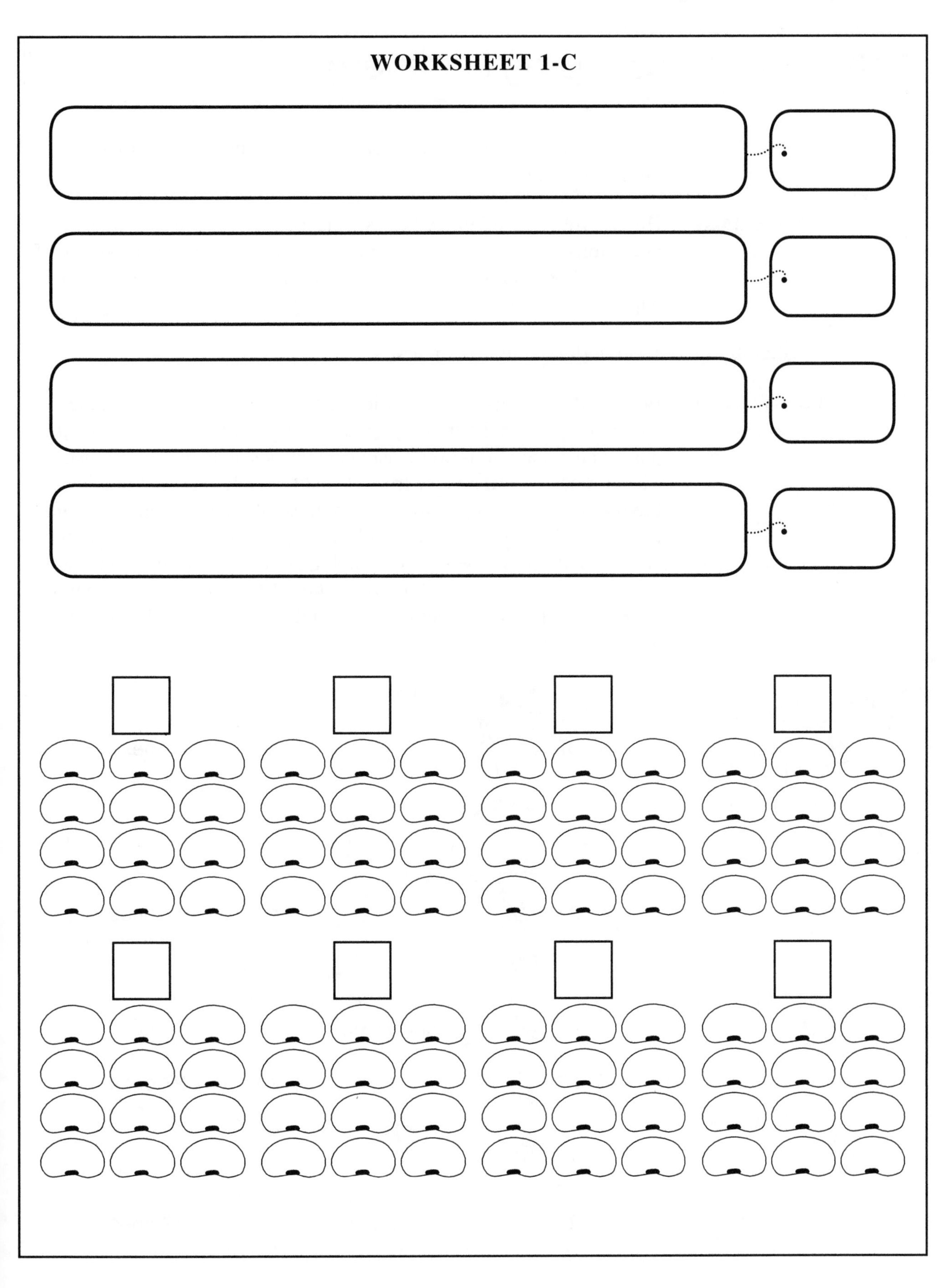

Activity 2: Regrouping Numbers

PURPOSE Develop number sense and place-value concepts by constructing numbers with regrouping.

MATERIALS Hundreds, tens, and units dice (see page A-27), one place-value operations board (page A-28), base-ten blocks (pages A-13 and A-15) or a set of beansticks consisting of ten sticks and loose beans, and paper for recording

GROUPING Work individually or in pairs.

GETTING STARTED Begin with the tens and units dice. One student rolls the dice; the second constructs the two-digit number represented by the dice on the place-value operations board using either rods and cubes or beansticks. The number should be constructed first by using the least number of blocks or beansticks possible. The highest place-value block should then be traded for 10 of the next lower order blocks. The new number constructed should then be recorded. Continue the trading until only units blocks are used. Each number should be recorded as shown in the example below. Students then switch roles. Repeat the activity with three dice.

Example:

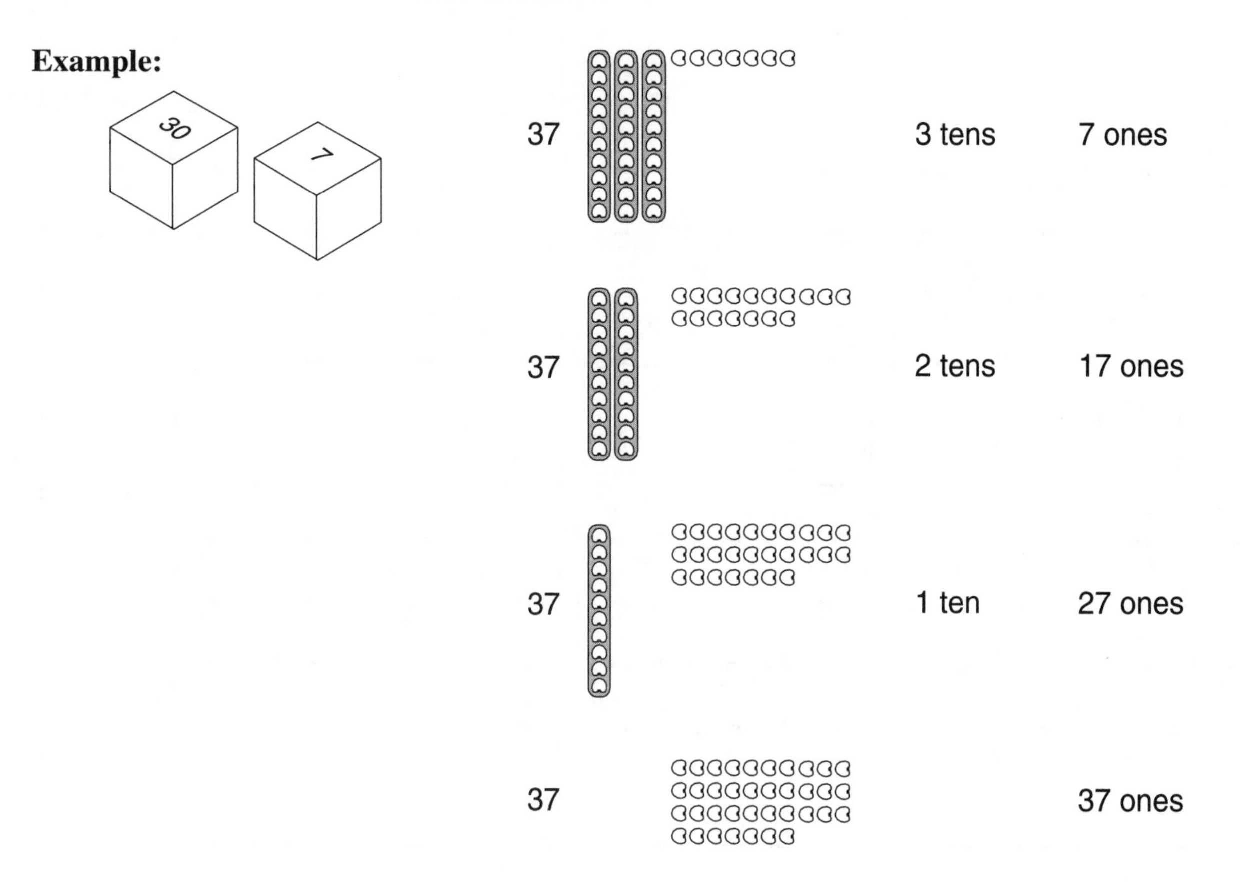

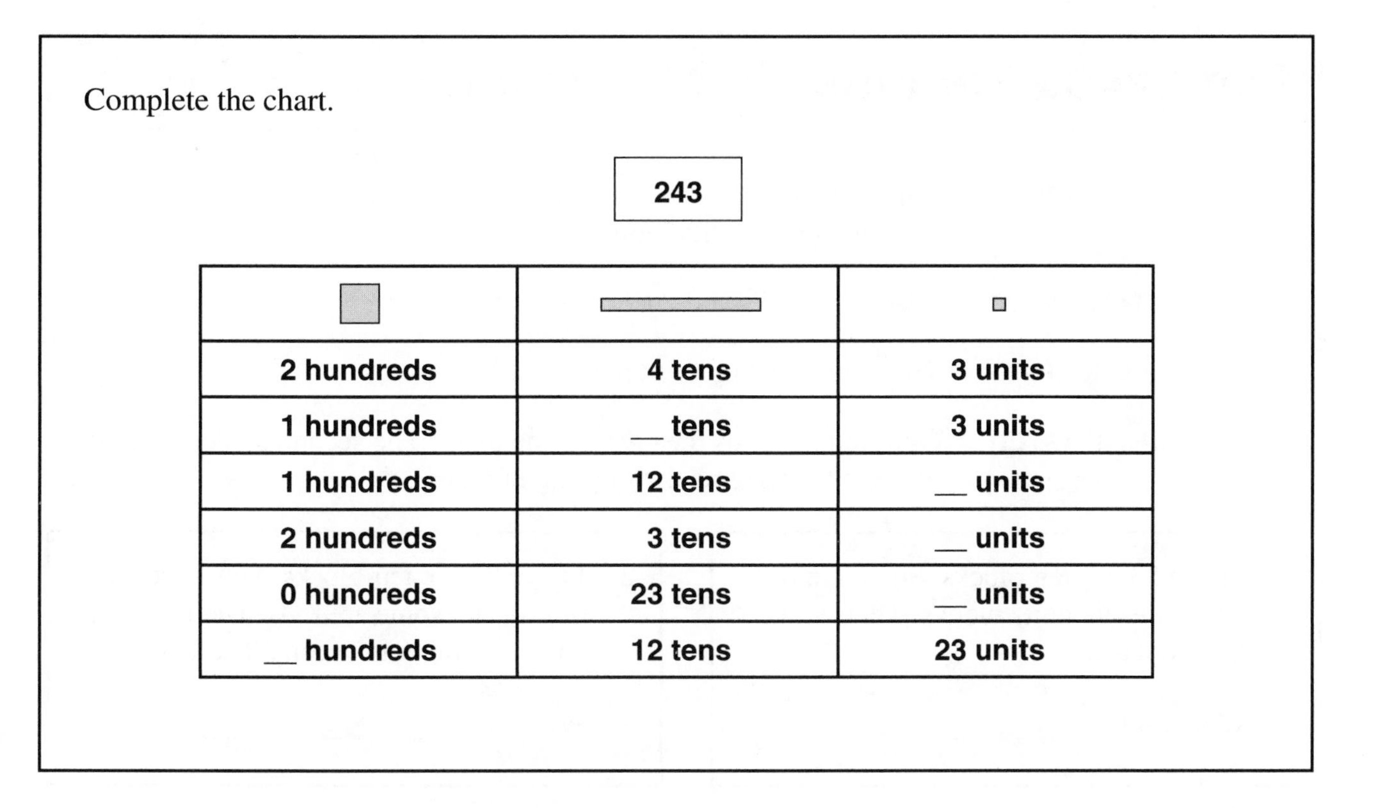

Complete the chart.

243

2 hundreds	**4 tens**	**3 units**
1 hundreds	**__ tens**	**3 units**
1 hundreds	**12 tens**	**__ units**
2 hundreds	**3 tens**	**__ units**
0 hundreds	**23 tens**	**__ units**
__ hundreds	**12 tens**	**23 units**

EXTENSIONS Duplicate several charts like the one shown below. Fill in some of the parts and then have your partner complete the chart.

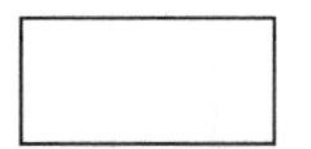

__ hundreds	**__ tens**	**__ units**
__ hundreds	**__ tens**	**__ units**
__ hundreds	**__ tens**	**__ units**
__ hundreds	**__ tens**	**__ units**
__ hundreds	**__ tens**	**__ units**
__ hundreds	**__ tens**	**__ units**

Activity 3: Find the Missing Numbers

PURPOSE Reinforce number sense and place-value concepts in a problem-solving situation.

MATERIALS Base-ten blocks and clue cards

GROUPING Work individually or in pairs.

GETTING STARTED One student reads the clues to the other. The second student uses the blocks to determine the missing number.

1. I have 5 base-ten blocks. Some are rods and some are units blocks. Their value is less than 20.

   Who am I? _______________

2. I have 8 base ten-blocks. Some are units blocks and some are rods. Their value is an odd number between 50 and 60.

   Who am I? _______________

3. I have 4 base-ten blocks. Some are rods and some are units blocks. Their value is between 30 and 40.

   Who am I? _______________

4. I have 6 base-ten blocks. Some are flats, some are rods, and some are units blocks. I am a palindrome.

   Who am I? _______________

5. I have 2 base-ten blocks.

   Who am I? _______________

6. I have 3 base-ten blocks. None of them are flats.

   Who am I? _______________

7. I have 4 base-ten blocks. None of them are flats.

   Who am I? _______________

8. I have 4 base-ten blocks. Only one of them is a flat.

   Who am I? _______________

EXTENSIONS Create additional clue cards for your partner to solve.

Activity 4: A Visit to Fouria

PURPOSE Use a model to reinforce place-value concepts, to introduce the base-four numeration system, and to develop understanding of the regrouping process.

MATERIALS Red, blue, and white chips (10 of each color) and a die

GROUPING Work in pairs.

GETTING STARTED While on an Intergalactic Numismatics Tour, you encounter a meteor shower and are forced to make an unscheduled stop on the planet Fouria. The monetary system used on Fouria consists of three coins: a white coin (worth $1 in our money), a red coin, and a blue coin. The red coin is equivalent in value to four white coins, and the blue coin is equivalent to four red coins.

Unlike its sister planet, Ufouria, Fouria turns out to be a rather dull place to visit. To help pass the time in the waiting area, you and a fellow passenger play a coin trading game. The rules of the game are as follows:

- Players alternate turns.
- On each turn, a player rolls the die and places that number of white Fourian coins in the White column on his or her Coin Trading Game Sheet.
- Whenever possible, a player must trade four white coins for one red coin and/or four red coins for one blue coin.
- Coins must always be placed in the appropriately labeled column, and no more than three of any of the coins may be in any column at the end of a turn.
- The first player to get two blue coins is the winner.

Make a Coin Trading Game Sheet and play the game with a partner. At the end of each turn, record the number of each color coin on your game sheet in a table like the one at the right.

Coin Trading Game Sheet		
Red	Blue	White

Turn	Number Rolled	Result		
		B	R	W
1				
2				
3				
4				
5				
6				
7				
8				
9				
10				

THE COIN TRADING GAME REVISITED

1. A third passenger has been watching you play. She suggests it is more challenging to start the game with three blue coins and to remove the number of white coins equal to the number rolled on each turn. The first player to remove all the coins from the playing board is the winner. Your playing partner is confused. "How can you remove white coins when there aren't any on the board?" he asks. Explain how this can be done.

2. Play this new version of the game with a partner. Again, record the result of each of your moves in a table like the one on the preceding page.

---

1. You become bored with the games and go to the spaceport newsstand to buy something to read. Glancing at the cover of a Fourian magazine, you notice that the price is given as 123_4. At the checkout stand, the clerk explains that this means one blue coin, two red coins, and three white coins. How many of each color coin does each of the following prices represent?

   a. 231_4 ______________________ b. 102_4 ______________________

   c. 13_4 ______________________ d. 20_4 ______________________

2. How would Fourians write each of the following prices?

   a. 1 blue, 2 red, 1 white __________ b. 2 red, 3 white __________

   c. 2 red ______________________ d. 2 blue, 3 white __________

3. Back in the waiting area, you begin leafing through a magazine you purchased. You note that the first page is numbered 1_4, but when you get to the fourth page, you are surprised to find it numbered 10_4. Fill in the blanks below to show how the remaining pages of the magazine would be numbered.

   1_4 ____ ____ 10_4 ____ ____ ____ ____ ____ ____

   ____ ____ ____ ____ 33_4 ____ ____ ____ ____ ____

   ____ ____ ____ ____ ____ ____ ____ ____ ...

   ____ ____ 333_4 ____ ____ ____ ____ ____ ____ ____

1. After reading for a while, you decide to have dinner. The price of your meal was 121_4. When you go to the cashier to pay for your meal, you realize you don't have any Fourian money with you. "No problem," the cashier says. "You may pay with dollars." What is the cost of your meal in dollars?
2. a. On your way back to the waiting area, you stop at the newsstand to buy a souvenir. It costs 1312_4. How many dollars is this?

   b. You give the cashier two $100 bills. How much Fourian money should you get back in change?

1. Back in the waiting area, you find a mathematics book left behind by a Fourian student. Flipping through the book, you come across the examples shown below. Explain what the small 1 in the second example means and how it was obtained.

$$\begin{array}{r} 2\,3\,1_4 \\ +\,1\,0\,2_4 \\ \hline 3\,3\,3_4 \end{array} \qquad \begin{array}{r} {\scriptstyle 1} \\ 2\,3_4 \\ +\,3\,3_4 \\ \hline 1\,2\,2_4 \end{array}$$

2. Use what you learned in the examples in Exercise 1 to find the following sums.

   a. $121_4 + 211_4$ b. $123_4 + 221_4$

3. A few pages later, you find the following examples. Explain what is being done in steps A, B, and C.

$$\begin{array}{r} 3\,3\,3_4 \\ -\,1\,2\,1_4 \\ \hline 2\,1\,2_4 \end{array} \qquad \begin{array}{r} {}^{2}\not{3}\,{}^{1}1_4 \leftarrow (A) \\ -\,1\,2_4 \\ \hline 1\,3_4 \\ \uparrow\;\uparrow \\ (C)\;(B) \end{array}$$

4. Use what you learned in the examples in Exercise 3 to find the following differences.

   a. $323_4 - 211_4$ b. $221_4 - 132_4$

EXTENSIONS

1. Explain how the place-value system used to write Fourian numerals is related to the place value system used to write base-ten numerals.
2. Explain how the regrouping used in Fourian addition and subtraction is related to the regrouping used in regular (base ten) addition and subtraction.

Activity 5: It All Adds Up

PURPOSE Develop an understanding of addition of multi-digit numbers and regrouping.

MATERIALS Place-value operations board (page A-28), base-ten blocks (pages A-13 and A-15) or beansticks, place-value dice, and paper for recording. Instructions for constructing beansticks and place-value dice can be found on page A-27.

GROUPING Work individually or in pairs.

GETTING STARTED The first student rolls the dice and constructs the two-digit number on the operations board using the correct base-ten blocks. (See the example.) The number should also be recorded on a student record sheet. The second student then rolls the dice and constructs the number beneath that of the first student as shown below. Record the number below the first on the record sheet and draw lines below the two representations. The blocks are then moved together to represent the sum of the two numbers. Record the representation of the sum. (If a column has 10 or more blocks, then a 10-for-1 trade should be made.)

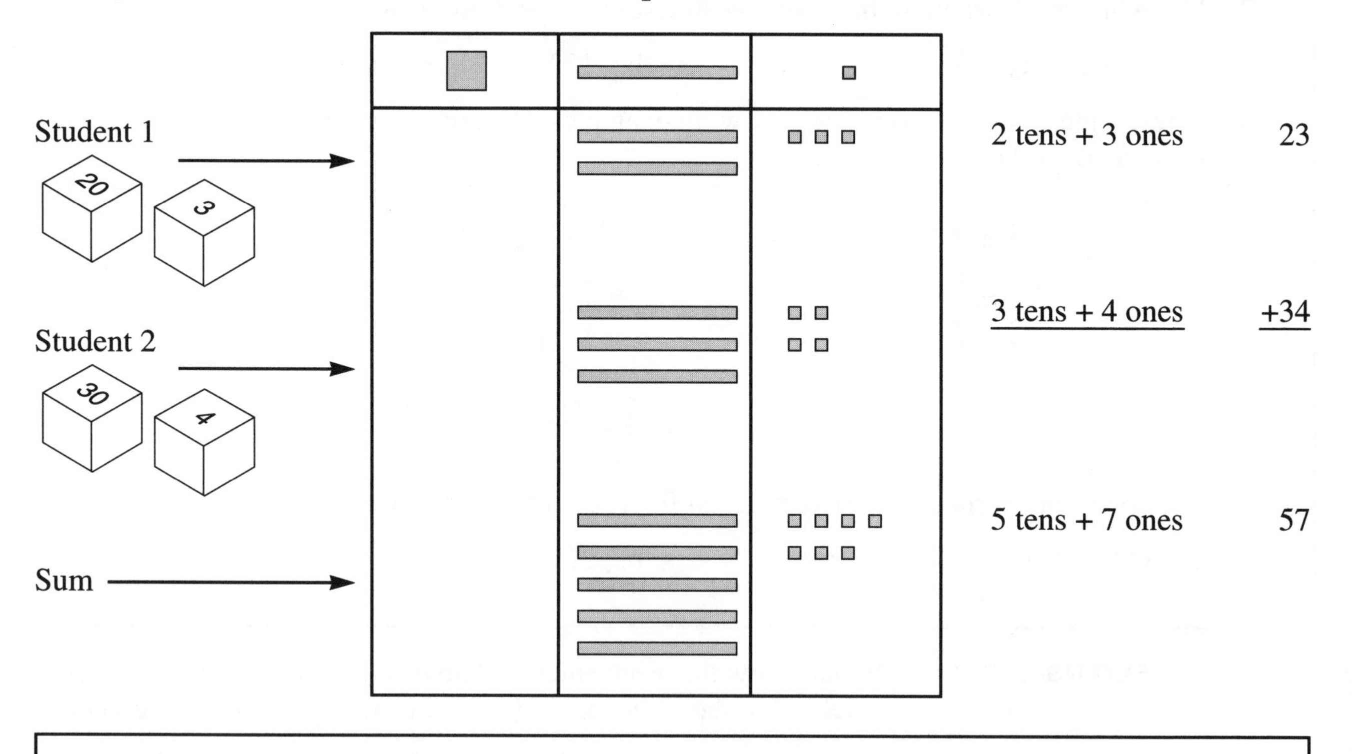

In each of the problems you completed and recorded, mark those that required you to make a trade in order to determine the correct sum.

Activity 6: What's the Difference?

PURPOSE Use two different models to develop an understanding of subtraction of multi-digit numbers with regrouping.

MATERIALS Place-value operations board, base-ten blocks or beansticks, two sets of place-value dice, and paper for recording

GROUPING Work individually or in pairs.

GETTING STARTED For the *take away* model of subtraction, each student rolls a set of dice. The person with the larger number constructs the two-digit number on the operations board using the correct number of base-ten blocks. The student with the smaller number removes (*takes away*) the number of blocks represented by the two-digit number on the dice. If necessary, make a 10-for-1 trade. Record the representation of the difference.

Example 1:

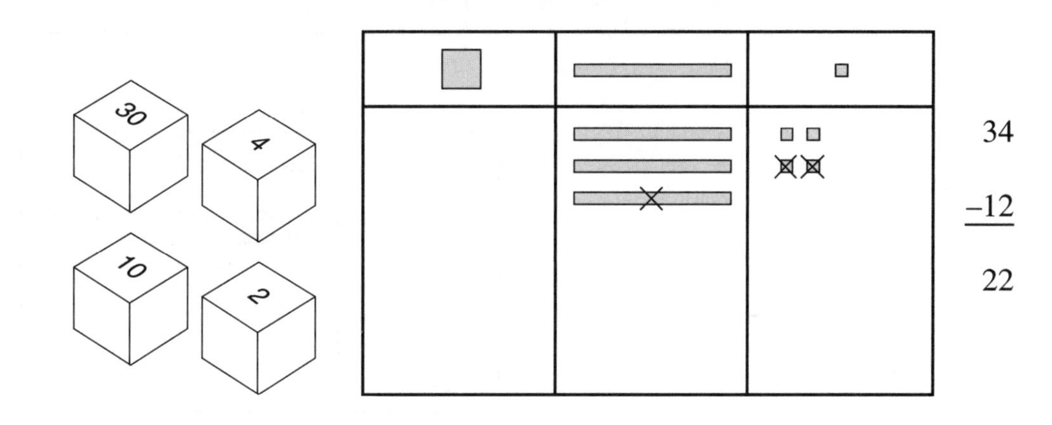

Record your work in the table below.

Larger Number	–	Smaller Number	=	Difference
	–		=	
	–		=	
	–		=	
	–		=	
	–		=	

For the *comparison* model of subtraction, each student rolls a set of dice. The person with the larger number constructs it at the top of the operations board using base-ten blocks. The other student constructs the smaller number below the first, as shown in Example 2. To make the comparison, stack the blocks representing the smaller number on top of those representing the larger number. If necessary, make a 10-for-1 trade. The blocks that are not covered represent the difference between the two numbers.

Example 2:

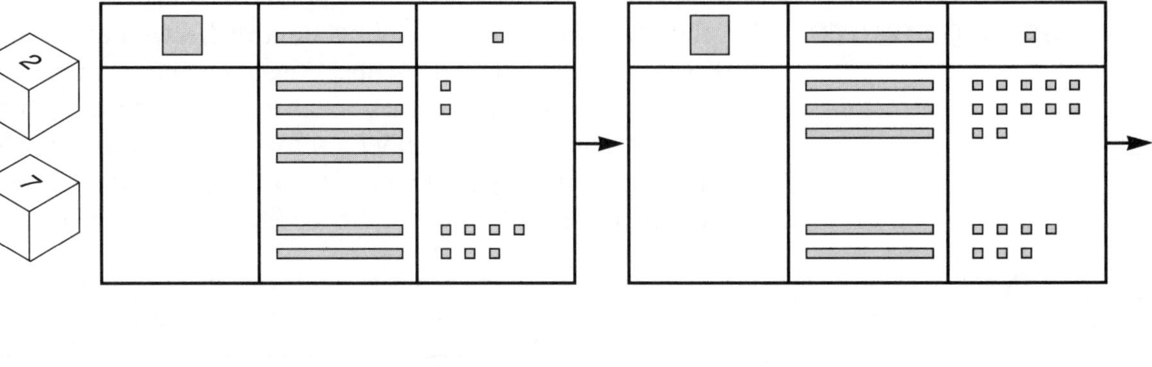

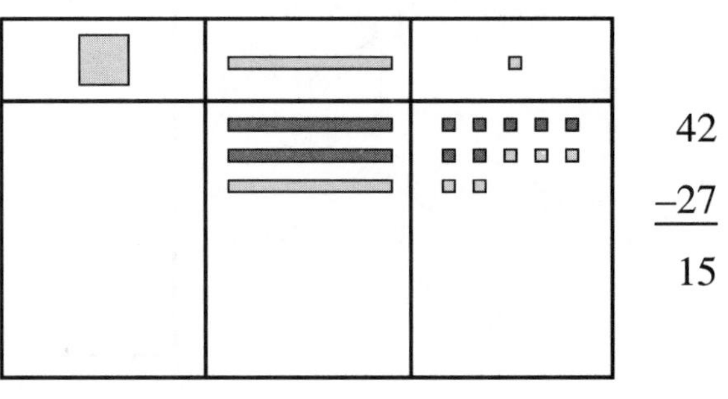

1. Roll the dice 10 times and determine the difference between the two numbers represented using the *take away* and *comparison* models. Record all problems as illustrated in the examples.

2. Write two word problems for each of the two models for subtraction, *take away* and *comparison*. Each problem should be a real-world application that is relevant to the life of an elementary student and correctly represents the designated model.

Activity 7: Multiplication Arrays

PURPOSE Develop the concept of multiplication as repeated addition.

MATERIALS Colored squares (page A-7), ceramic tiles, squares cut from construction paper, or square grid paper; a copy of the multiplication and division frame (page A-29), and paper for recording

GROUPING Work individually in pairs.

GETTING STARTED In a multiplication problem, two factors determine a product. Every multiplication fact can be illustrated as a rectangular array on a multiplication and division frame. The factor to the left of the frame determines the number of groups (rows). The number above the frame determines the number in each group (columns).

Use the **Multiplication and Division Frame** to construct each of the first 10 multiples of 4. Record your answers on paper as shown below.

factor
factor | product

Example:

$1 \times 4 \rightarrow$ 1 group of 4 $\rightarrow 4$
$1 \times 4 = 4$

$2 \times 4 \rightarrow$ 2 groups of 4 $\rightarrow 4 + 4$
$2 \times 4 = 8$

$3 \times 4 \rightarrow$ 3 groups of 4 $\rightarrow 4 + 4 + 4$
$3 \times 4 = 12$

1. What happens to the rectangular arrays and the products when the previous factors are used but their order is reversed, such as 4×1, 4×2, 4×3?

2. On square grid paper, construct a 3×4 rectangle and a 4×3 rectangle. Cut out the rectangles and place one on top of the other. What must be done to align them so that they match each other? What can you conclude from this? Is this always true?

Activity 8: Find the Missing Factor

PURPOSE Develop the operation of division as equal sharing or *partitioning*.

MATERIALS Colored squares, a copy of the multiplication and division frame, and one place-value die

GROUPING Work in pairs.

GETTING STARTED The following problem is an example of the *partitioning* model for division: "Tasha has 15 marbles. She wishes to give the same number of marbles to each of 5 friends. How many marbles does each person receive?" Use the multiplication and division frame to solve 15 ÷ 5, or $5\overline{)15}$. Record the factor (divisor) 5 to the left of the frame. Using 15 squares, determine how many squares must be placed in each of the 5 groups to form a rectangular array.

Example: a. Place a square in each group.

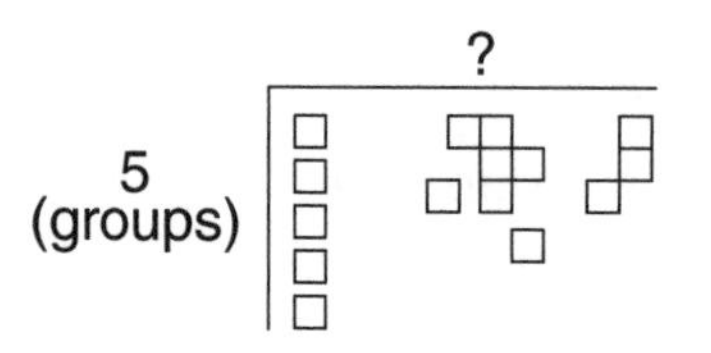

b. Now arrange the rest of the squares in groups to construct a rectangular array.

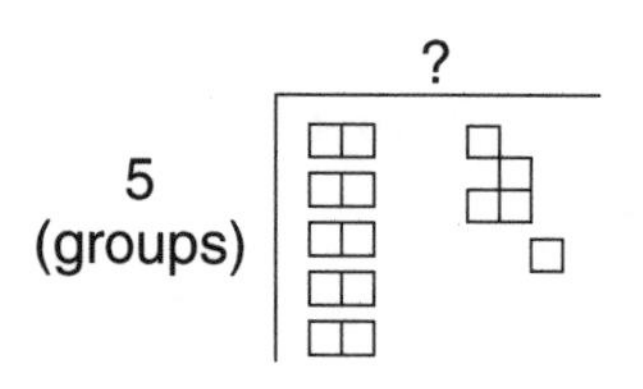

c. How many squares are there in each group (row)? ____________

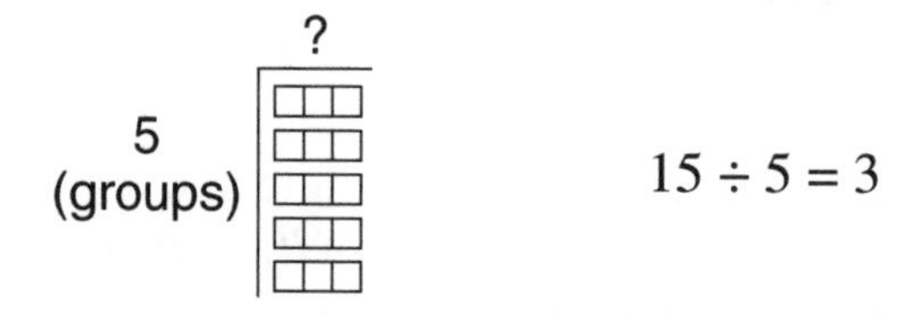

Student 1 rolls a die to determine a factor. Write the number to the left of the frame. Student 2 picks up a handful of squares (without counting) to determine the product (dividend) and then uses the factor to construct a rectangular array as in the previous example.

Example: The number of squares in the handful equals 15.

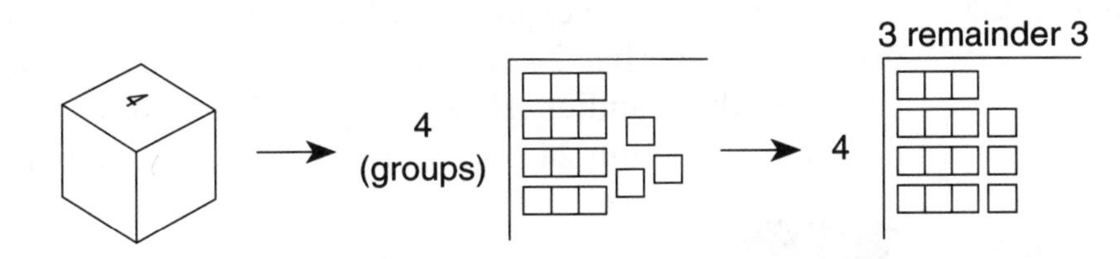

How many squares are in each of the four groups? ______

How many squares are remaining? ______

a. Repeat this activity 10 times, alternating turns.

b. Describe the relationship between the *partitioning* model for division and multiplication.

EXTENSIONS Use the *integer-division* key on your calculator to solve the following problems.

1. $64 \div 3$
2. $42 \div 2$
3. $116 \div 6$
4. $57 \div 4$
5. $94 \div 6$
6. $43 \div 8$
7. $81 \div 9$
8. $79 \div 5$
9. $157 \div 7$
10. If you use the [÷] key, $15 \div 4 = 3.75$. How would you use the calculator to determine the *whole number* remainder? Show the sequence of keys to be pressed.

Activity 9: How Many Cookies?

PURPOSE Develop the concept of division as repeated subtraction.

MATERIALS Colored squares and paper for recording

GROUPING Work individually or in pairs.

GETTING STARTED The following is an example of the repeated subtraction model of division: "Tyler has 24 candies in a jar and wants to decorate each cookie with 6 candies. How many cookies will he be able to decorate?" Solving 24 ÷ 6, or $6\overline{)24}$, is illustrated below.

Example:

$$\begin{array}{r} 6\overline{)24} \\ -6 \\ \hline 18 \end{array}$$ Remove 6 one time.

$$\begin{array}{r} -6 \\ \hline 12 \end{array}$$ Remove 6 two times.

$$\begin{array}{r} -6 \\ \hline 6 \end{array}$$ Remove 6 three times.

$$\begin{array}{r} -6 \\ \hline 0 \end{array}$$ Remove 6 four times.

24 ÷ 6 = 4

Write more word problems that illustrate the *repeated subtraction* model for division and solve them as shown above.

Solve the following problems using the *constant* feature of your calculator.

Example: 120 ÷ 20

[ON] [−] [2] [0] [Cons] [1] [2] [0] [Cons] [Cons] [Cons] [Cons] [Cons] [Cons]

[ON] [1] [2] [0] [−] [2] [0] [=] [=] [=] [=] [=] [=]

1. 96 ÷ 6
2. 318 ÷ 3
3. 107 ÷ 24
4. 858 ÷ 13
5. 94 ÷ 23
6. 152 ÷ 4
7. 297 ÷ 3
8. 432 ÷ 17

Explain how you determined the quotient and the remainder in each problem using the *repeated subtraction* model.

Activity 10: Multi-digit Multiplication

PURPOSE Develop the multiplication algorithm with multi-digit numbers and reinforce place-value concepts.

MATERIALS Base-ten blocks, dice, multiplication and division frame, and paper for recording

GROUPING Work individually or in pairs.

GETTING STARTED Roll the dice to generate two-digit numbers as shown. Construct these numbers on the top and left of the frame using blocks. Then construct a rectangle inside the frame with the appropriate base-ten blocks.

Example:

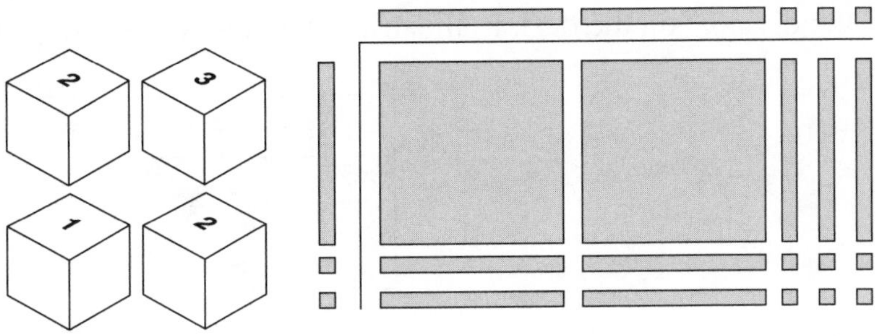

Count the blocks to determine the product.

2 flats = 200

7 rods = 70

6 units = 6

Product = 276

Record your work in expanded notation to reinforce place value in the factors and in the product.

$$\begin{array}{rcrcrcr} (20 & + & 3) & & & & \\ \times\,(10 & + & 2) & & & & \\ \hline 200 & + & 30 & & & & \\ & & 40 & + & 6 & & \\ \hline 200 & + & 70 & + & 6 & = & 276 \end{array}$$

Roll the place-value dice to generate factors for more multiplication problems. Solve the problems using the multiplication and division frame as illustrated in the example. Record all answers in expanded form.

EXTENSIONS Use the multiplication and division frame and base-ten blocks to illustrate multiplication of binomials. Let a flat equal x^2, a rod equal x, and a unit cube equal 1.

Example: $x + 2 =$ ▭ + ▫ ▫

Complete the following:

1. $(x + 2)(x + 3) =$
2. $(x + 4)(x + 5) =$
3. $(x + 1)(x + 7) =$
4. $(x + 6)(x + 8) =$

Activity 11: Multi-digit Division

PURPOSE Develop the division algorithm and reinforce the partitioning model and the place-value concepts associated with division.

MATERIALS Base-ten blocks, three place-value dice (hundreds, tens, and ones), place-value operations board, paper squares approximately 6 in. × 6 in., and paper for recording

GROUPING Work individually or in pairs.

GETTING STARTED Roll the three dice to generate a 3-digit number and construct it on the place-value operations board. Roll the ones die to determine the factor (divisor). Place a number of the paper squares (same number as the divisor) on the desk to illustrate the number of groups needed.

Example: $5\overline{)167}$

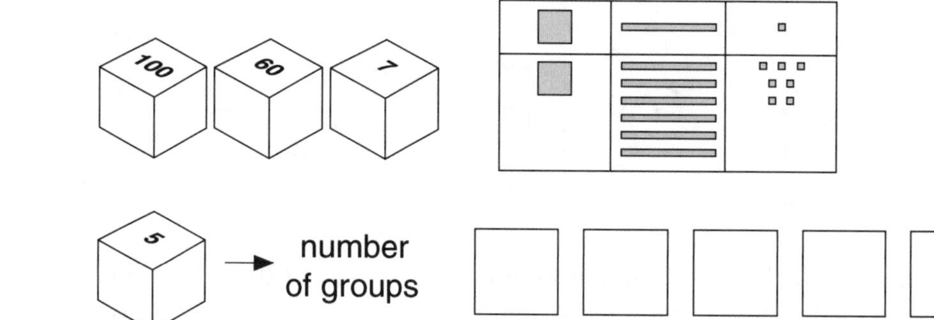

Solution:

a. Place an equal number of the highest place-value blocks (flats) on each square using the largest number of blocks possible. If this is not possible, record a 0 in the hundreds place above the division frame as shown. Make a 1-for-10 trade with the next lower place-value block (rods) and add 10 rods to that column. In some cases, it may be necessary to make more than one trade.

b. There are now 16 rods. Place an equal number of rods on each square using the largest number of rods possible. The rods placed on each square represent what whole number? ________

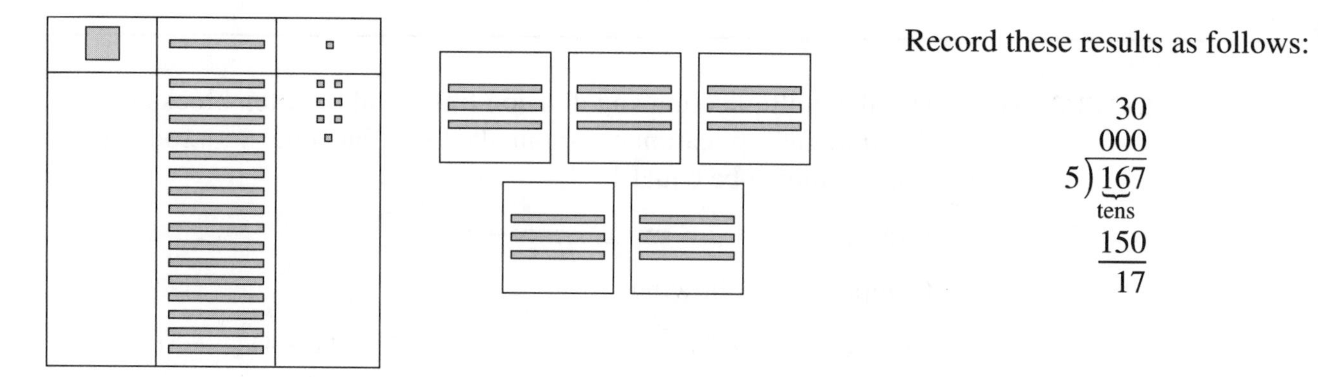

c. Proceed as shown on the previous page. Make 1-for-10 trades and share an equal number of units blocks among the squares.

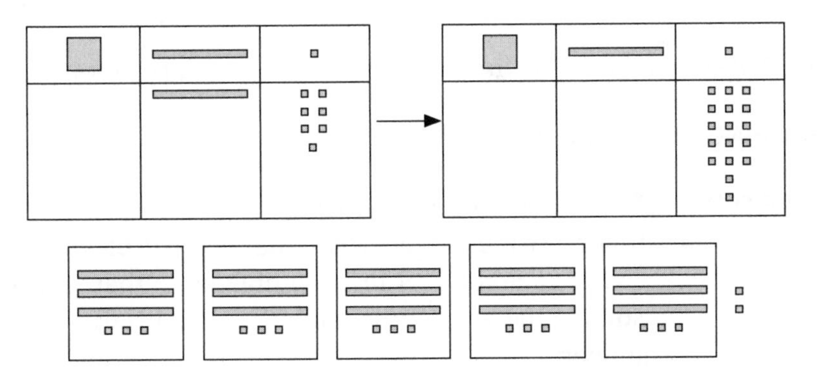

The units blocks placed on each square represent what whole number? ______

Record the results of the sharing as shown:

```
      3
     30
    000
 5)167
    150
     17
   units
     15
      2
```

How many blocks remain? ______ Can they be shared equally among the squares? ______

Record the final results as follows:

```
      3
     30
    000   = 33 remainder 2
 5)167
    150
     17
     15
      2
```

Generate a factor (divisor) and a product (dividend) for six problems using the place-value dice. Use base-ten blocks and the method described above to model multi-digit division. Record the results as illustrated in the example.

Activity 12: Estimate to Calculate

PURPOSE Develop estimation strategies for computation.

GROUPING Work individually.

GETTING STARTED Do **not** perform the following computations. Use an estimation strategy (for example, rounding, compatible numbers, etc.) and rewrite each number in the problems so that you could easily estimate the answer mentally. Explain briefly why you chose each number.

EXPLANATION

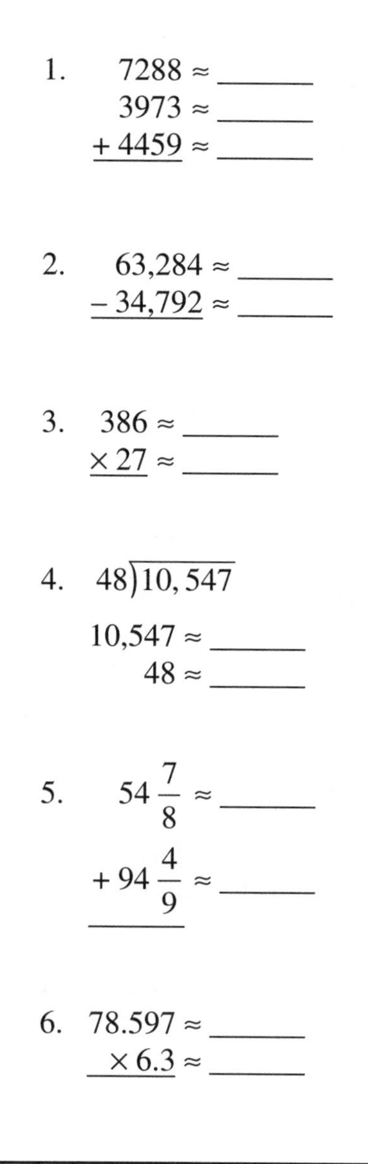

Activity 13: Target Number

PURPOSE Apply the guess and check problem-solving strategy to reinforce the inverse relationship between multiplication and division and to develop estimation skills.

MATERIALS One calculator for each pair of students

GROUPING Work in pairs.

GETTING STARTED Construct several tables like the one shown below. Use only whole numbers and use the calculator to find only the product or quotient as described in each game. Estimated factors must be determined mentally.

TARGET = ______			
Turn	**Factors**	**Product**	**Difference**
1			
2			
3			
		TOTAL	

GAME 1

- Players choose a target number, and each player records it in a table.
- On alternate turns, each player chooses two factors (other than 1 and the target), and records them in the table. Then the player multiplies the factors and records the product.
- If the product equals the target number, the player wins. If not, the player records the absolute value of the difference between the product and the target number.
- If neither player has won after three turns, the person with the smallest total for the three differences is the winner.

GAME 2

- Player 1 chooses a target number. Player 2 chooses a constant factor and an acceptable range for the answer, for example, ± 20.
- On alternate turns, each player chooses another factor and multiplies it by the constant factor to obtain a product.
- If the absolute value of the difference between the product and the target number is less than or equal to the range chosen by the players, the player wins.
- If the absolute value of the difference is greater than the range, the players alternate turns until one player obtains a product within the specified range of the target number.

SAMPLE GAME

Target Product = 650 **Constant Factor** = 19 **Range** = ±8

	Constant Factor		Estimated Factor		Product	| Difference |
Player 1	19	×	37	=	703	53
Player 2	19	×	30	=	570	80
Player 1	19	×	33	=	627	23
Player 2	19	×	34	=	646	4

Player 2 wins!

EXTENSIONS

1. Repeat **Game 1** using division. On each turn, the player chooses a dividend and a divisor and divides them. If the quotient equals the target number, the player wins. If not, the player records the difference between the quotient (rounded to the nearest whole number) and the target number in the table.

2. Repeat **Game 2** using division. Player 1 chooses the target quotient. Player 2 chooses the constant divisor and an acceptable range for the answer.

SAMPLE GAME 2

Target Quotient = 26 **Constant Divisor** = 14 **Range** = ±1

	Estimated Dividend		Constant Divisor		Quotient	| Difference |
Player 1	320	÷	14	≈	22.8	3.2
Player 2	380	÷	14	≈	27.1	1.1
Player 1	360	÷	14	≈	25.7	0.3

Player 1 wins!

Activity 14: Largest and Smallest

PURPOSE Use the guess and check problem-solving strategy to develop number and operation sense and to reinforce the concept of place value.

MATERIALS A calculator and paper for recording

GROUPING Work individually.

1. Arrange the digits 1, 2, 3, 4, and 5 to make a two-digit number and a three-digit number. Each digit may be used only once. Use a calculator to multiply the numbers. Try several different arrangements of the digits to determine the arrangement that results in the **largest** product.

2. Repeat the problem and arrange the digits so that you obtain the **smallest** product.

3. Analyze the results and determine a rule so that given any five digits $a < b < c < d < e$, $a \neq 0$, you can form a two-digit number and a three-digit number, multiply them, and guarantee that you will obtain the **largest** or the **smallest** product. Explain the reasoning for your answer on the basis of place-value concepts and the partial products obtained.

EXTENSIONS Write a rule for arranging seven digits into a four-digit number and a three-digit number to obtain the **largest** and the **smallest** product. Does the rule that you determined for the arrangement of five or seven digits also work for six digits or eight digits? Explain.

Chapter Summary

"Number sense refers to an intuitive feeling for numbers and their uses and interpretations; an appreciation for various levels of accuracy when figuring; the ability to determine arithmetical errors; and a common sense approach to using numbers.

—*Developing Number Sense*
Addenda Series, Grades 5–8

The development of number sense, operation sense, and a basic understanding of algorithms for operations is a focal point of the mathematics curriculum described in the *Curriculum and Evaluation Standards for School Mathematics.*

"The greatest revisions to be made in the teaching of computation include the following: fostering a solid understanding of, and proficiency with, simple calculations; abandoning the teaching of tedious calculations using paper-pencil algorithms in favor of exploring more mathematics; fostering the use of a wide variety of computation and estimation techniques suited to different mathematics settings; . . . and providing students ways to check the reasonableness of computations. . ."

—*Curriculum and Evaluation Standards for School Mathematics*

Activities 1–4 are examples of a developmentally appropriate curriculum design that proceeds from the concrete (beansticks) to the representational (pictures) to the abstract (numeral). For example, in Activity 2 you constructed two- and three-digit numbers; the beansticks provided the concrete model; representing the sticks or base-ten blocks on paper involved the pictorial representation of the number; and finally, the number itself was represented abstractly as a numeral.

Activities 5 and 6 develop the algorithms for addition and subtraction of whole numbers. Numbers were modeled on the place-value operations board; 10-for-1 or 1-for-10 trading demonstrated the regrouping necessary in the base-ten system for addition and subtraction, respectively. Recording each step of the activity on the Student Record Sheet helped develop a clear understanding of the addition and subtraction algorithm as well as the associated place-value concepts.

The *comparison* model for subtraction was emphasized, since it is the best way to explain the concepts of "how many more (greater) than?" and "how many less (fewer) than?" The model also illustrates the concept of subtraction as the inverse of addition. When the blocks representing the number to be subtracted are placed on top of the blocks of the larger number (see the example in Activity 6), the remaining blocks represent the number that must be added to the smaller number to obtain the larger. This concrete comparison of the two numbers promoted an understanding of the quantity of one number in relation to another.

Activities 7–11 developed the algorithms for multiplication and division and also illustrated the concept of inverse operations. The two words *factor* and *product* were stressed in both of these operations rather than the traditional *multiplier*, *multiplicand*, *dividend*, *divisor*, and *quotient*. This focus helped to reinforce the inverse relationship between the operations.

The final two activities reinforced all the principles of the earlier ones in a problem-solving setting using the calculator. The estimation and mental arithmetic strategies used in Activity 12 are important features of the Target Number activity. Investigation of the partial products obtained from multiplying the various arrangement of the digits in Activity 14 developed a deeper understanding of place-value concepts and lead to the correct arrangement of digits needed to obtain the largest or smallest product.

The inverse relationship between the four basic operations and the connections between multiplication as repeated addition and division as repeated subtraction are illustrated in the diagrams below.

	Once	Many
Join	+	×
Separate	−	÷

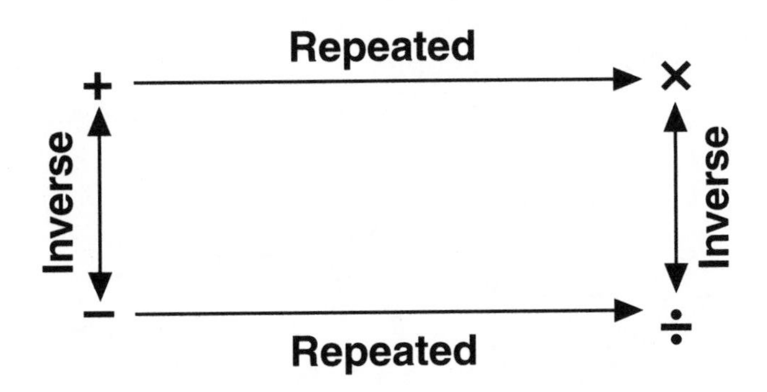

Chapter 4
Integers and Number Theory

"Number theory offers many rich opportunities for explorations that are interesting, enjoyable, and useful. These explorations have payoffs in problem solving, in understanding and developing other mathematical concepts, in illustrating the beauty of mathematics, and in understanding the human aspects of the historical development of number.

As they begin to understand the meaning of operations and develop a concrete basis for validating symbolic processes and situations, students should design their own algorithms and discuss, compare, and evaluate them with their peers and teacher. Students should analyze the way the various algorithms work and how they relate to the meaning of the operation and to the numbers involved."

—*Curriculum and Evaluation Standards for School Mathematics*

Many everyday situations cannot be adequately described without the use of both positive and negative numbers. Profit and loss, temperatures above and below 0°F, elevations above and below sea level, and deposits and withdrawals are just a few examples. This chapter introduces negative numbers by extending your knowledge of whole numbers to the set of integers.

In Activities 1 and 3, ● represents a proton, and ○ represents an electron. Protons and electrons are subatomic particles. Protons have a positive electrical charge of one unit, and electrons have a negative electrical charge of one unit. Because protons and electrons have opposite charges, when a proton and an electron are paired together, they neutralize each other; that is, the pair has an electrical charge of zero. You will use concrete representations for integers, like the charged-particle model, and your understanding of the operations with whole numbers to develop algorithms for the integer operations.

Number theory is primarily concerned with the study of the properties of the whole numbers. Number theory topics, such as multiples, factors, prime numbers, prime factorizations, least common multiples, and greatest common factors, are an integral part of the elementary mathematics curriculum. These concepts are used extensively when working with rational numbers. In this chapter, you will explore topics from number theory using concrete models and apply the concepts in problem-solving situations.

Activity 1: Charged Particles

PURPOSE Investigate a concrete model for integers.

MATERIALS Two different colored chips (or squares), 15 of each color

GROUPING Work individually or in groups of 2–3.

GETTING STARTED Use colored chips to represent protons and electrons and construct two different models that represent each integer. Sketch your models in the following boxes.

Examples:

The set at the right represents the number $^{+}2$.

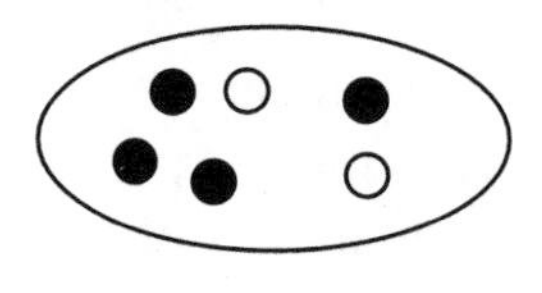

If the protons and electrons are paired, 2 protons are left over. The net electrical charge is $^{+}2$.

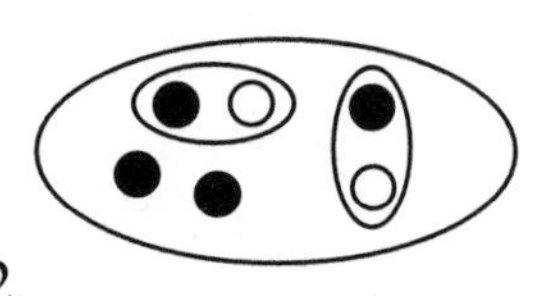

The set at the right represents the number $^{-}3$.

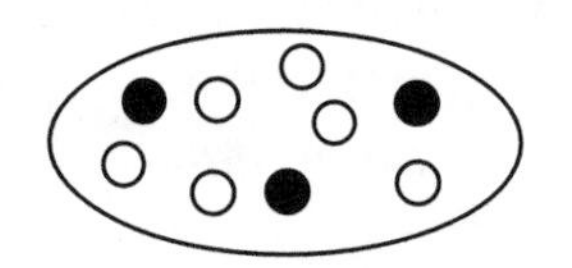

If the protons and electrons are paired, 3 electrons are left over. The net electrical charge is $^{-}3$.

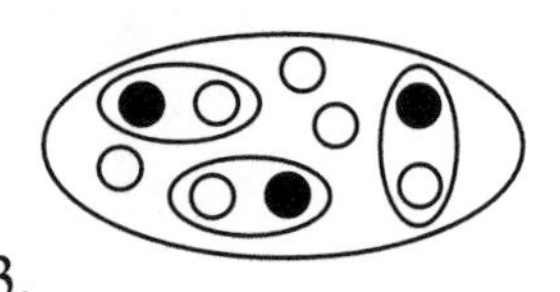

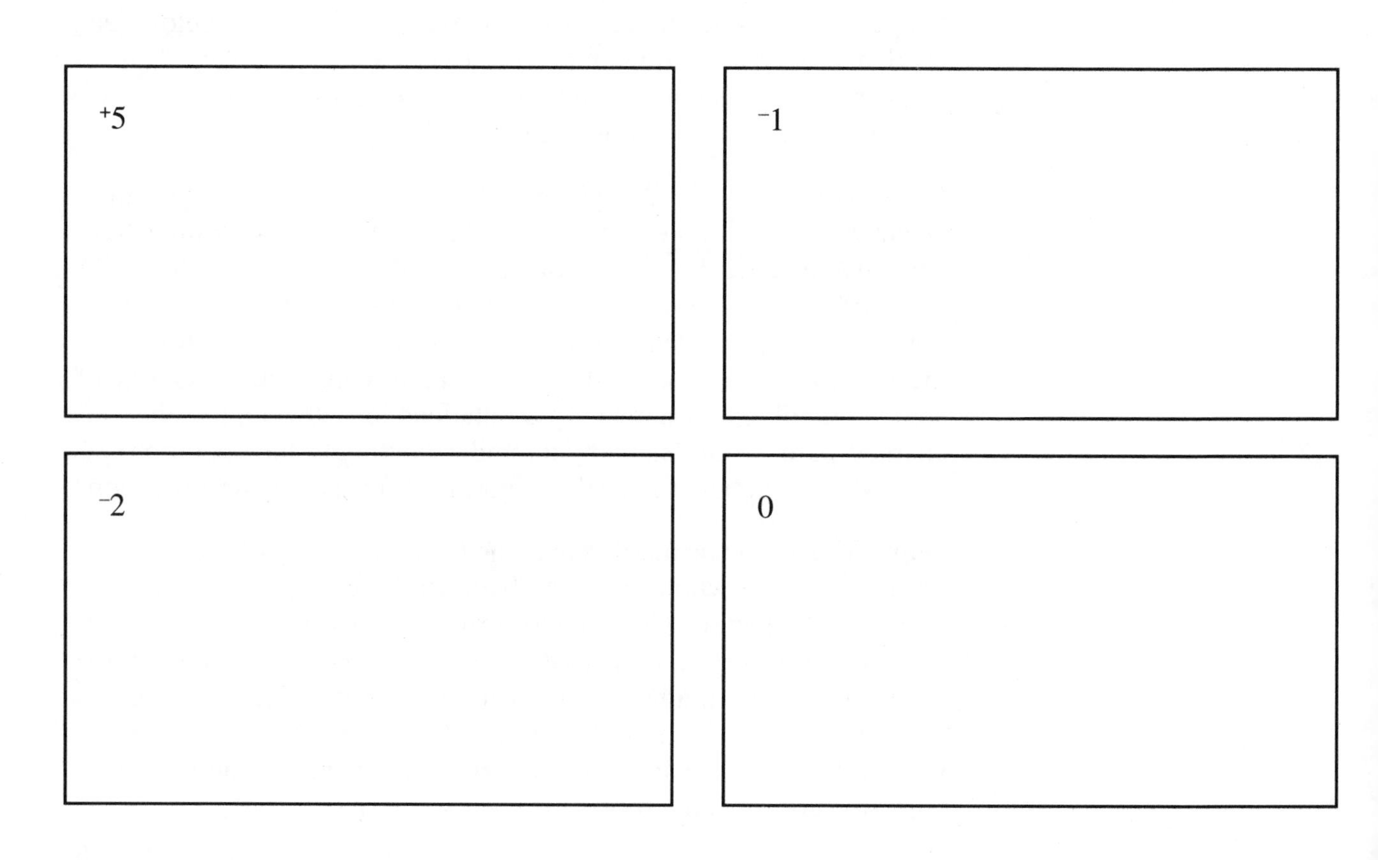

Activity 2: Coin Counters

PURPOSE Investigate a concrete model for integers and use it to discover algorithms for integer addition and subtraction.

MATERIALS A paper cup, 10 pennies, and a game marker

GROUPING Work in groups of 2–3.

GETTING STARTED

- At the beginning of game, each player places a game marker on zero on a number line like the one below.
- Players alternate turns.
- On your turn, place 6 pennies in the cup, cover the opening with your hand, shake the cup thoroughly, and drop the coins onto the table. Each HEAD means you move your marker to the right one unit; each TAIL means you move it to the left one unit.
- The first player to go past $^{+}10$ or $^{-}10$ is the winner. If there is no winner after ten turns, the player closest to $^{+}10$ or $^{-}10$ wins.

$^{-}10$ $^{-}9$ $^{-}8$ $^{-}7$ $^{-}6$ $^{-}5$ $^{-}4$ $^{-}3$ $^{-}2$ $^{-}1$ 0 $^{+}1$ $^{+}2$ $^{+}3$ $^{+}4$ $^{+}5$ $^{+}6$ $^{+}7$ $^{+}8$ $^{+}9$ $^{+}10$

Play the game twice. When you have finished, answer the following questions.

1. Did your group find a way to quickly determine where to place your marker after a coin toss? Explain.
2. If you were to represent the number of HEADS with an integer, would you use a positive or a negative integer?
3. If you were to represent the number of TAILS with an integer, would you use a positive or a negative integer?
4. Did your marker ever end up an odd number of units away from where it was at the start of your turn? Explain.
5. Did your marker ever end a turn in the same place it started the turn? Explain.
6. Use coins to construct two different representations for each integer. You may use more or less than 6 coins in a model.

   a. $^{+}4$ b. $^{-}3$ c. 0

You have seen how coins can be used to represent integers. Coins can also be used to model addition of integers. Think of the HEADS as a positive integer and the TAILS as a negative integer. For example, tossing 2 HEADS and 4 TAILS is the same as adding $^{+}2 + {}^{-}4$.

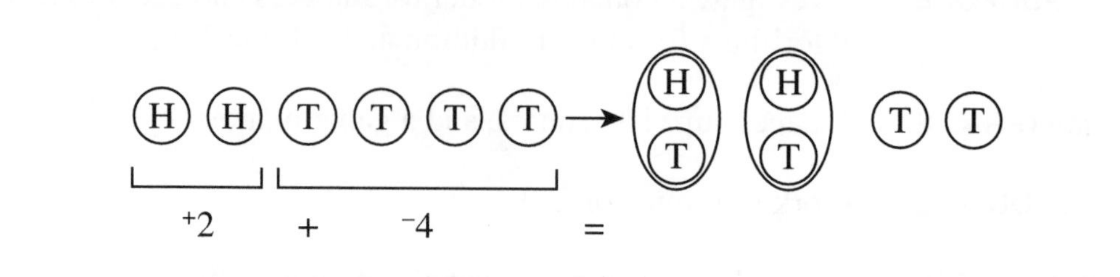

1. a. Why do the paired coins cancel each other out?

   b. If you tossed this combination of coins, how would you move your marker?

   c. What integer is represented by the combination of coins?

   d. Complete the equation: $^{+}2 + {}^{-}4 =$ ______.

2. Use coins to find the following sums. Make a sketch of your work. You may use more than 6 coins.

   a. $^{+}1 + {}^{-}5$ b. $^{+}6 + {}^{-}4$ c. $^{+}3 + {}^{-}3$ d. $^{-}5 + {}^{-}2$

Use the coin model to answer the following questions.

3. a. Is the sum of two negative numbers positive or negative?

   b. How can you determine the sum of two negative numbers without using coins?

4. When is the sum of a positive and a negative number

   a. equal to 0? b. positive? c. negative?

5. How can you determine the sum of a positive and a negative number without using coins?

6. Use your rules from Exercises 3 and 5 to compute the following:

   a. $^{-}17 + {}^{+}25$ b. $^{+}13 + {}^{-}7$ c. $^{-}36 + {}^{-}19$ d. $^{-}11 + {}^{+}11$

Activity 3: Subtracting Integers

PURPOSE Use the charged-particle model to develop a rule for subtracting integers.

MATERIALS Two different colored chips (or squares), 15 of each color

GROUPING Work individually or in groups of 2–3.

GETTING STARTED Use the colored chips to represent protons and electrons.

The following examples illustrate subtraction of integers using the charged-particle model.

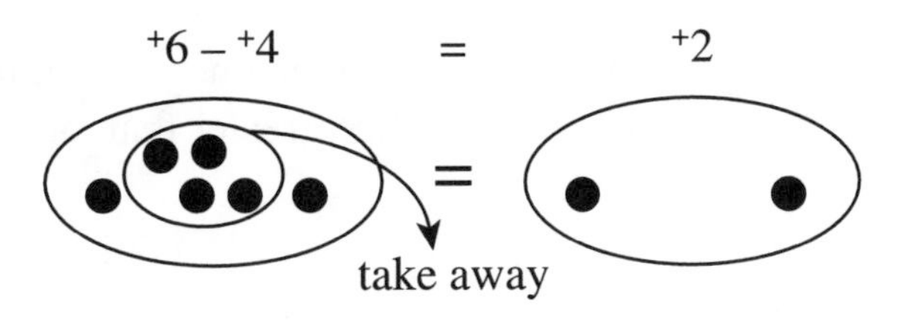

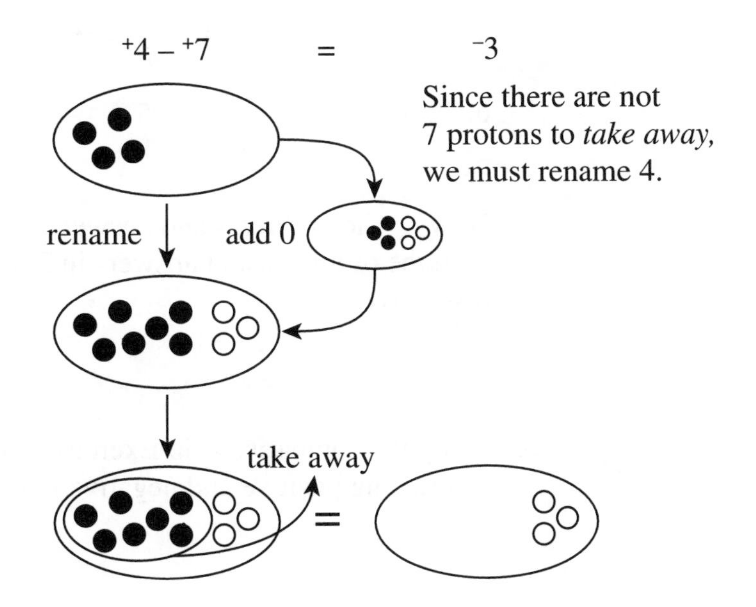

1. How could you rename $^{-}2$ to compute the difference $^{-}2 - ^{+}5$ using the charged-particle model?

2. Use the charged-particle model and colored chips to compute the following differences. Make a drawing to illustrate what you did in each problem.

   a. $^{+}5 - {}^{+}9$ b. $^{+}3 - {}^{-}4$ c. $^{-}2 - {}^{+}5$

   d. $^{-}6 - {}^{-}5$ e. $^{-}5 - {}^{+}3$ f. $^{-}4 - {}^{-}8$

3. Use the results from Exercise 2 in answering the following questions.

   a. When you subtract a positive integer from another integer, is the difference greater than or less than the original integer?

   b. When you subtract a negative integer from another integer, is the difference greater than or less than the original integer?

4. Determine the following:

   a. $^{+}5 + {}^{-}9$ b. $^{+}3 + {}^{+}4$ c. $^{-}2 + {}^{-}5$

   d. $^{-}6 + {}^{+}5$ e. $^{-}5 + {}^{-}3$ f. $^{-}4 + {}^{+}8$

5. How do the problems and answers in Exercise 4 (a)–(f) compare with the problems and answers in Exercise 2 (a)–(f), respectively?

6. Study the comparisons in Exercise 5 to help write a rule for subtracting positive and negative integers.

7. Use your rule from Exercise 6 to determine the following:

   a. $^{-}17 - {}^{-}25$ b. $^{+}13 - {}^{-}7$

   c. $^{-}36 - {}^{-}19$ d. $^{-}11 - {}^{+}11$

Activity 4: Addition Patterns

PURPOSE Develop a rule for addition of integers by exploring patterns.

GROUPING Work individually or in pairs.

GETTING STARTED Fill in the missing entries in each list of problems and observe the patterns. Use the completed lists to answer the questions.

4	+	3	=	7
4	+	2	=	6
4	+	1	=	_____
4	+	0	=	_____
4	+	$^{-}1$	=	_____
4	+	_____	=	_____
4	+	_____	=	_____
4	+	_____	=	_____
4	+	_____	=	_____

1. When is the sum of a positive number and a negative number

   a. positive?

   b. zero?

   c. negative?

2. Write a rule for adding a positive number and a negative number.

$^{-}4$	+	2	=	$^{-}2$
$^{-}4$	+	1	=	$^{-}3$
$^{-}4$	+	0	=	_____
$^{-}4$	+	_____	=	_____
$^{-}4$	+	_____	=	_____
$^{-}4$	+	_____	=	_____
$^{-}4$	+	_____	=	_____

Observe the pattern in the set of problems at the left and write a rule for the addition of two negative numbers.

EXTENSIONS The following is a problem situation that illustrates addition of two positive integers. "Annie's allowance is $5.00 more than Michael's. Michael's allowance is $10.00 per week. What is Annie's allowance?" Write problem situations that illustrate the addition of (1) a positive and a negative integer and (2) two negative integers. Write two problems for each case.

Activity 5: Subtraction Patterns

PURPOSE Develop a rule for subtraction of integers by exploring patterns.

GROUPING Work individually or in pairs.

GETTING STARTED In each of the following sets of problems, observe the patterns of the numbers and fill in the missing entries.

1.							2.						
	4	–	0	=	4	______		3	–	4	=	$^{-}1$	______
	4	–	1	=	3	______		2	–	4	=	$^{-}2$	______
	4	–	2	=	____	______		1	–	4	=	____	______
	4	–	3	=	____	______		0	–	4	=	____	______
	4	–	____	=	____	______		____	–	4	=	____	______
	4	–	____	=	____	______		____	–	____	=	____	______
	4	–	____	=	____	______		____	–	____	=	____	______
	____	–	____	=	____	______		____	–	____	=	____	______

3.							4.						
	4	–	3	=	1	______		$^{-}4$	–	3	=	$^{-}7$	______
	4	–	2	=	2	______		$^{-}4$	–	2	=	____	______
	4	–	1	=	____	______		$^{-}4$	–	1	=	____	______
	4	–	____	=	____	______		$^{-}4$	–	____	=	____	______
	4	–	____	=	____	______		____	–	____	=	____	______
	4	–	____	=	____	______		____	–	____	=	____	______
	____	–	____	=	____	______		____	–	____	=	____	______
	____	–	____	=	____	______		____	–	____	=	____	______

5. For each of the subtraction problems above, write a related addition problem using numbers that have the same absolute value as those in the given problem.

   Examples: $4 - 5 = 4 + {}^{-}5$ $\quad$ ${}^{-}4 - 1 = {}^{-}4 + {}^{-}1$

6. Write a rule for the subtraction of integers.

EXTENSIONS Write problem situations that illustrate (1) subtraction of a negative integer from a positive integer and (2) subtraction of a negative integer from a negative integer. Write two problems for each case.

Activity 6: Multiplication and Division Patterns

PURPOSE Develop rules for multiplication and division of integers by exploring patterns.

GROUPING Work individually or in pairs.

GETTING STARTED In each of the following sets of problems, observe the patterns of the numbers and fill in the missing entries.

1.

4	×	3	=	12
4	×	2	=	____
4	×	1	=	____
4	×	____	=	____
4	×	____	=	____
____	×	____	=	____
____	×	____	=	____
____	×	____	=	____

2.

4	×	5	=	20
3	×	5	=	15
____	×	5	=	10
____	×	5	=	____
____	×	5	=	____
____	×	____	=	____
____	×	____	=	____
____	×	____	=	____

3. Write a rule for multiplying a positive number and a negative number.

4.

$^{-}3$	×	2	=	____
$^{-}3$	×	1	=	____
$^{-}3$	×	0	=	____
$^{-}3$	×	____	=	____
$^{-}3$	×	____	=	____
____	×	____	=	____
____	×	____	=	____
____	×	____	=	____

5.

4	×	$^{-}6$	=	____
3	×	$^{-}6$	=	____
____	×	$^{-}6$	=	$^{-}12$
____	×	____	=	$^{-}6$
____	×	____	=	____
____	×	____	=	____
____	×	____	=	____
____	×	____	=	____

6. Write a rule for multiplying two negative numbers.

Recall that multiplication and division are *inverse operations.*

Example: $12 \div 4 = 3$, since $3 \times 4 = 12$.

More generally, $A \div B = C$ means that $C \times B = A$.

$^{-}12 \div {}^{+}4 = ?$ Think: $^{+}4 \times ? = {}^{-}12$ So $? = {}^{-}3$.

Use the inverse relationship between multiplication and division to compute the quotients of several pairs of integers. Use the results to write a rule for division of integers.

Activity 7: A Square Experiment

PURPOSE Develop the concepts of prime, composite, and square numbers using a geometric model.

MATERIALS 30 squares per group

GROUPING Work individually or in groups of 3–4.

GETTING STARTED Use squares or graph paper to form all the rectangular arrays possible with each different number of squares. Record your results in Table 1.

Examples: Number of Squares 1 2 3

Rectangular Arrays

Note: is not a rectangular array.

When an array is described by its dimensions, the figure has an altitude of 1 unit and a base of 2 units and is labeled 1×2. The figure is labeled 2×1.

TABLE 1

Number of Squares	Dimensions of the Rectangular Arrays	Total Number of Arrays
1		
2		
3	1×3, 3×1	2
4		
5		
6	1×6, 6×1, 3×2, 2×3	4
7		
8		
9		
10		
11		
12		

1. Use the results from Table 1 to complete Table 2.

TABLE 2

Number of Squares That Produced:			
A	B	C	D
Only One Array	Only Two Arrays	More Than Two Arrays	An Odd Number of Arrays

2. Suppose you have 24 squares.

   a. How many rectangular arrays can be made?

   b. In which column(s) in Table 2 would you place 24?

   c. What are the factors of 24?

3. a. What are the factors of 16?

   b. How many rectangular arrays can you make with 16 squares?

   c. In which column(s) of Table 2 would you place 16?

4. Look at the data in Table 1 and Table 2. How is the number of factors of a given number related to the number of rectangular arrays?

5. a. Why is it that the numbers in column D of Table 2 produce an odd number of arrays?

   b. What are the next three numbers that would be placed in column D?

6. What is the mathematical name for the numbers in

   a. column B?

   b. column C?

   c. column D?

7. Which numbers can be placed in two lists? Why?

8. Can any numbers be placed in three lists? If so, which ones?

9. Write each of the following composite numbers as a product of primes.

   a. 28   b. 42

   c. 150   d. 231

10. a. Can every composite number be written as a product of primes? Explain your reasoning.

    b. If two people write the same number as a product of primes,

       i. how would their factorizations be alike?

       ii. how might the factorizations be different?

Activity 8: A Sieve of Another Sort

PURPOSE Use a "sieve" to investigate primes, composites, multiples, and prime factorizations.

MATERIALS Orange, red, blue, green, and yellow colored pencils or crayons

GROUPING Work individually.

GETTING STARTED Eratosthenes, a Greek mathematician, invented the "sieve" method for finding primes over 2200 years ago. This activity explores a variation of Eratosthenes' sieve.

As you discovered in Activity 7, *one* is neither prime nor composite. To show this, mark an X through 1.

The first prime number is 2. Color the diamond in which 2 is located **orange**. Use **red** to color the upper-left corner of the key and the upper-left corner of all squares containing multiples of 2. Any number with a corner colored will fall through the sieve.

What was the first multiple of 2 that fell through the sieve? _____

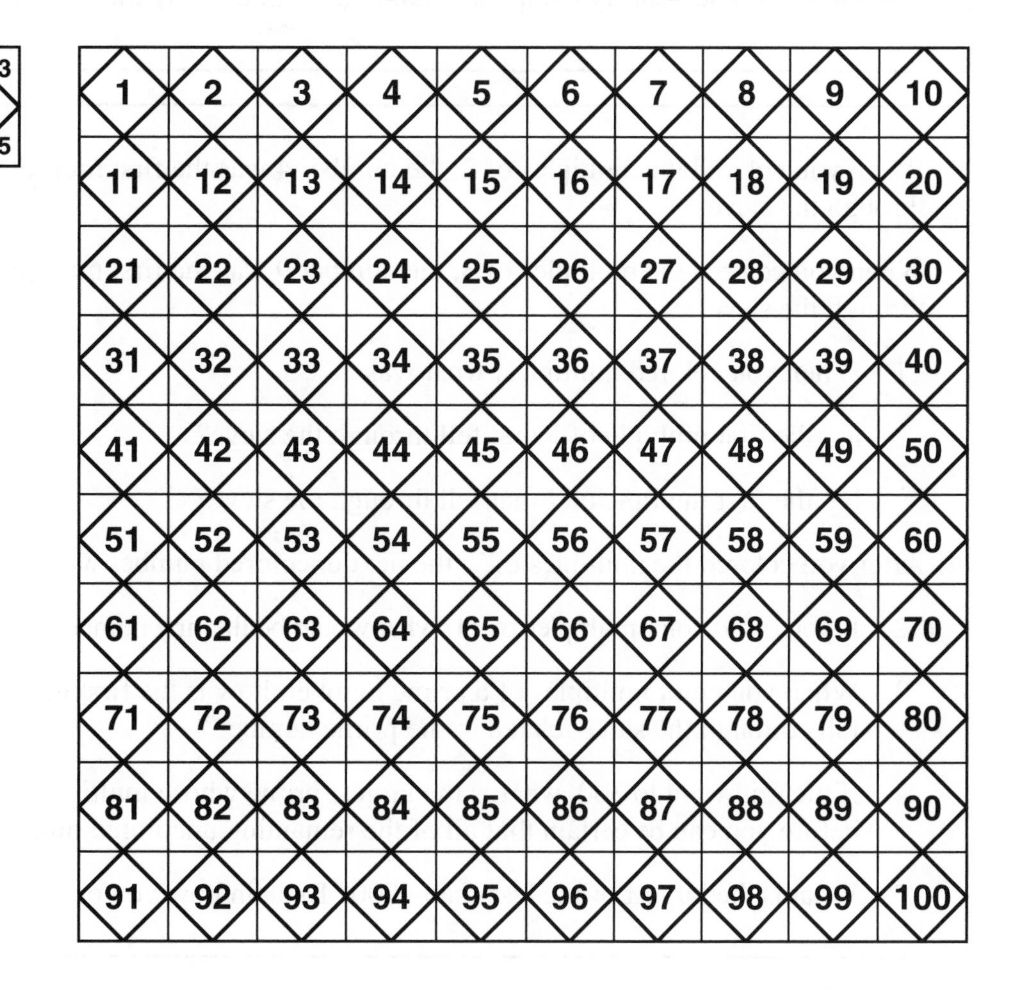

The next uncolored number is 3. Color the diamond surrounding the 3 **orange**. Use **blue** to color the upper-right corner of the key and the upper-right corner of all squares containing multiples of 3.

What was the first multiple of 3 that fell through the sieve? _____

Repeat this process for 5 and 7. Color the diamonds surrounding the numbers **orange**. Use **green** to color the lower-right corners of the key and of all squares containing multiples of 5. Use **yellow** to color the lower-left corners of the key and of all the squares containing multiples of 7. Note the first multiples of 5 and of 7 that fell through the sieve.

Finally, use **orange** to color the diamond surrounding all the numbers in the grid that are in squares with no corners colored. These numbers are all primes.

1. How do you know that 2, 3, 5, and 7 are prime numbers?
2. How can you tell that 2, 3, 5, and 7 are prime numbers from the way the sieve is colored?
3. How can you identify composite numbers from the way the sieve is colored?

When you colored the multiples of 2, the number 4 was the first multiple of 2 that fell through the sieve.

When you colored the multiples of 3, the number 9 was the first multiple of 3 that fell through the sieve.

1. When you colored multiples, what was
   a. the first multiple of 5 that fell through the sieve?
   b. the first multiple of 7 that fell through the sieve?

After you colored the multiples of 7, the next uncolored number was 11.

2. If you could color multiples of 11, what would be the first number to fall through the sieve?
3. When you color multiples of a prime number, how is the first multiple of the prime that falls through the sieve related to the prime number?
4. If the grid went to 300, what is the largest prime whose multiples must be colored before you can be certain that all of the remaining uncolored numbers are prime?
5. What is the largest prime less than 1000? Explain how you obtained your answer.

The sieve can be used for more than finding primes.

1. List the numbers that are colored with the code for 2 and for 3.

2. The numbers in Exercise 1 are multiples of 2 and 3 at the same time. What numbers are they?

3. How could you use the color code to find

   a. the multiples of 14?

   b. the multiples of 30?

The sieve can also help you find the prime factorization of a number.

Example: From the sieve, you find that the prime factors of 72 are 2 and 3. $72 = 2 \times 3 \times 12 \leftarrow$ 12 is not prime

2 is prime
↓

Again from the sieve, the prime factors of 12 are 2 and 3.

$$72 = 2 \times 3 \times (2 \times 3 \times 2)$$
$$= 2 \times 2 \times 2 \times (3 \times 3)$$

Since 2 is a prime, you are done. $72 = 2^3 \times 3^2$

1. Find the prime factorization of

   a. 54 b. 84 c. 100

1. Pairs of prime numbers like 3 and 5 that differ by 2 are called *twin primes*. List all the twin primes less than 100.

2. What is the longest string of consecutive composite numbers on the grid?

3. Several of the numbers on the grid are divisible by three different primes. What is the smallest number that is divisible by four different primes?

Activity 9: Interesting Numbers

PURPOSE Review and apply the concepts of prime number, odd, even, perfect cube, and perfect square.

GROUPING Work individually.

GETTING STARTED The lyrics of an old song assert that "One is the loneliest number." This makes 1 interesting, but all numbers have some property that makes them interesting.

121

I'm a perfect square, since I'm equal to 11×11, but there is something else interesting about me. What is it?

1. Find three more perfect squares.

A *palindrome* is a number that is unchanged when its digits are reversed.

2. Find three more palindromes.

64

I'm equal to $4 \times 4 \times 4$. This means I'm a perfect cube, but I'm also a ________________. This makes me doubly interesting!

1. Find three more perfect cubes.
2. What is the smallest multi-digit number that is both a perfect cube and a palindrome?

13

I'm prime, since my only factors are 1 and 13, but I have another property that not all primes have. What is it? **HINT:** Reverse my digits.

1. How many palindromes between 100 and 200 are primes?
2. Find three prime numbers such that when their digits are reversed, the result is also a prime number.

Use the rating scale at the right to investigate some interesting three-digit numbers.

Example: 169

Perfect Square (169 = 13 × 13)	7 points
Sum of Digits Greater Than 14 (1 + 6 + 9 = 16)	4 points
Odd Number	2 points
Three Factors (1, 13, and 169)	3 points
Interest Rating	16 points

INTEREST RATING	
Prime Number	15 points
Perfect Cube	10 points
Perfect Square	7 points
Sum of Digits Greater Than 14	4 points
Even Number	3 points
Odd Number	2 points
Each Factor	1 point

1. Choose any three-digit number. Find its interest rating using the scale.

   Number _______
   Interest Rating _______

2. Can you find a three-digit number with an interest rating greater than 30 points? Explain.

3. What is the greatest possible interest rating for a three-digit prime number? Explain.

4. a. Explain why a number that is greater than 1 and both a perfect cube and a perfect square would have an interest rating of at least 24 points.

   b. Are there any three-digit numbers that are both perfect cubes and perfect squares?

   c. If so, what are their interest ratings?

5. Try to find the three-digit number with the highest interest rating.

Activity 10: Arrays Anyone?

PURPOSE Use geometric models to develop the concepts of greatest common factor and least common multiple.

MATERIALS Two different colored squares, 30 of each color

GROUPING Work individually or in groups of 3–4.

GETTING STARTED Arrange the given numbers of squares (we'll call them red squares and blue squares) into a rectangular array. The array must satisfy the following conditions:

a. Each column of the array may contain only red or only blue squares.

b. Each column must contain the same number of squares.

The goal is to find the array that has the greatest possible number of squares in each column.

Example: 6 red squares
9 blue squares

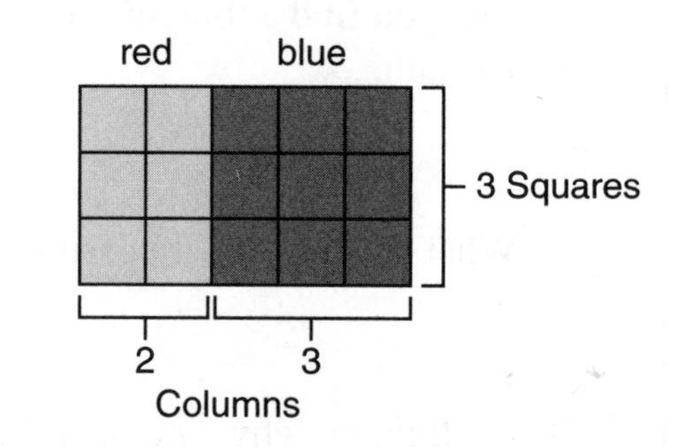

Number of Red Squares	Number of Blue Squares	Number of Columns of Red Squares	Number of Columns of Blue Squares	Number of Squares in Each Column
6	9	2	3	3
6	8			
16	24			
16	9			
24	30			

1. a. List the factors of 16.

   b. List the factors of 24.

   c. List the common factors of 16 and 24.

   d. What is the *greatest common factor* (GCF) of 16 and 24?

2. a. List the factors of 30.

   b. List the common factors of 24 and 30.

   c. What is the GCF of 24 and 30?

3. How do the *greatest common factors* found in Exercises 1 and 2 compare to the **Number of Squares in Each Column** of the corresponding array?

---

1. a. List the first ten multiples of 16.

   b. List the first ten multiples of 24.

   c. List the common multiples of 16 and 24.

   d. What is the *least common multiple* (LCM) of 16 and 24?

2. a. List the first ten multiples of 30.

   b. List the common multiples of 24 and 30.

   c. What is the LCM of 24 and 30?

3. How do the *least common multiples* found in Exercises 1 and 2 compare to the product of the **Number of Columns of Red Squares, Number of Columns of Blue Squares,** and the **Number of Squares in Each Column** of the corresponding array?

Activity 11: GCF and LCM Revisited

PURPOSE Develop algorithms for calculating greatest common factors and least common multiples by extending the array model to Venn diagrams.

GROUPING Work individually or in groups of 3–4.

GREATEST COMMON FACTORS

Example: Find the GCF of 24 and 30.

Prime Factorizations: $24 = 2^3 \times 3$ $30 = 2 \times 3 \times 5$

Array Model

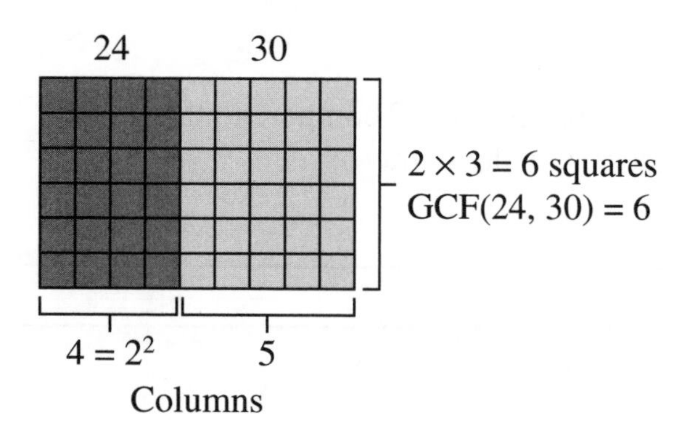

Venn Diagram

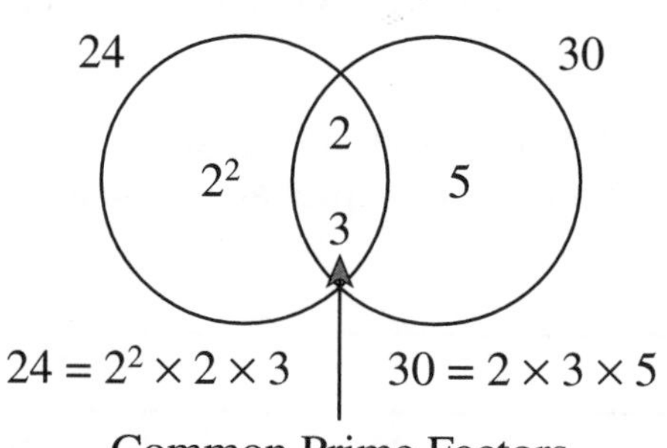

$24 = 2^2 \times 2 \times 3$ $30 = 2 \times 3 \times 5$

Common Prime Factors

GCF(24, 30) = $2 \times 3 = 6$

Find the prime factorizations of each number. Then place the prime factors in the appropriate part of the Venn diagram and find the GCF of the numbers.

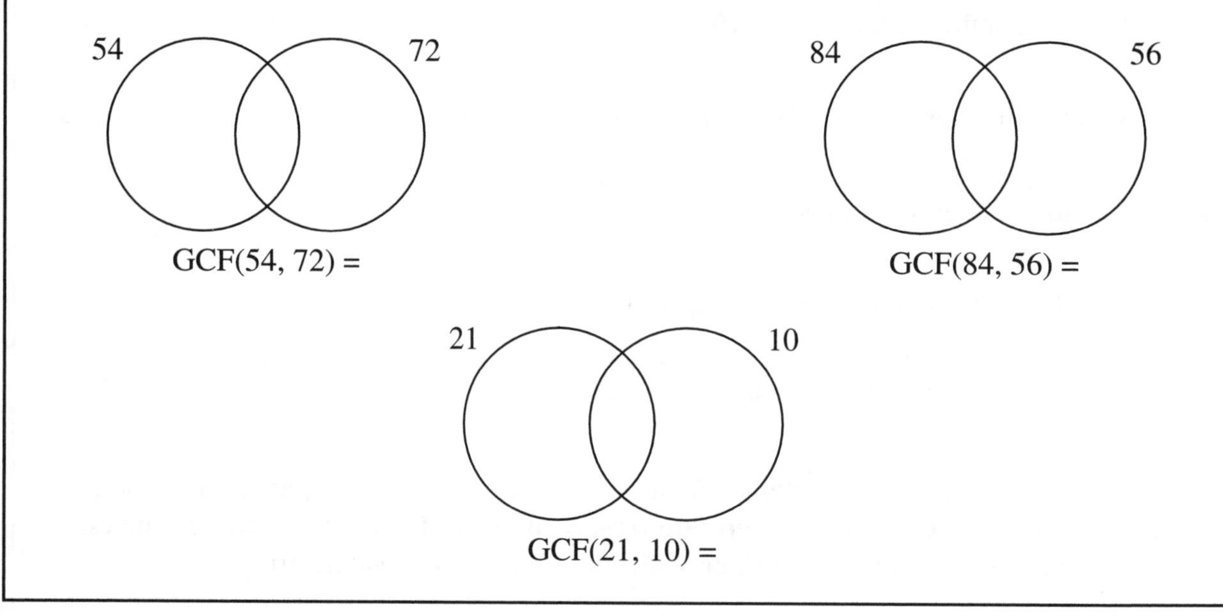

LEAST COMMON MULTIPLES

Example: Find the LCM of 24 and 30.

Prime Factorizations: $24 = 2^3 \times 3$ $30 = 2 \times 3 \times 5$

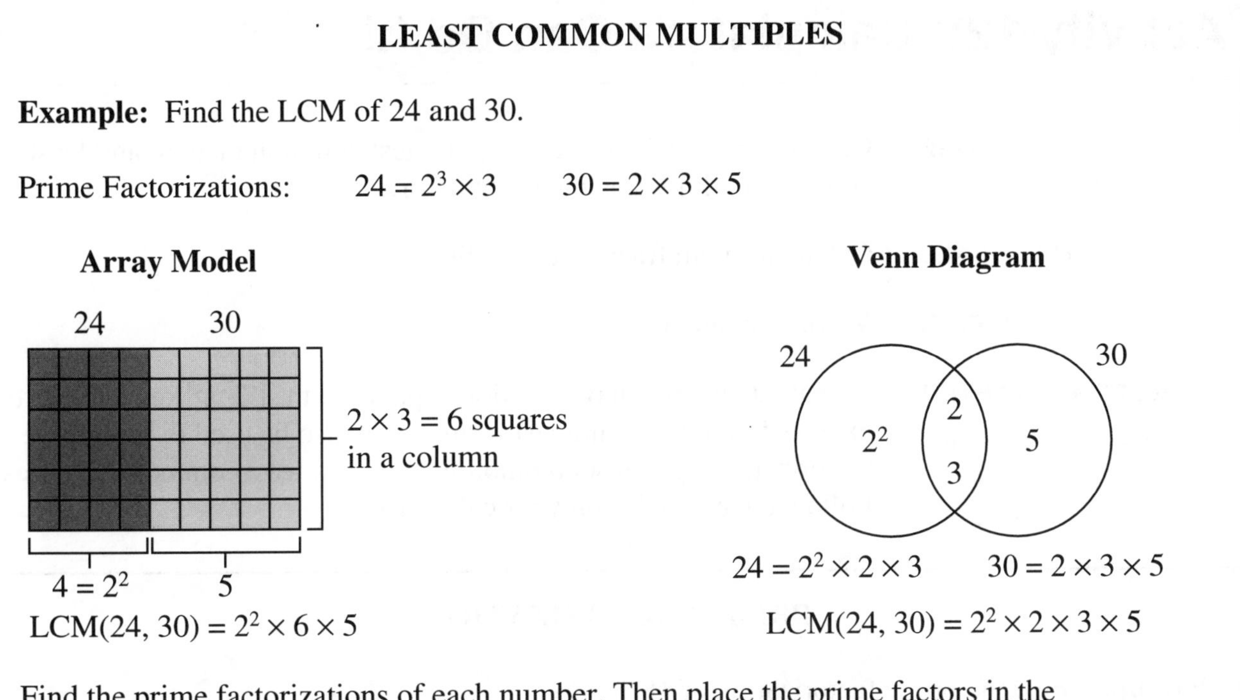

Find the prime factorizations of each number. Then place the prime factors in the appropriate part of the Venn diagram and find the LCM of the numbers.

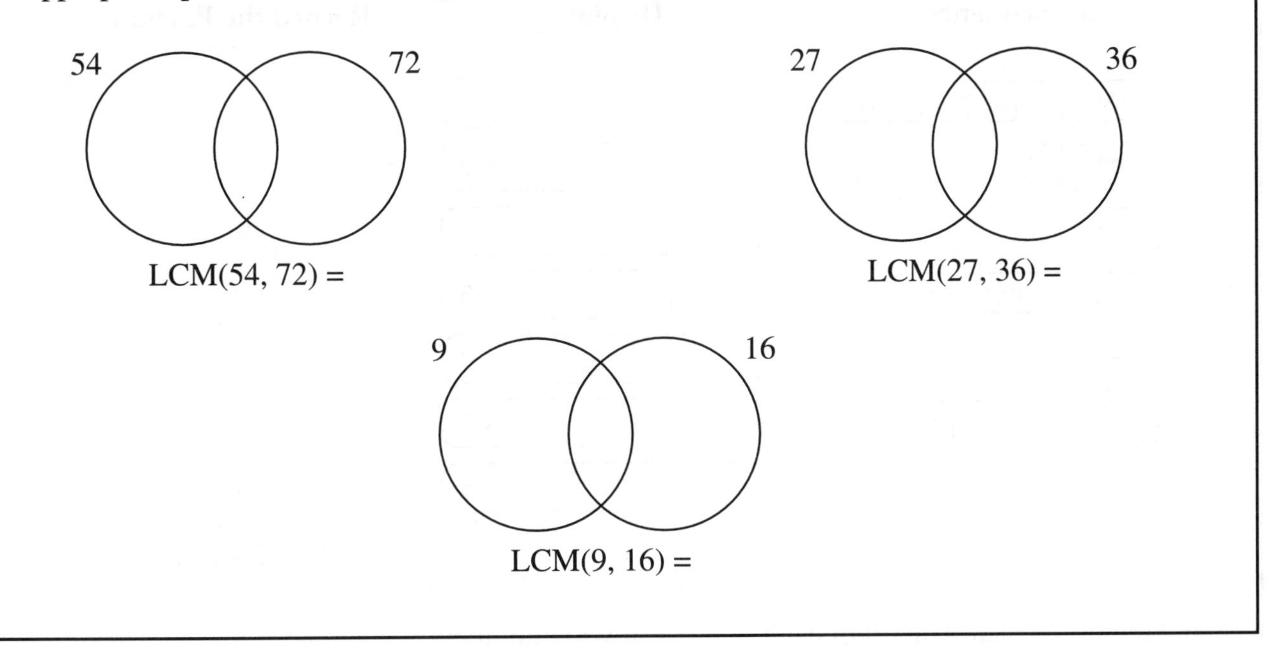

EXTENSIONS

1. a. Find the following products.

   LCM(24, 30) × GCF(24, 30) LCM(54, 72) × GCF(54, 72)

   b. What can you conclude about the product of the LCM and GCF of two numbers?

2. Use Venn diagrams to find the GCF and LCM of 18, 45, and 60.

Activity 12: Calculators Can Do It!

PURPOSE Calculate prime factorizations, greatest common factors, and least common multiples using a calculator.

MATERIALS Calculator with fraction capability

GROUPING Work individually.

GETTING STARTED On calculators that have fraction capability, the **Simp** key is used to express fractions in simplest form. It can also be used to find prime factorizations, greatest common factors, and least common multiples. Follow the examples on your calculator.

PRIME FACTORIZATION

Example: Find the prime factorization of 18.

Key Sequence	Display	Record the Factors
ON/AC 1 8 / 1 8	18/18	
Simp =	N/D → n/d 9/9	
x ↔ y	2	2
x ↔ y Simp =	N/D → n/d 3/3	
x ↔ y	3	3
x ↔ y Simp =	1/1	
x ↔ y	3	3

$18 = 2 \times 3 \times 3$

Find the prime factorization of each of the following numbers.

60 102 210 924

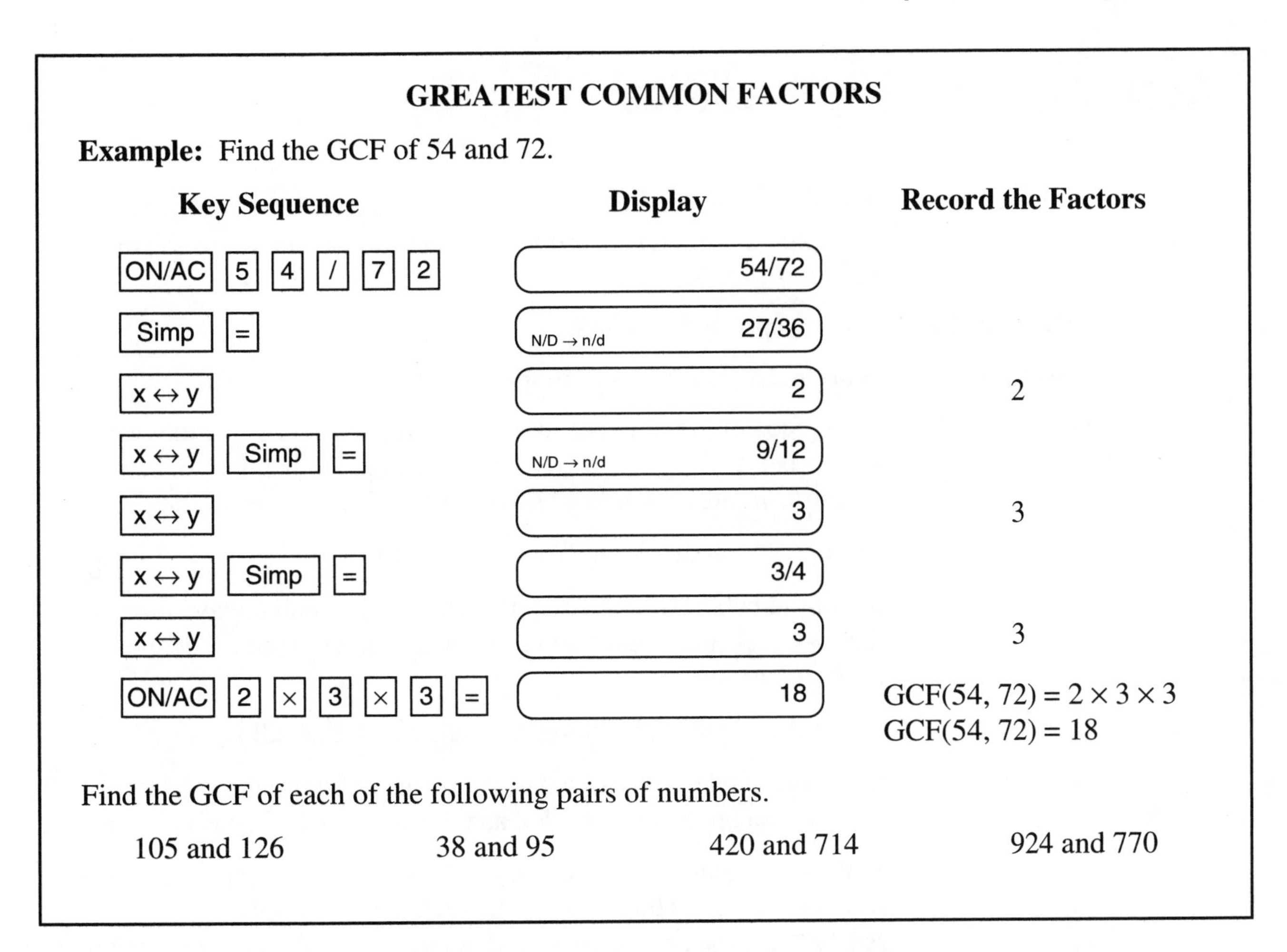

GREATEST COMMON FACTORS

Example: Find the GCF of 54 and 72.

Key Sequence	Display	Record the Factors
ON/AC 5 4 / 7 2	54/72	
Simp =	N/D → n/d 27/36	
x ↔ y	2	2
x ↔ y Simp =	N/D → n/d 9/12	
x ↔ y	3	3
x ↔ y Simp =	3/4	
x ↔ y	3	3
ON/AC 2 × 3 × 3 =	18	GCF(54, 72) = 2 × 3 × 3 GCF(54, 72) = 18

Find the GCF of each of the following pairs of numbers.

105 and 126 38 and 95 420 and 714 924 and 770

LEAST COMMON MULTIPLES

Example: Find the LCM of 54 and 72.

In Activity 11, you discovered that for any two numbers a and b,

$$\text{LCM}(a, b) \times \text{GCF}(a, b) = a \times b.$$

Or, $\text{LCM}(a, b) = a \times b \div \text{GCF}(a, b)$.

GCF(54, 72) = 18

$$\text{LCM}(54, 72) = \underset{(a \times b)}{54 \times 72} \div \underset{\text{GCF}(54,\,72)}{18}$$
$$= 216$$

Find the LCM of each of the following pairs of numbers.

105 and 126 38 and 95 420 and 714 924 and 770

Activity 13: Pool Factors

PURPOSE Apply the concepts of greatest common factor, least common multiple, and relatively prime numbers in a geometric problem situation.

MATERIALS Graph paper and straightedge

GROUPING Work individually or in groups of 3–4.

GETTING STARTED On a piece of graph paper, draw several pool tables like the one shown below, but with different dimensions. Label the pockets *A, B, C,* and *D* in order, starting with the lower-left pocket as shown.

Place a ball on the dot in front of pocket *A*.

Shoot the ball as indicated by the arrows. The ball always travels on the diagonals of the grid and rebounds at an angle of 45 degrees when it hits a cushion.

Count the number of squares through which the ball travels.

Count the number of *hits,* that is, the number of times the ball hits a cushion, the initial hit at the dot, and the hit as the ball goes into a pocket.

In the table, enter the dimensions of each pool table, the number of squares through which the ball travels, and the number of hits. Analyze the data in the table and determine a rule that predicts the number of squares and the number of hits, given the dimensions of any pool table.

Height	Base	Number of Hits	Number of Squares
4	6		
5	7		
3	2		

Pool Table

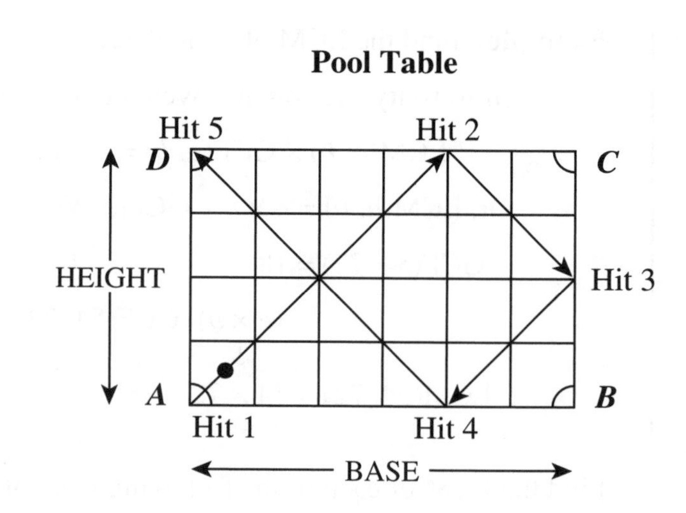

EXTENSIONS Add a column headed **Final Pocket** to the table. In this column, for each pool table, record the letter of the pocket into which the ball finally fell. Use this data to find a rule that will predict which pocket the ball will fall into for any pool table.

Activity 14: How Many Factors?

PURPOSE Use patterns to discover a rule for determining the number of factors of any number.

GROUPING Work individually or in pairs.

Find the prime factorization (using exponential form) and the total number of factors for each number in the table.

Number	2	3	4	8	9
Prime Factorization					
Number of Factors					

Each of the numbers has only one prime factor. How is the total number of factors of each number related to the exponent in its prime factorization?

Use your answer to find a number that has 7 factors. _____ 11 factors. _____

Check your predictions by listing the factors of each number.

Find the prime factorization (using exponential form) and the total number of factors for each number in the table.

Number	6	24	60	72	100
Prime Factorization					
Number of Factors					

Each of the numbers in the table has more than one prime factor. How is the total number of factors of each number related to the exponents in its prime factorization?

Find the number of factors of 360 using the exponents in the prime factorization and by listing.

State a rule for determining the number of factors of any number from its prime factorization.

Use your answer to find a number that has 20 factors. Check your answer by listing the factors.

Chapter Summary

In your early mathematical experiences you probably thought about whole numbers as representing simple quantities, such as six marbles or five pencils. Thus the whole number 6 could be modeled by a set containing six objects. But when your understanding of whole numbers was extended to the integers in Activities 1–6, your concept of a number changed.

Integers do represent quantities. However, when you think about an integer, you usually think of it as representing not just a quantity, but also a direction. Thus you think of $^{+}5$ and $^{-}5$ as opposites, as a \$5 profit and a \$5 loss, or as 5 more than zero and 5 less than zero. This interpretation distinguishes integers from whole numbers and is reflected in the models used to represent integers. On a number line, $^{+}5$ and $^{-}5$ are both located 5 units from zero, but $^{-}5$ is to the left of zero and $^{+}5$ is to the right. When modeled as particles, $^{+}5$ and $^{-}5$ in their simplest forms are represented by the same number of particles, but the particles have opposite charges. These ideas were explored in Activities 1–3.

The extension of the whole numbers to the integers required not only that you alter your concept of a number, but also that you modify your interpretations of the operations with numbers. Addition could still be thought of in terms of the union of sets. However, because the objects in the sets might be opposites, you found that in some cases you had to pair the opposites in order to find the sum. That is,

$$\begin{aligned} {}^{+}6 + {}^{-}8 &= {}^{+}6 + ({}^{-}6 + {}^{-}2) \\ &= ({}^{+}6 + {}^{-}6) + {}^{-}2 \\ &= {}^{-}2 \qquad \text{since } {}^{+}6 + {}^{-}6 = 0. \end{aligned}$$

Similarly, subtraction could still be thought of as *taking away*. But in some cases, the objects in the sets representing the two numbers in the subtraction problem were opposites. In these cases, it was necessary to rename the number being subtracted before you could *take away*. The renaming was accomplished by adding zero. For example,

$$\begin{aligned} {}^{+}6 - {}^{-}8 &= {}^{+}6 + 0 - {}^{-}8 \\ &= {}^{+}6 + ({}^{+}8 + {}^{-}8) - {}^{-}8 \\ &= {}^{+}6 + {}^{+}8 + ({}^{-}8 - {}^{-}8) \\ &= {}^{+}6 + {}^{+}8 \qquad \text{since } {}^{-}8 - {}^{-}8 = 0. \end{aligned}$$

As a result, you discovered that subtraction can be interpreted as adding the opposite.

These changes in the meanings of addition and subtraction and the resulting algorithms were explored in Activities 2 and 3. The results were verified in Activities 4 and 5 by examining patterns.

In Activity 6, patterns were analyzed to discover an algorithm for multiplying integers. The algorithm was extended to division by applying the inverse relationship between multiplication and division to find missing factors.

In Activities 7 and 8, you discovered that the positive integers can be classified by how many factors they have. *Prime numbers* have exactly two factors; *composite numbers* have more than two; and *one*, which is in a class of its own, has exactly one factor.

Initially, this classification of the positive integers probably seemed rather arbitrary and pointless. By the end of Activity 8, you discovered that every integer greater than one can be expressed as a product of prime numbers, and that this product is unique except for the order of the factors. This means that, in a sense, the prime numbers are the building blocks from which all of the integers greater than one are constructed. This result is so important that it is known as the *Fundamental Theorem of Arithmetic.*

In Activities 7–9, you also learned how to find the prime factorization of a number and explored other interesting classifications: *squares, cubes, palindromes,* and *emirps* (prime numbers, like 13, that are also prime when their digits are reversed). In Activity 14, you discovered that the prime factorization of a number can be used to find the total number of factors of a given number. If you add 1 to each of the exponents in the prime factorization of the number, then the number of factors is the product of the sums. This is just one of the many applications of prime factorization.

Activities 10–13 were devoted to the study of factors and multiples of numbers. The concepts of *least common multiple* (LCM) and *greatest common factor* (GCF) and procedures for calculating them were developed using arrays and Venn diagrams. Least common multiples and greatest common factors will be used extensively in your study of the rational numbers.

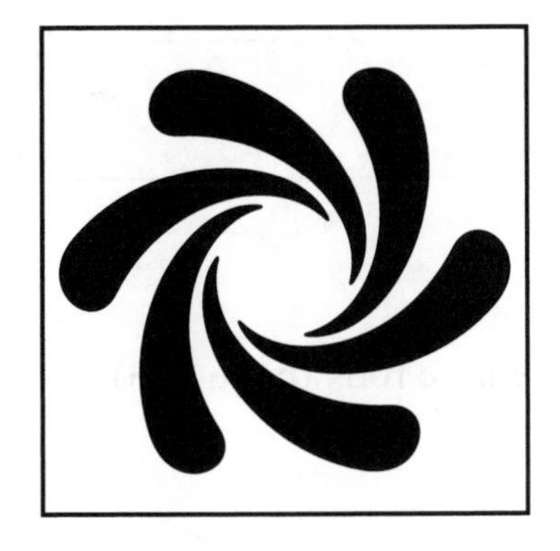

Chapter 5 Rational Numbers as Fractions

"The transition from whole numbers to fractions and decimals can be difficult for students. Although they may multiply the numerators and then the denominators, for example, they often do not understand why a similar procedure does not work in adding fractions. Concrete or representational models can help students clarify these anomalies."

—*Curriculum and Evaluation Standards for School Mathematics*

Other than whole-number computation, no topic in the elementary mathematics curriculum demands more time than the study of fractions. Yet, despite the years of study, most students enter high school with a poor concept of fractions and an even poorer understanding of operations with fractions. It is not surprising then that adults, when asked about their knowledge of fractions, often respond, *"Yours is not to reason why, just invert and multiply."*

Rational numbers should be taught as the natural extension of the whole numbers. The fraction $\frac{3}{4}$ can be viewed as the solution to the problem of dividing 3 dollars among 4 people: $3 \div 4$. For students to understand this connection between whole numbers and fractions, teaching about fractions and their operations must be grounded in concrete models, as was the instruction regarding whole numbers. A firm concept of fractions and a feeling for their magnitude must be established before they can be compared and ordered meaningfully. This must precede computation.

Given a firm grasp of the concept of a fraction, students can develop operation sense and a deeper understanding of the algorithms for operations with fractions.

This chapter provides activities that firmly establish the concept of fractions. All operations are explored through a variety of concrete models and reinforced by representing the models pictorially.

Activity 1: What Is a Fraction?

PURPOSE Develop the concept of a fraction.

MATERIALS Pattern blocks (pages A-7–A-11) and Cuisenaire rods (page A-3)

GROUPING Work individually or in pairs.

Use pattern blocks to solve the following problems.

1. The trapezoid is what fractional part of the hexagon? ________
2. The blue parallelogram is what fractional part of the hexagon? ________
3. The triangle is what fractional part of the hexagon? ________
4. The triangle is what fractional part of the blue parallelogram? ________
5. The triangle is what fractional part of the trapezoid? ________

What fractional part of each figure is *shaded*? *unshaded*?

1. shaded ________ unshaded ________
2. shaded ________ unshaded ________
3. shaded ________ unshaded ________
4. shaded ________ unshaded ________

Use Cuisenaire rods to solve the following:

1. If the orange rod = 1, each rod is what fractional part of the orange rod?

   a. red ________ b. green ________

   c. yellow ________ d. purple ________

2. If the purple rod = 1, each rod is what fractional part of the purple rod?

   a. brown ________ b. orange ________

   c. dark green ________ d. black ________

---

1. If the red rod = $\frac{1}{2}$, which rod = 1? ____________
2. If the red rod = $\frac{1}{3}$, which rod = 1? ____________
3. If the white rod = $\frac{1}{5}$, which rod = 1? ____________
4. If the white rod = $\frac{1}{4}$, which rod = $1\frac{3}{4}$? ____________
5. If the red rod = $\frac{1}{2}$, which rod = $1\frac{1}{2}$? ____________
6. If the red rod = $\frac{1}{3}$, which rod = $1\frac{2}{3}$? ____________

Explain how you used the rods to arrive at your answers.

Construct a shape similar to this one with two trapezoids and one blue parallelogram.

1. Given that the shape = 1, what pattern block(s) would you use to represent each of the following fractions?

   a. $\frac{1}{4}$ ______ b. $\frac{3}{4}$ ______ c. $\frac{1}{8}$ ______

Fill in the same shape using one red block, two blue blocks, and one green block.

2. What fraction is represented by each of the following?

   a. a blue block ______ b. a red block ______ c. a green block ______

Activity 2: Equivalent Fractions

PURPOSE Develop the concept of a fraction using concrete models and a problem-solving approach.

MATERIALS Cuisenaire rods (page A-3), pattern blocks (pages A-7–A-11), and fraction strips (page A-30)

GROUPING Work individually or in pairs.

For the following problems, use Cuisenaire rods to construct the trains.

1. Make all of the possible one-color trains the same length as a dark green rod and complete the following:

   a. $\frac{\text{light green}}{\text{dark green}} = \frac{1}{2} = \frac{\ }{6}$

   b. $\frac{\text{red}}{\text{dark green}} = \frac{1}{\ } = \frac{\ }{6}$

   c. $\frac{\text{purple}}{\text{dark green}} = \frac{\ }{\ } = \frac{\ }{\ }$

   d. $\frac{\text{dark green}}{\text{dark green}} = \frac{\ }{6} = \frac{\ }{\ }$

2. Make all of the possible one-color trains the same length as a brown rod and complete each of the following:

   a. $\frac{\text{purple}}{\text{brown}} = \frac{\ }{\ } = \frac{\ }{\ } = \frac{\ }{\ }$

   b. $\frac{\text{dark green}}{\text{brown}} = \frac{\ }{\ } = \frac{\ }{\ }$

   c. $\frac{\text{red}}{\text{brown}} = \frac{\ }{\ } = \frac{\ }{\ }$

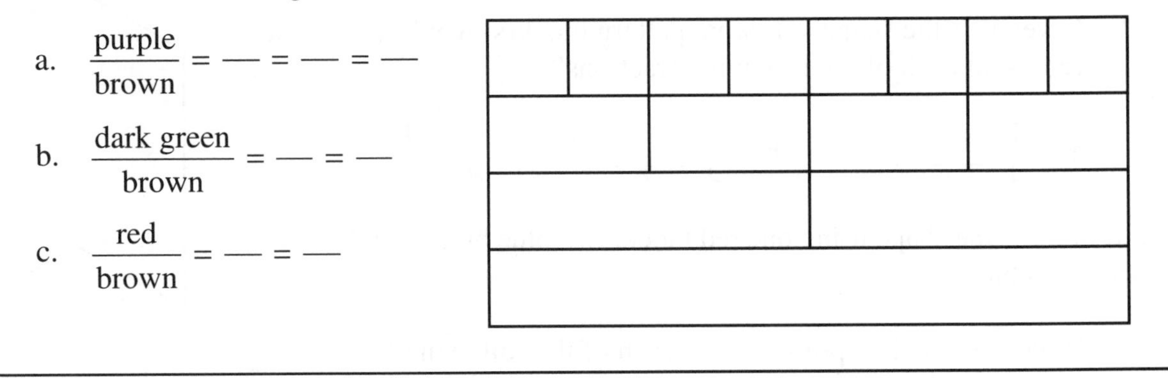

Use pattern blocks to construct a shape similar to the star and complete the following:

If the star shape = 1, then

a. $\dfrac{\text{trapezoid}}{\text{star}} = \dfrac{}{12} = \dfrac{1}{}$

b. $\dfrac{\text{2 blue parallelograms}}{\text{star}} = \dfrac{}{6} = \dfrac{1}{} = \dfrac{}{12}$

c. $\dfrac{\text{hexagon}}{\text{star}} = \dfrac{6}{} = \dfrac{}{6} = \dfrac{1}{}$

Use the fraction strips to find strips that can be folded into parts so that the resulting strip is equal in length to the fraction given in each problem. Folds may be made **only** on the lines on the strips.

Write the name of the equivalent fractions in the space provided.

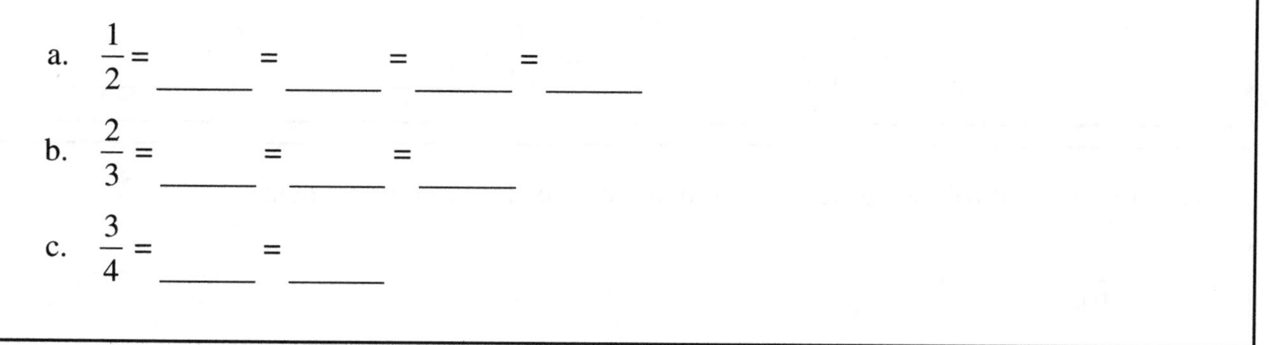

a. $\frac{1}{2}$ = ______ = ______ = ______ = ______

b. $\frac{2}{3}$ = ______ = ______ = ______

c. $\frac{3}{4}$ = ______ = ______

EXTENSIONS Make all the one-color trains that are equal in length to (a) the blue rod and (b) the orange rod. Write all of the equivalent fractions that can be illustrated with (a) the blue rod equal to 1 and (b) the orange rod equal to 1.

Activity 3: How Big Is It?

PURPOSE Develop the ability to estimate the magnitude of a fraction.

MATERIALS A deck of fraction cards (pages A-31 and A-32) and a copy of the fraction sorting board (page A-33)

GROUPING Work individually or in pairs.

GETTING STARTED Use these rules to complete the problems that follow.

A fraction is close to

1 if the numerator and denominator are approximately the same size.

$\mathbf{\frac{1}{2}}$ if the denominator is about twice as large as the numerator.

0 if the numerator is very small compared to the denominator.

Determine if each of the following fractions is *close to* 1, *close to* $\frac{1}{2}$, or *close to* 0.

a. $\frac{5}{6} \approx$ ______ b. $\frac{11}{13} \approx$ ______ c. $\frac{1}{15} \approx$ ______ d. $\frac{8}{9} \approx$ ______

e. $\frac{2}{13} \approx$ ______ f. $\frac{6}{13} \approx$ ______ g. $\frac{2}{51} \approx$ ______ h. $\frac{33}{35} \approx$ ______

i. $\frac{4}{9} \approx$ ______ j. $\frac{4}{100} \approx$ ______ k. $\frac{7}{12} \approx$ ______ l. $\frac{6}{7} \approx$ ______

1. Complete the following fractions so that they are close to but less than $\frac{1}{2}$.

a. $\frac{\quad}{100}$ b. $\frac{\quad}{25}$ c. $\frac{\quad}{9}$ d. $\frac{\quad}{14}$

e. $\frac{7}{\quad}$ f. $\frac{3}{\quad}$ g. $\frac{11}{\quad}$ h. $\frac{8}{\quad}$

2. Complete the following fractions so that they are close to but less than 1.

a. $\frac{\quad}{27}$ b. $\frac{\quad}{12}$ c. $\frac{\quad}{75}$ d. $\frac{\quad}{8}$

e. $\frac{9}{\quad}$ f. $\frac{3}{\quad}$ g. $\frac{11}{\quad}$ h. $\frac{95}{\quad}$

THE FRACTION SORTING GAME

This is a game for two players. Cut out the fraction cards and shuffle them. One student places each card in the appropriate column on the fraction sorting board and justifies each placement to the other player. Reshuffle the deck and reverse the roles.

Activity 4: Fraction War

PURPOSE Reinforce estimation and comparison of fractions in a game format.

MATERIALS A deck of playing cards (remove the face cards), the fraction arrays (page A-34), and the fractions game board (page A-35)

GROUPING Work in pairs.

GETTING STARTED Rules for Fraction War.

Example:

FRACTIONS GAME BOARD

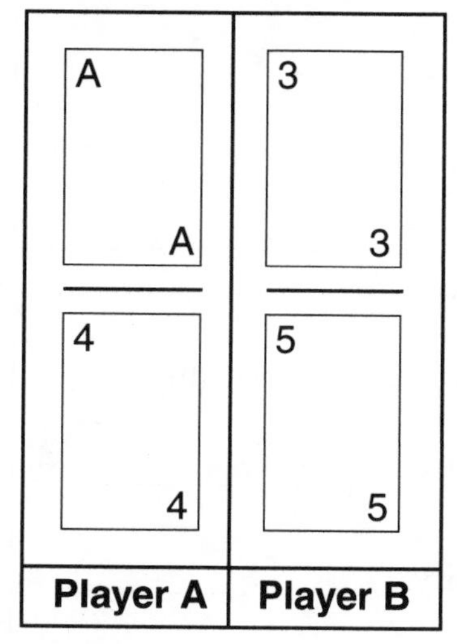

Player A is the winner for a round in Game a: *the smaller fraction*

1. Shuffle the cards and deal them face down to the players. Players choose a goal for each game from the following list. Each player turns up two cards and forms a fraction on the fractions game board according to the rules of the chosen game. The ace represents 1.

   a. Form a fraction by placing the card with the smaller number in the numerator. Player with the smaller fraction is the winner.

   b. Form a fraction by placing the card with the smaller number in the numerator. Player with the larger fraction is the winner.

   c. Form a fraction by placing the card with the smaller number in the numerator. Player with the fraction whose value is closest to $\frac{1}{2}$ is the winner.

   d. Form a fraction by placing the card with the larger number in the numerator. Player with the larger fraction is the winner.

   e. Place the first card in the numerator and the second in the denominator. Player with the fraction whose value is closest to 2 is the winner.

   f. Each player decides where to place each card. Player with the fraction whose value is closest to 1 is the winner.

2. The winner of each round collects the four cards and places them face up at the bottom of his/her pile of cards. If the fractions formed are equivalent, each player turns over two additional cards and forms a new fraction. The winner of the round gets all eight cards.

3. When the players have played all the face-down cards, the player with the most face-up cards is the winner of the game. Reshuffle the cards and choose a different goal for a new game.

Activity 5: What Comes First?

PURPOSE Develop an understanding of comparing and ordering fractions.

MATERIALS Cuisenaire rods, pattern blocks, fraction arrays (page A-34)

GROUPING Work individually.

Use Cuisenaire rods to build all possible one-color trains that are the same length as a brown rod, and complete the following:

a. Which is larger, $\frac{5}{8}$ or $\frac{1}{2}$? ______ Complete the inequality: ______ > ______

b. Which is smaller, $\frac{3}{8}$ or $\frac{1}{4}$? ______ Complete the inequality: ______ < ______

c. Which is larger, $\frac{7}{8}$ or $\frac{3}{4}$? ______ Complete the inequality: ______ > ______

Use pattern blocks to construct a star shape (see Activity 2) and complete the following:

a. Which is larger, $\frac{5}{12}$ or $\frac{1}{2}$? ______ Complete the inequality: ______ < ______

b. Which is smaller, $\frac{2}{3}$ or $\frac{7}{12}$? ______ Complete the inequality: ______ > ______

c. Which is larger, $\frac{5}{6}$ or $\frac{3}{4}$? ______ Complete the inequality: ______ < ______

Use the fraction array to order the following fractions.

a. $\frac{1}{2}, \frac{3}{5}, \frac{4}{7}$ ________ > ________ > ________

b. $\frac{2}{3}, \frac{3}{4}, \frac{7}{8}$ ________ > ________ > ________

c. $\frac{5}{12}, \frac{2}{5}, \frac{3}{7}$ ________ < ________ < ________

d. $\frac{5}{6}, \frac{11}{12}, \frac{4}{5}$ ________ < ________ < ________

Activity 6: Adding and Subtracting Fractions

PURPOSE Develop the operations of addition and subtraction of fractions using concrete models.

MATERIALS Pattern blocks and fraction strips (page A-30)

GROUPING Work individually.

If the yellow hexagon = 1, then the red trapezoid = $\frac{1}{2}$, the blue parallelogram = $\frac{1}{3}$, and the green triangle = $\frac{1}{6}$. Use pattern blocks to solve the following:

a. 1 red + 3 green = 1 red + 1 red = 3 green + 3 green =

$\frac{1}{2} + \frac{3}{6} = \frac{1}{2} + \frac{1}{2} = \frac{3}{6} + \frac{3}{6} =$ ______

b. 1 red + 1 blue = ___ green + ___ green = ______

$\frac{1}{2} + \frac{1}{3} =$ ______ + ______ = ______

c. 1 blue + 1 green = ______ + ______ = ______

$\frac{1}{3} + \frac{1}{6} =$ ______ + ______ = ______ = ______

d. 1 red – 1 blue = ______ – ______ = ______

$\frac{1}{2} - \frac{1}{3} =$ ______ – ______ = ______

e. 1 red – 1 green = ______ – ______ = ______

$\frac{1}{2} - \frac{1}{6} =$ ______ – ______ = ______ = ______

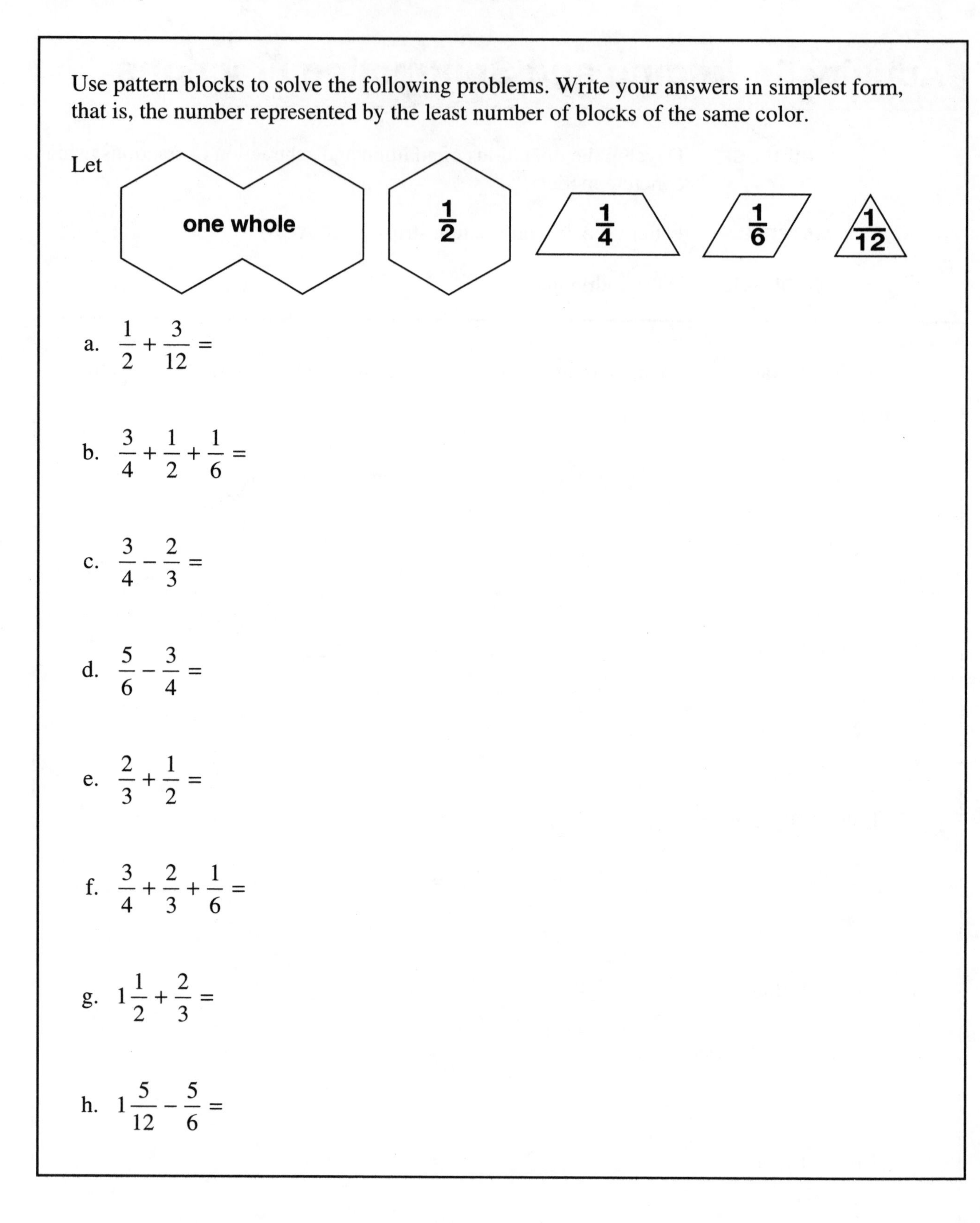

Use pattern blocks to solve the following problems. Write your answers in simplest form, that is, the number represented by the least number of blocks of the same color.

Let

a. $\frac{1}{2} + \frac{3}{12} =$

b. $\frac{3}{4} + \frac{1}{2} + \frac{1}{6} =$

c. $\frac{3}{4} - \frac{2}{3} =$

d. $\frac{5}{6} - \frac{3}{4} =$

e. $\frac{2}{3} + \frac{1}{2} =$

f. $\frac{3}{4} + \frac{2}{3} + \frac{1}{6} =$

g. $1\frac{1}{2} + \frac{2}{3} =$

h. $1\frac{5}{12} - \frac{5}{6} =$

When adding fractions using fraction strips, you must fold the strips to show only the fractions that are needed. Strips are placed as shown in the following figures. A longer strip must then be found that has fold lines in common with the two fractions.

Example for Addition: $\frac{1}{3} + \frac{1}{4} =$

$$\frac{1}{3} \quad + \quad \frac{1}{4}$$

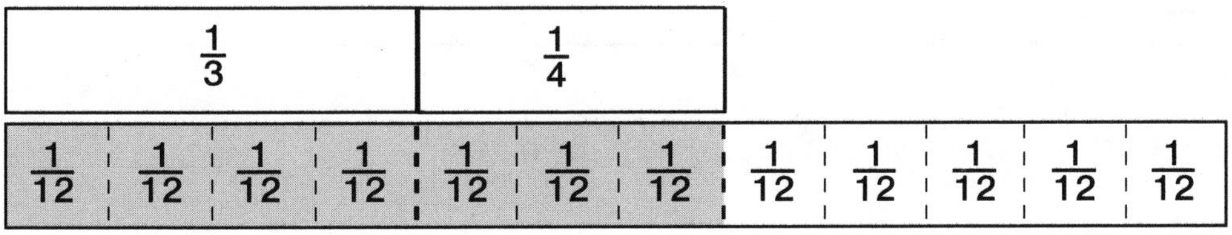

$\frac{1}{3}$	$\frac{1}{4}$

$\frac{1}{12}$	$\frac{1}{12}$	$\frac{1}{12}$	$\frac{1}{12}$	$\frac{1}{12}$	$\frac{1}{12}$	$\frac{1}{12}$	$\frac{1}{12}$	$\frac{1}{12}$	$\frac{1}{12}$	$\frac{1}{12}$	$\frac{1}{12}$

$$\frac{4}{12} + \frac{3}{12} = \frac{7}{12}$$

Example for Subtraction: $\frac{2}{3} - \frac{1}{4} =$

$$\frac{2}{3} \quad - \quad \frac{1}{4} \quad =$$

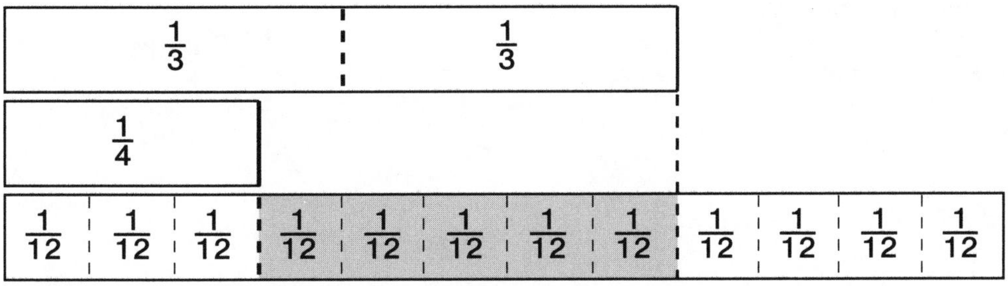

$\frac{1}{3}$	$\frac{1}{3}$

$\frac{1}{4}$

$\frac{1}{12}$	$\frac{1}{12}$	$\frac{1}{12}$	$\frac{1}{12}$	$\frac{1}{12}$	$\frac{1}{12}$	$\frac{1}{12}$	$\frac{1}{12}$	$\frac{1}{12}$	$\frac{1}{12}$	$\frac{1}{12}$	$\frac{1}{12}$

$$\frac{8}{12} - \frac{3}{12} = \frac{5}{12}$$

Use your fraction strips to solve the following problems.

a. $\frac{1}{2} + \frac{3}{5} =$

b. $\frac{7}{8} - \frac{1}{4} =$

c. $\frac{7}{10} - \frac{2}{5} =$

d. $\frac{5}{12} + \frac{1}{3} =$

e. $\frac{2}{3} + \frac{3}{4} =$

f. $\frac{3}{4} - \frac{1}{6} =$

Activity 7: Multiplying Fractions

PURPOSE Use pattern blocks and paper folding to develop an algorithm for multiplying fractions.

MATERIALS Pattern blocks and paper for folding

GROUPING Work individually.

Example: $\frac{2}{3}$ of $\frac{1}{4}$ means two of three equal parts of $\frac{1}{4}$.

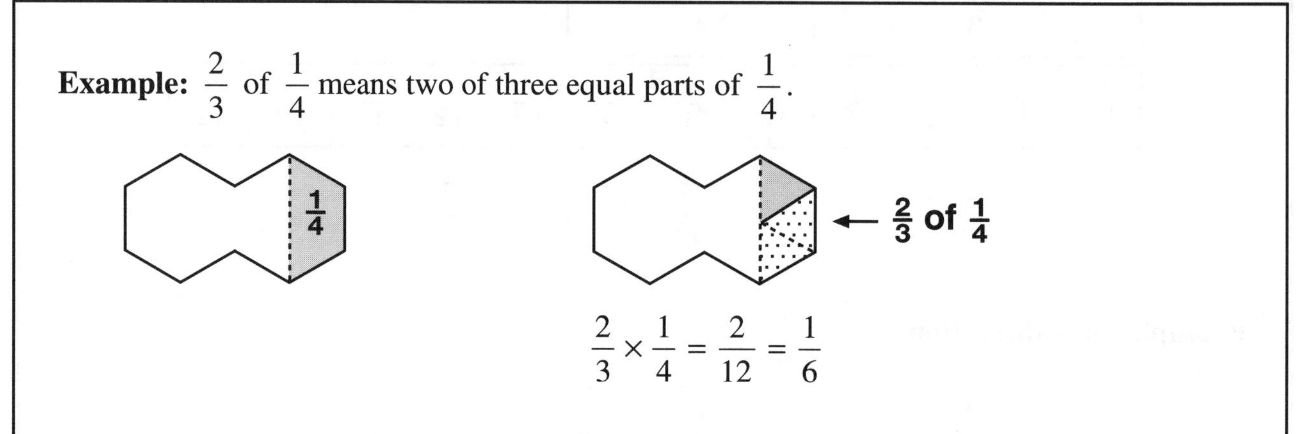

$$\frac{2}{3} \times \frac{1}{4} = \frac{2}{12} = \frac{1}{6}$$

Place pattern blocks on Figure A to solve the following. Record your solution both pictorially and numerically.

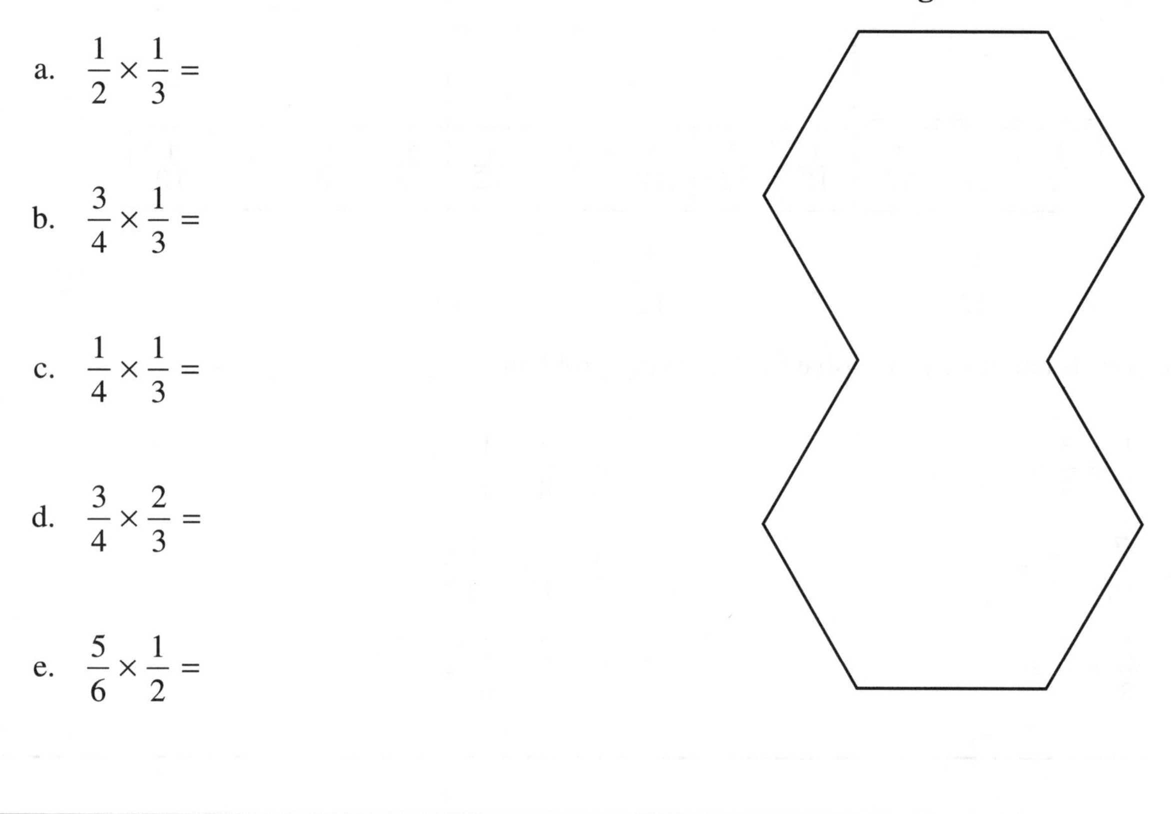

a. $\frac{1}{2} \times \frac{1}{3} =$

b. $\frac{3}{4} \times \frac{1}{3} =$

c. $\frac{1}{4} \times \frac{1}{3} =$

d. $\frac{3}{4} \times \frac{2}{3} =$

e. $\frac{5}{6} \times \frac{1}{2} =$

Example: $\frac{1}{2}$ of $\frac{2}{3}$ means one of the two equal parts of two thirds.

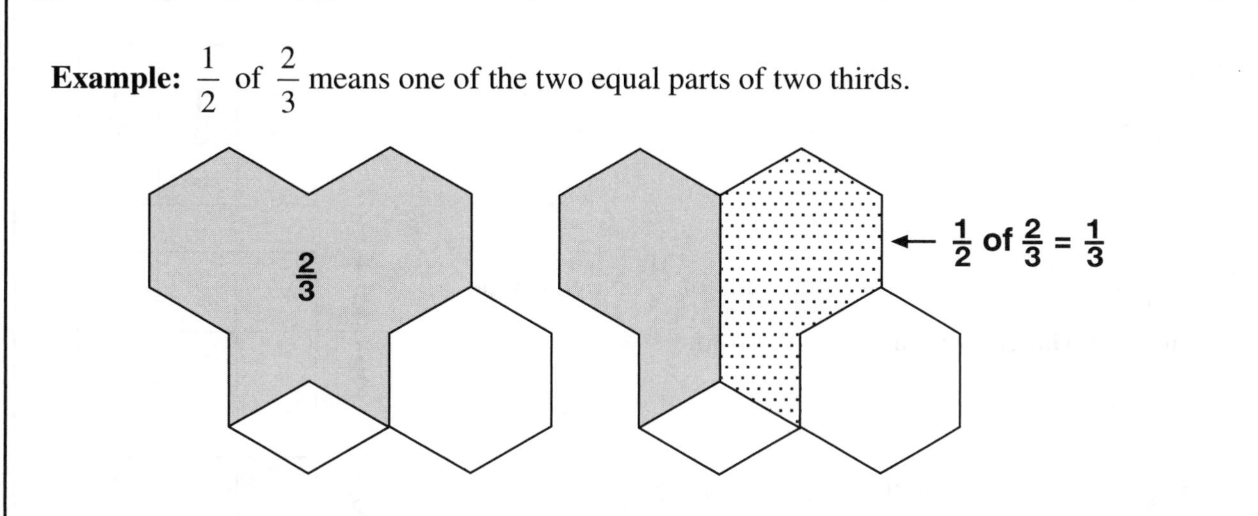

Construct a figure similar to the one shown above, and solve the following. Record each step of your solutions both pictorially and numerically.

a. $\frac{3}{4} \times \frac{1}{6} =$

b. $\frac{3}{8} \times \frac{2}{3} =$

c. $\frac{7}{12} \times \frac{1}{2} =$

d. $\frac{5}{8} \times \frac{1}{3} =$

$\frac{1}{3}$ of 1 means one of the three equal parts of 1. Divide a piece of paper into thirds with vertical folds.

$\frac{1}{2}$ of $\frac{1}{3}$ means one of the two equal parts of $\frac{1}{3}$. Now, divide the thirds into halves with a horizontal fold.

$\frac{2}{3}$ of $\frac{1}{2}$ means two of the three equal parts of $\frac{1}{2}$.

Fold sheets of paper to solve the following. Record each step of your solution pictorially and numerically.

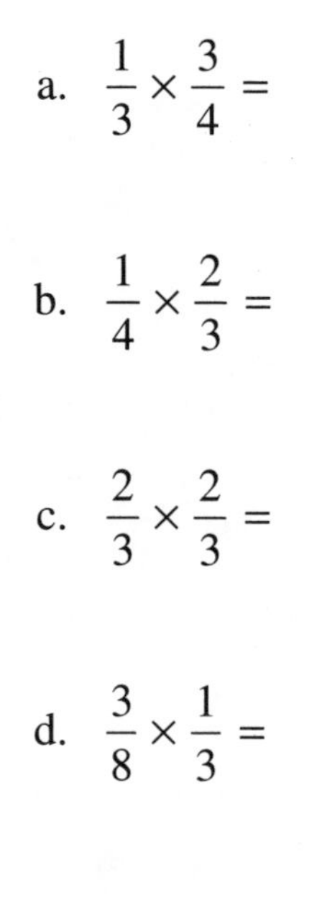

a. $\frac{1}{3} \times \frac{3}{4} =$

b. $\frac{1}{4} \times \frac{2}{3} =$

c. $\frac{2}{3} \times \frac{2}{3} =$

d. $\frac{3}{8} \times \frac{1}{3} =$

EXTENSIONS From what you have observed in this activity, write a rule for multiplication of fractions.

Activity 8: Dividing Fractions

PURPOSE Use pattern blocks to illustrate the repeated subtraction model for dividing fractions.

MATERIALS Pattern blocks (pages A-7–A-11)

GROUPING Work individually.

GETTING STARTED Recall the use of the multiplication and division frame for the division of whole numbers.

Example: $3\overline{)6}$ can mean how many groups of 3 are there in 6? $3\overline{)\boxplus}^{2}$

In the following example [double hexagon] represents 1.

Example: $1 \div \frac{1}{2}$ means: How many groups of $\frac{1}{2}$ are there in 1?

2

$\frac{1}{2}$ | one

Complete each sentence and use your pattern blocks to solve the following problems.

a. $\frac{1}{3} \div \frac{1}{6}$ means ________________________________

$\frac{1}{3} \div \frac{1}{6} =$

b. $\frac{1}{2} \div \frac{1}{4}$ means ________________________________

$\frac{1}{2} \div \frac{1}{4} =$

c. $\frac{5}{6} \div \frac{5}{12}$ means ________________________________

$\frac{5}{6} \div \frac{5}{12} =$

d. $\frac{3}{4} \div \frac{1}{4}$ means ________________________________

$\frac{3}{4} \div \frac{1}{4} =$

e. $\frac{3}{2} \div \frac{3}{4}$ means ________________________________

$\frac{3}{2} \div \frac{3}{4} =$

To model the problem $\frac{1}{2} \div \frac{1}{3}$, let

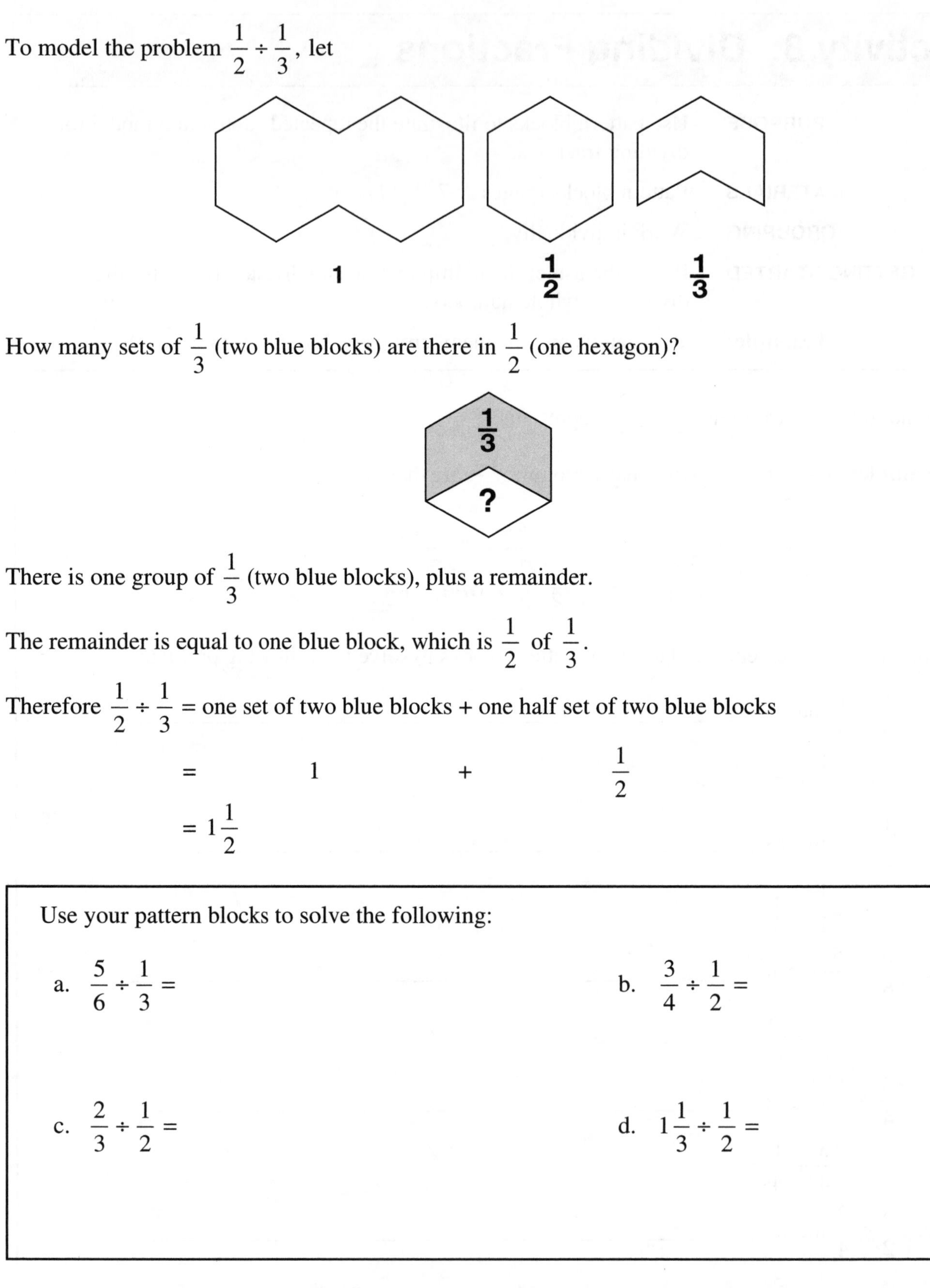

How many sets of $\frac{1}{3}$ (two blue blocks) are there in $\frac{1}{2}$ (one hexagon)?

There is one group of $\frac{1}{3}$ (two blue blocks), plus a remainder.

The remainder is equal to one blue block, which is $\frac{1}{2}$ of $\frac{1}{3}$.

Therefore $\frac{1}{2} \div \frac{1}{3}$ = one set of two blue blocks + one half set of two blue blocks

$$= 1 + \frac{1}{2}$$

$$= 1\frac{1}{2}$$

Use your pattern blocks to solve the following:

a. $\frac{5}{6} \div \frac{1}{3} =$

b. $\frac{3}{4} \div \frac{1}{2} =$

c. $\frac{2}{3} \div \frac{1}{2} =$

d. $1\frac{1}{3} \div \frac{1}{2} =$

EXTENSIONS Describe how you would use fraction strips to solve the previous problems. Draw at least one illustration of your method.

Activity 9: Square Fractions

PURPOSE Reinforce the concept of fractions and equivalent fractions, and illustrate operations with fractions using geometric models.

MATERIALS A sheet of paper 20 cm square (colored construction paper works well), and scissors (optional)

GROUPING Work individually or as a class activity directed by the instructor.

GETTING STARTED Work through each section of the activity in order, following the folding and cutting directions carefully. Fold and crease the paper sharply so that it will tear cleanly if scissors are not used. Questions in each section can be used as a script if the activity is directed by the instructor. Answers should be discussed and reasoning explained.

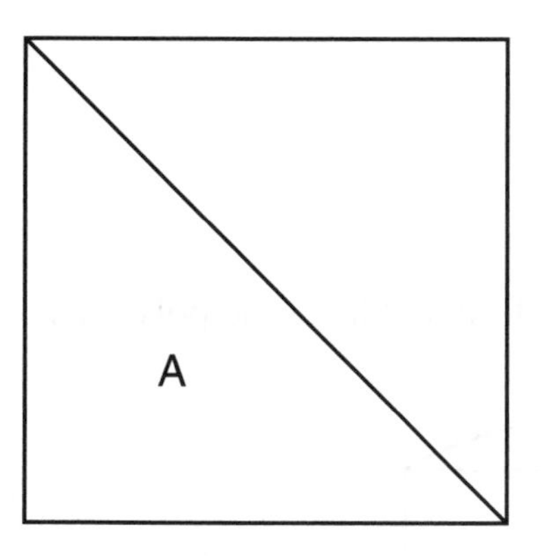

1. Fold the square as shown and cut or tear along the fold to divide the square into two congruent parts.

   Each polygon is what fraction of the original square? ________

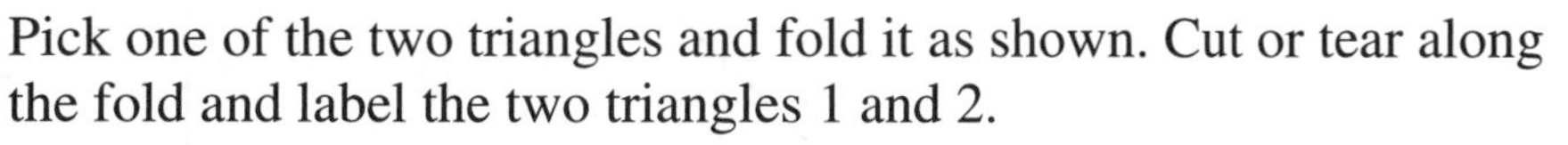
Pick one of the two triangles and fold it as shown. Cut or tear along the fold and label the two triangles 1 and 2.

2. Triangle 1 is what fractional part of

   a. triangle A? ______

   b. the original square? ______

   Explain your answers.

In the remaining large triangle, fold the vertex of the right angle to the midpoint of the longest side. Cut along the fold and label the polygons B and 3 as shown.

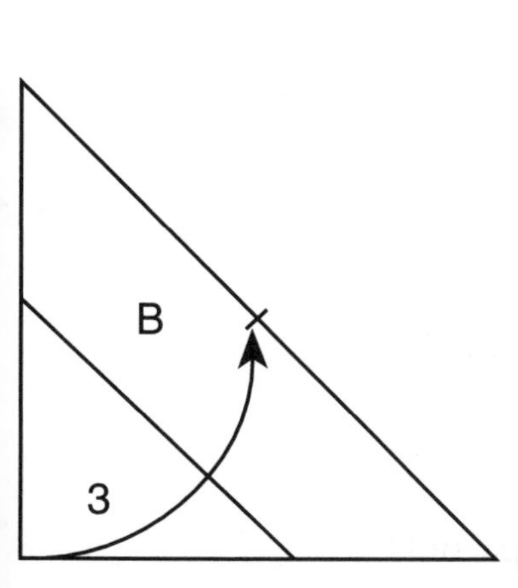

3. Triangle 3 is what fractional part of

   a. triangle 1? ______

   b. triangle A? ______

   c. trapezoid B? ______

Fold one of the endpoints of the longest side of trapezoid B to the midpoint of that side as shown. Cut along the fold and label the polygons C and 4.

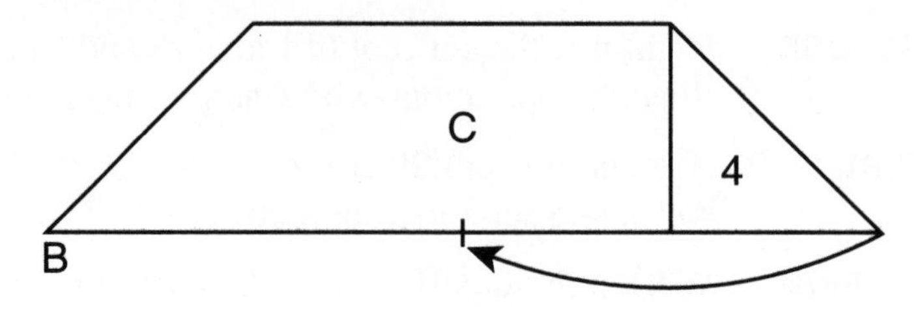

4. Triangle 4 is what fractional part of

   a. triangle 3? ______

   b. triangle A? ______

   c. trapezoid C? ______

Fold trapezoid C as shown. Cut along the fold and label the polygons D and 5.

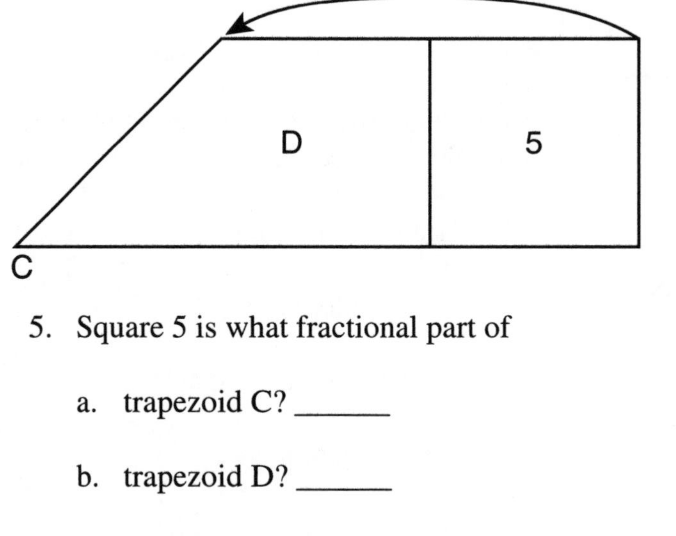

5. Square 5 is what fractional part of

   a. trapezoid C? ______

   b. trapezoid D? ______

   c. triangle 3? ______

   d. original square? ______

6. Trapezoid C is what fractional part of

   a. triangle A? ______ b. square 5? ______

7. Trapezoid D is what fractional part of

   a. triangle 1? ______ b. trapezoid C? ______

Fold trapezoid D as shown. Cut along the fold and label the two polygons 6 and 7.

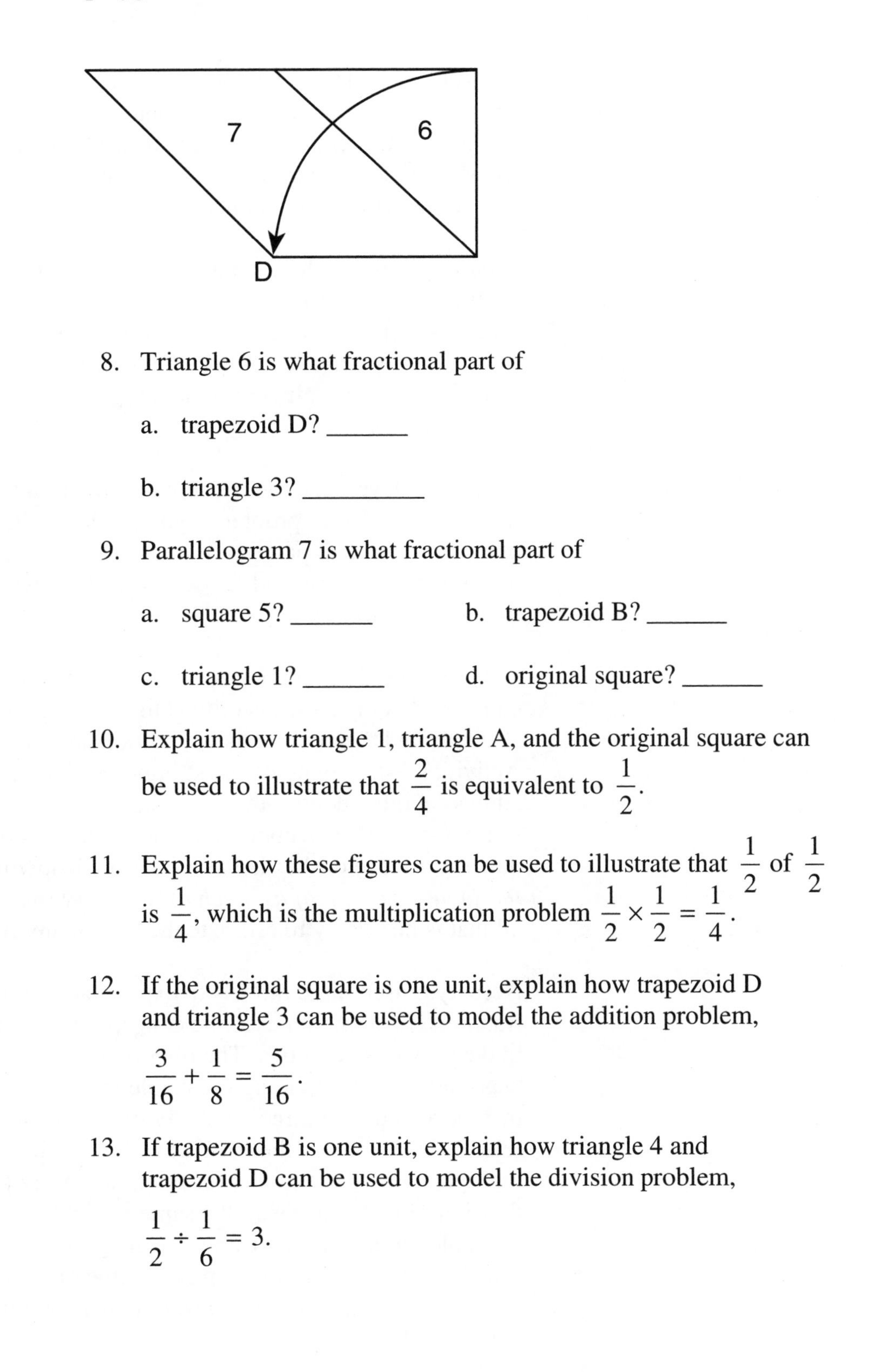

8. Triangle 6 is what fractional part of

   a. trapezoid D? ______

   b. triangle 3? _________

9. Parallelogram 7 is what fractional part of

   a. square 5? ______ b. trapezoid B? ______

   c. triangle 1? ______ d. original square? ______

10. Explain how triangle 1, triangle A, and the original square can be used to illustrate that $\frac{2}{4}$ is equivalent to $\frac{1}{2}$.

11. Explain how these figures can be used to illustrate that $\frac{1}{2}$ of $\frac{1}{2}$ is $\frac{1}{4}$, which is the multiplication problem $\frac{1}{2} \times \frac{1}{2} = \frac{1}{4}$.

12. If the original square is one unit, explain how trapezoid D and triangle 3 can be used to model the addition problem,

    $\frac{3}{16} + \frac{1}{8} = \frac{5}{16}$.

13. If trapezoid B is one unit, explain how triangle 4 and trapezoid D can be used to model the division problem,

    $\frac{1}{2} \div \frac{1}{6} = 3$.

Chapter Summary

The activities in this chapter emphasized the development of a conceptual understanding of rational numbers (fractions) and their operations. Concrete materials and structured lessons were used to show how the operations with fractions are an extension of the operations with whole numbers.

A curriculum developmentally appropriate to students is one of the central themes of the NCTM *Standards*. This implies that lessons progress from the concrete to the representational (pictorial) to the abstract. The activities in this chapter illustrated one aspect of such a curriculum through careful development of the operations with fractions.

Activity 1 developed the concept of a fraction through a variety of models. In different problems, you (a) determined the fractional part of a whole, (b) compared two areas or the length of two strips to determine a fraction, and (c) determined the whole unit, given the fractional part.

Activity 2 used several models to introduce equivalent fractions. The concept of equivalence is critical to the understanding of ordering rationals and the operations of addition and subtraction of fractions. Activities 3–5 may be the most important in the chapter. They address another central theme of the *Standards*: developing number sense. Once the concept of a fraction is understood, then one can develop a sense of the size of a fraction. Knowing terms like *about the same size as, half as much as,* and *very small as compared to* is all that is necessary to estimate the size of any fraction.

Activity 6 used three models to explore addition and subtraction of fractions. Each model used the concept of equivalence as developed in the previous activities. The importance of common denominators was connected to dividing the whole into equivalent parts that could then be added or subtracted. Activity 7 illustrated the language relationship between the word *of* and multiplication of rationals through a geometric model for fractions. Activity 8 developed division of fractions using the same model that was used for division of whole numbers. That is, how many groups of one factor (the divisor) are there in the product (the dividend). By modeling division this way, you can come to understand the familiar rule "invert and multiply."

Chapter 6
Exponents and Decimals

"Students should come to understand and appreciate mathematics as a coherent body of knowledge rather than a vast, perhaps bewildering, collection of facts and rules. Understanding this structure promotes students' efficiency in investigating the arithmetic of fractions, decimals, integers, and rationals through the unity of common ideas. It also offers insights into how the whole number system is extended to the rational number system and beyond."

—*Curriculum and Evaluation Standards for School Mathematics*

With the advent of the hand calculator, many educators argued that decimals should be taught prior to fractions and that decimals should receive much greater attention. However, children have many everyday experiences with common fractions such as "half an apple," or "divide the candy bar among three friends," etc. These common connections to simple fractions precede experiences with decimals, which are usually associated with money.

Instruction related to decimals and computation with decimals will have its greatest impact when it is based on the same models and understanding as fractions and whole numbers. The activities in this chapter extend and reinforce the models which have been used previously in the study of whole numbers and fractions. Decimal numbers will be modeled with base-ten blocks
in a variety of ways.

Example: *If a rod = 1, then the small cube = 0.1.*
If a flat = 1, then a rod = 0.1, and the small cube = 0.01

These models reinforce the concept of a decimal number being part of a whole.

Number and operation sense and estimation skills developed in the chapter on whole numbers also will be reviewed and applied in the development of algorithms for multiplication of decimals. These common threads further illustrate how whole number concepts extend to decimals.

Activity 1: Paper Powers

PURPOSE Develop an understanding of exponents and exponential change.

MATERIALS Sheets of newsprint, rulers, and calculators (a scientific model is best for this activity)

GROUPING Work individually or in pairs. This activity may also be done as a class activity directed by the instructor.

1. How many times do you think you can fold a sheet of newsprint if you continue to fold the result in half each time?
2. Fold the sheet in half as many times as you can. After each fold, count the number of layers and record the result for the **Number of Layers** in **Standard Form** in the table. What happens to the number of layers after each fold?
3. Now record the **Number of Layers** in **Factored Form** and in **Exponential Form** in the table.

Folds	1	2	3	4	5	6	7	8
No. Layers (Std. Form)	2							
No. Layers (Fact. Form)		2×2						
No. Layers (Exp. Form)	2^1							
Approx. Height (cm)							1.5	

4. How does your estimate for the number of folds in Exercise 1 compare to the actual number of folds you were able to make?
5. If a large sheet of newsprint is folded in half seven times as described above, the height of the stack is approximately 1.5 cm. Use this number to determine the missing entries in the **Approx. Height of the Stack** row of the table.
6. Use your calculator to extend the table and determine the number of layers needed to approximate your height.

   Your height (cm) ______ No. of layers ______ Height of the stack ______
7. If a sheet could be folded 30 times, do you think the stack would reach the top of the World Trade Center? ______ (Y/N) an orbiting satellite? ______ (Y/N) the moon? ______ (Y/N)

   Use the pattern in the table to determine the height of the stack after 30 folds. Describe how your answer compares to the height of the World Trade Center and to the distance of an orbiting satellite and the moon from the Earth.

Activity 2: The King's Problem

PURPOSE Apply problem-solving strategies in an extended activity.

GROUPING Work individually or in groups of 2–3.

GETTING STARTED Legend has it that when the inventor of the game of chess explained the game to his king, the king was so delighted that he asked the man what gift he would like as a reward.

"My wants are simple," the man replied. "If you but give me one grain of rice for the first square on the playing board, two for the second, four for the third, and so on for all sixty-four squares, doubling the number of grains each time, I will be satisfied."

1. Suppose the king agreed to the request.

1		3		5		7	
	10		12		14		16
17		19		21		23	
	26		28		30		32
33		35		37		39	
	42		44		46		48
49		51		53		55	
	58		60		62		64

   a. How many grains of rice would the inventor receive? **HINT:** How would the number of grains of rice on the seventh square compare to the total number of grains on the first six squares?

   b. How would the total number of grains of rice on the black squares compare to the total number of grains on the white squares?

2. a. How much would the number of grains of rice you found in Exercise 1(a) weigh?

   b. How many bushels of rice would this be?

   c. How large would a building need to be to hold the rice? (Make a sketch of the building and label its dimensions.)

   d. At today's prices, what would be the retail value of the rice?

3. Consult a recent world almanac.

   a. Does the United States produce enough rice in one year to satisfy the inventor's request? Explain.

   b. Is enough rice produced in the world in one year to satisfy the inventor's request? Explain.

Activity 3: What's My Name?

PURPOSE Develop the relationship between fractions and decimals.

MATERIALS Base-ten blocks (pages A-13 and A-15)

GROUPING Work individually.

GETTING STARTED Let 1 rod = 1, and 1 small cube = 0.1.

1. Write the fraction and the decimal for the shaded part of each figure. If necessary, use the cubes to determine the decimal part.

	Fraction	Decimal
a. 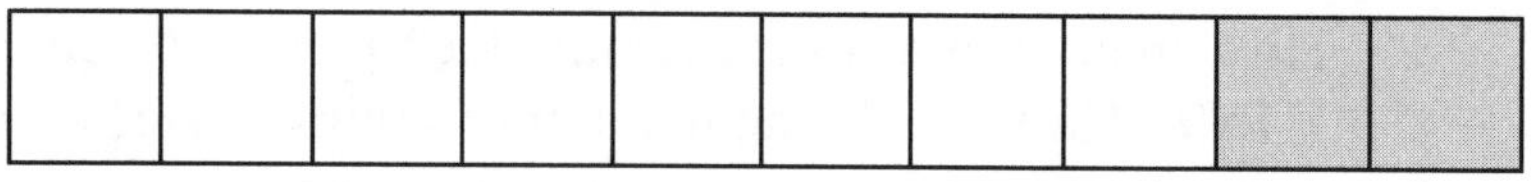	______	______
b. 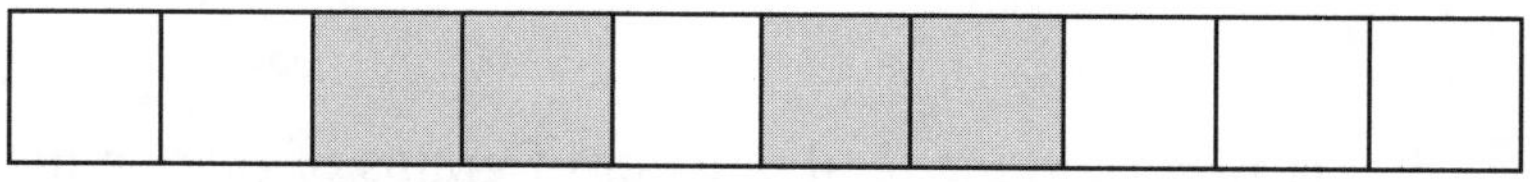	______	______
c. 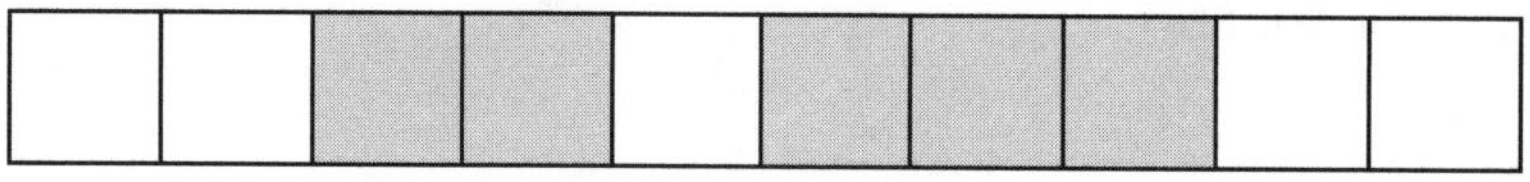	______	______
d. 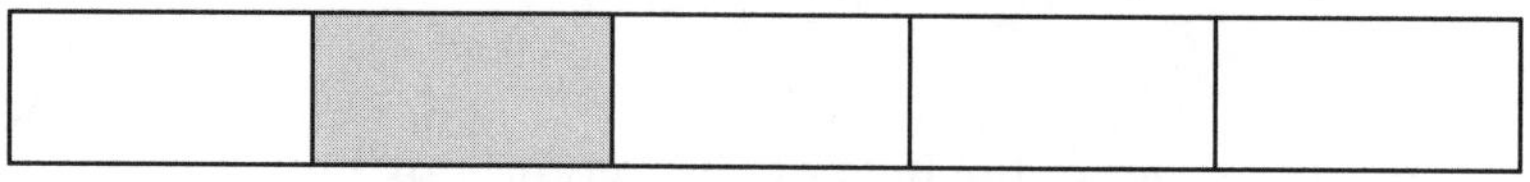	______	______
e. 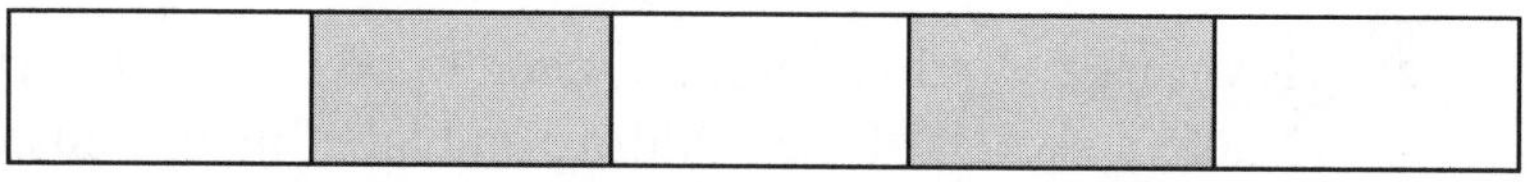	______	______
f. 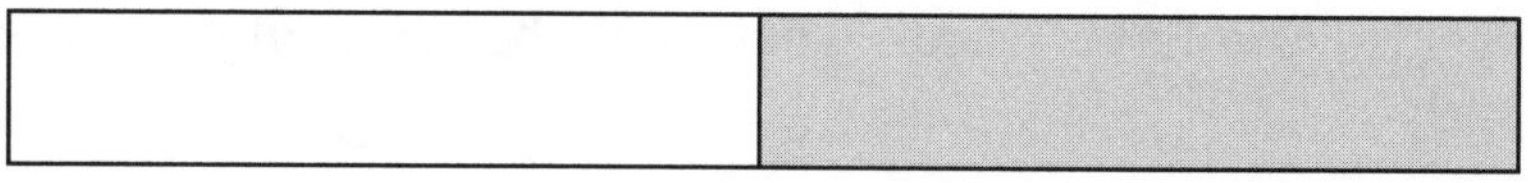	______	______

2. Match the letters of the problems that have equal decimal answers.

In the following problems, a flat = 1, a rod = 0.1, and a small cube = 0.01. Write the fraction and the decimal for the shaded part of each figure.

	Fraction	Decimal
a.	______	______
b.	______	______
c.	______	______

Shade in the 100 grid that represents a flat to show the correct fraction or decimal and fill in the missing number in each problem.

	Fraction	Decimal
a.	$\frac{63}{100}$	______
b.	______	0.07
c.	$\frac{14}{100}$	______

Activity 4: Who's First?

PURPOSE Develop an understanding of comparing and ordering decimal numbers.

MATERIALS Base-ten blocks, place-value operations board (page A-28), and a colored chip

GROUPING Work individually or in pairs.

GETTING STARTED Represent the decimal point by placing the chip between the large cube (1) and the flat (0.1). Construct each of the numbers on your place-value operations board and record the results on a student record form.

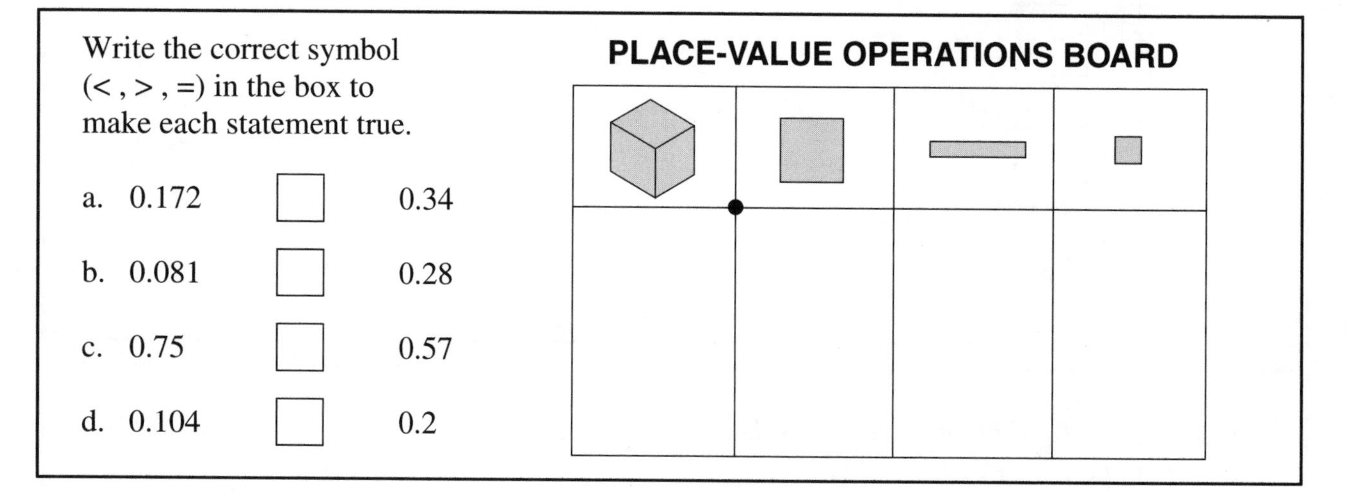

Write the correct symbol (< , > , =) in the box to make each statement true.

a. 0.172 ☐ 0.34

b. 0.081 ☐ 0.28

c. 0.75 ☐ 0.57

d. 0.104 ☐ 0.2

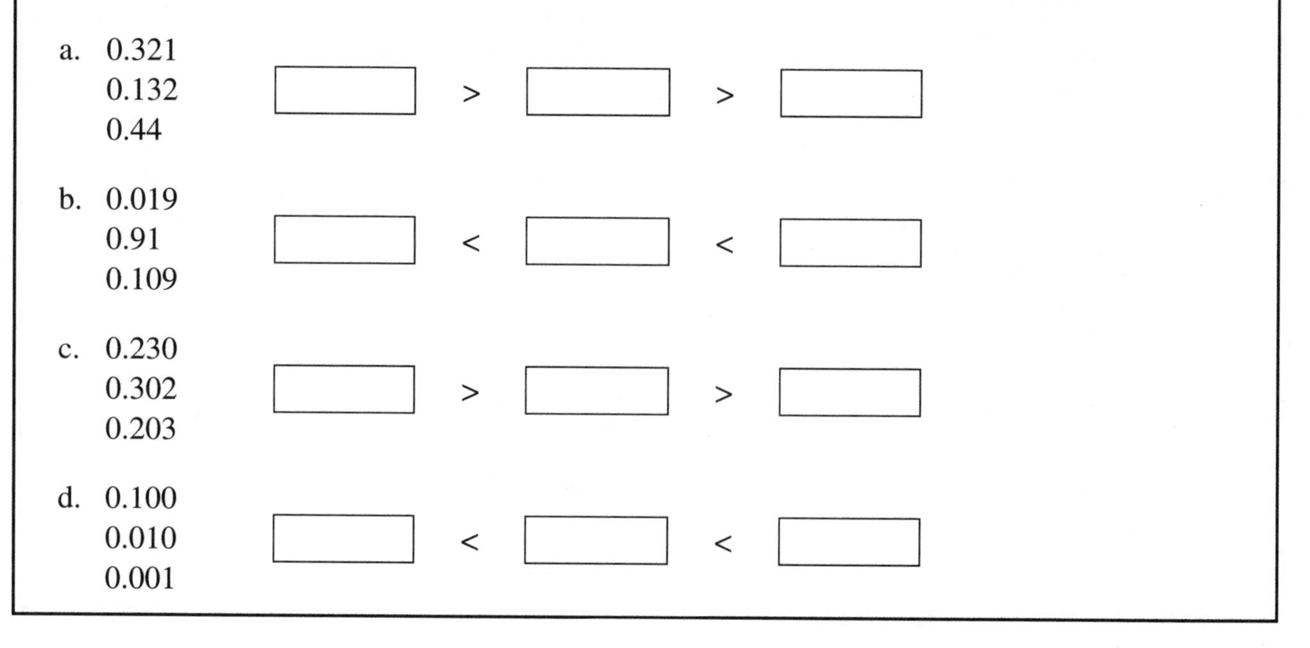

In each of the following problems, place the decimals in the correct boxes to make the statement true.

a. 0.321, 0.132, 0.44 ☐ > ☐ > ☐

b. 0.019, 0.91, 0.109 ☐ < ☐ < ☐

c. 0.230, 0.302, 0.203 ☐ > ☐ > ☐

d. 0.100, 0.010, 0.001 ☐ < ☐ < ☐

Activity 5: Race for the Flat

PURPOSE Develop an algorithm for addition of decimals.

MATERIALS Base-ten blocks and number cubes, labeled as shown

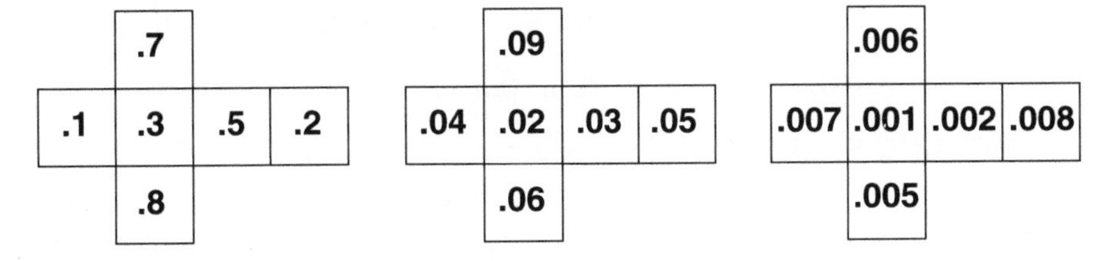

GROUPING Work in pairs.

GETTING STARTED In this game, the flat = 1, the rod = 0.1, and the cube = 0.01. Players alternate rolling the tenths and hundredths dice. After each roll, a player collects a number of base-ten blocks equivalent to the corresponding numbers on the dice and places them on his or her flat. On each roll after the first, a player adds the new blocks to those on the flat, trading when necessary. Each addition must be recorded and validated by the blocks on the flat as shown below. The winner is the first player to cover the flat exactly. On any turn, a player may choose to roll only one die.

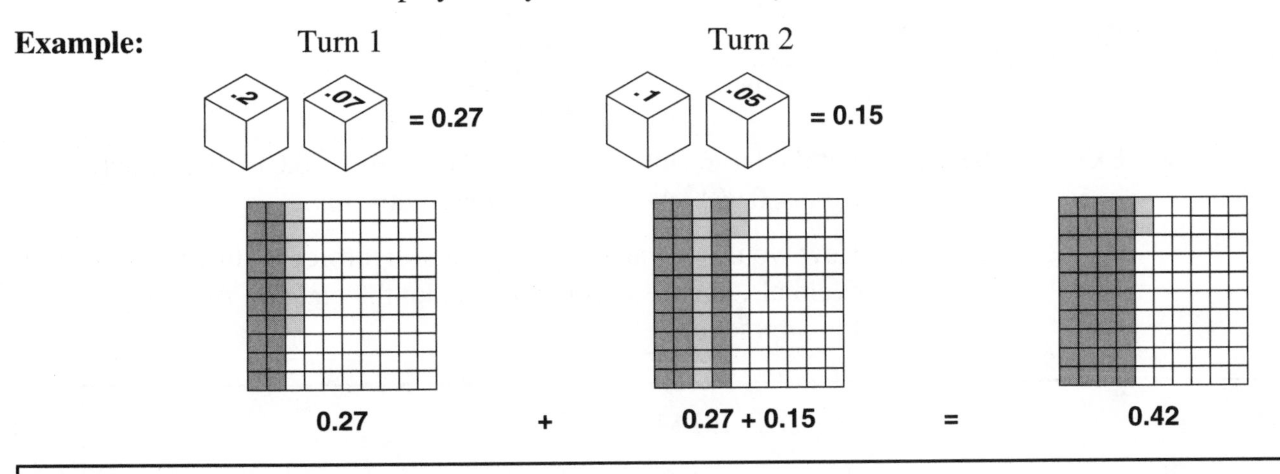

Play another game in which a player does not have to fill the flat exactly. Choose a goal for the game from one of the following:

a. The winner is the person whose number is
b. The winner is the person whose number is

Players roll the dice and add blocks to the flat as above. For Goal (a), a player's strategy will determine when to stop rolling the dice. For Goal (b), players stop adding blocks on the roll that covers the flat, with or without some remainder.

At the conclusion of a game, players compare the decimal numbers represented by all of their blocks. The winner is the person whose number meets the goal of the game.

Activity 6: Empty the Board

PURPOSE Develop an algorithm for subtraction of decimals.

MATERIALS Base-ten blocks, tenths and hundredths decimal dice (see Activity 5), a colored chip, and the place-value operations board

GROUPING Work in pairs.

GETTING STARTED In this activity, a flat = 1, a rod = 0.1, and the small cube = 0.01. Each player arranges a stack of five flats on his/her place-value operations board. Players alternate turns rolling the dice. After each roll, a player removes blocks equivalent to the sum of the numbers on the dice, making trades when necessary. After each roll, a player must record the subtraction problem that was modeled by removing the blocks. The first player to empty the entire board is the winner.

Example:

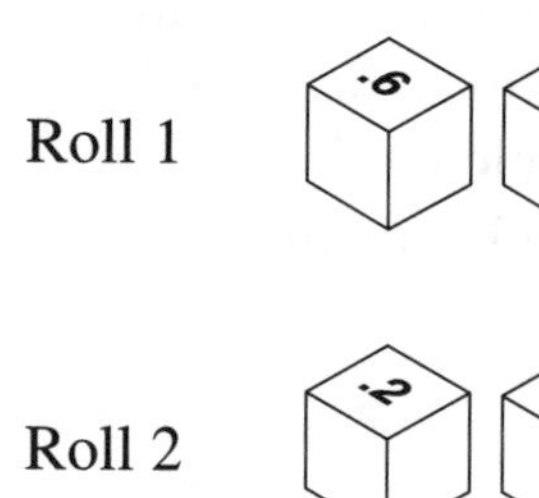

$$\begin{array}{r} 5.00 \\ -\ .69 \\ \hline 4.31 \end{array}$$

$$\begin{array}{r} 4.31 \\ -\ .23 \\ \hline 4.08 \end{array}$$

PLACE-VALUE OPERATIONS BOARD

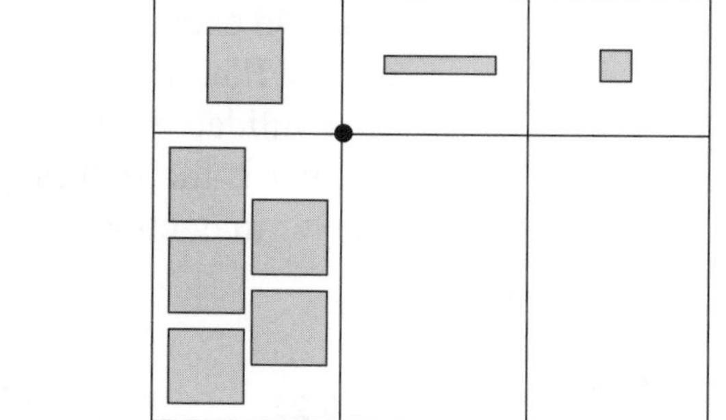

EXTENSIONS

1. Let the large cube = 1, a flat = 0.1, a rod = 0.01, and a small cube = 0.001. Also use the thousandths die.
2. How would you adapt this game to reinforce addition of decimals, and how would you adapt the game in Activity 5 for subtraction?

DECIMAL PUZZLE

In each of the given numbers, the decimal point may be placed in front of the first digit, behind the last digit, or between any two digits. Place the decimal point in each number so that the sum of the resulting numbers is 100.

$$\begin{array}{r} 68 \\ 959 \\ 481 \\ 1734 \\ 2279 \\ +4011 \\ \hline 100 \end{array}$$

Activity 7: Decimal Arrays

PURPOSE Develop an algorithm for multiplication of decimals.

MATERIALS Base-ten blocks

GROUPING Work individually or in pairs.

GETTING STARTED Recall from Activity 7 in Chapter 5 on multiplication of fractions that 0.2×0.3 means 0.2 of 0.3. In this activity, a flat = 1, a rod = 0.1, and the small cube = 0.01.

Example: To determine 0.2×0.3, place a flat on your paper and label the factors as shown.

0.3

0.2

Place three rods vertically on the flat as shown to model 0.3.

0.3

0.2

Place two rods horizontally as shown to model 0.2.

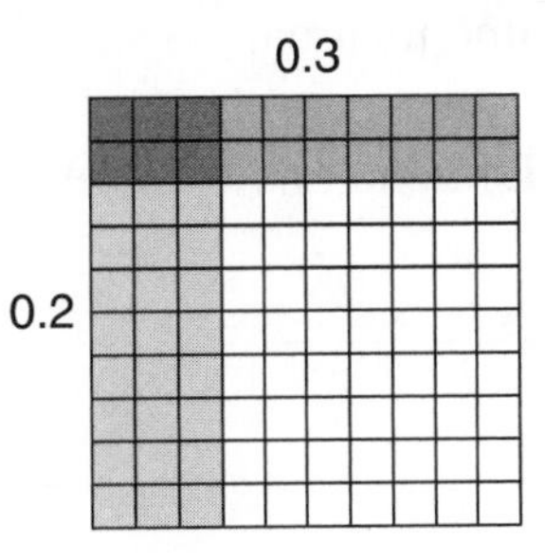

The product of 0.2×0.3 is determined by the rectangular array formed by the overlapping parts of the rods. The six squares in the array represent 0.06 (*6 parts of 100*).

So, 0.2×0.3 or 0.2 of 0.3 is 0.06.

Use base-ten blocks to solve these multiplication problems. Record your results on the 100 grids by labeling the dimensions (factors) and shading in the correct number of rods. Find the number of squares in the overlapping area to determine each product.

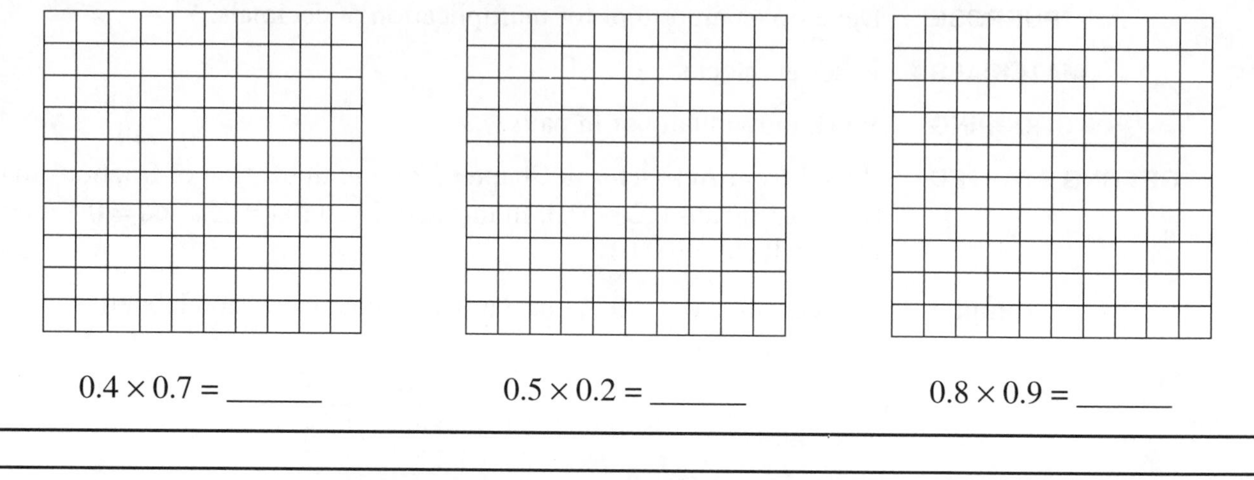

$0.4 \times 0.7 =$ ______ $0.5 \times 0.2 =$ ______ $0.8 \times 0.9 =$ ______

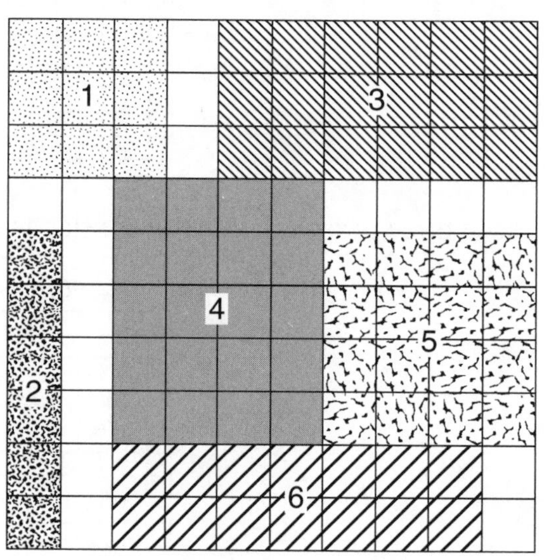

The 100 grid above represents a flat that is equal to 1. Write the dimensions for each shaded rectangle and determine its area.

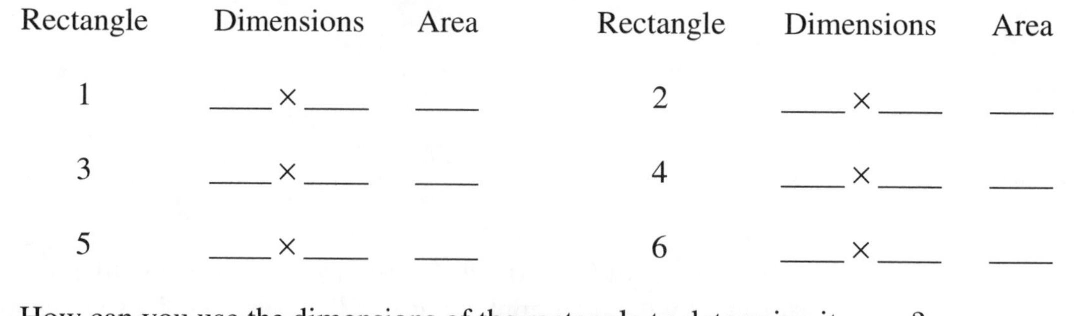

Rectangle	Dimensions	Area	Rectangle	Dimensions	Area
1	____ × ____	____	2	____ × ____	____
3	____ × ____	____	4	____ × ____	____
5	____ × ____	____	6	____ × ____	____

How can you use the dimensions of the rectangle to determine its area?

Activity 8: Decimal Multiplication

PURPOSE Use base-ten blocks to extend multiplication of decimals.

MATERIALS Base-ten blocks and the multiplication and division frame (page A-25)

GROUPING Work individually.

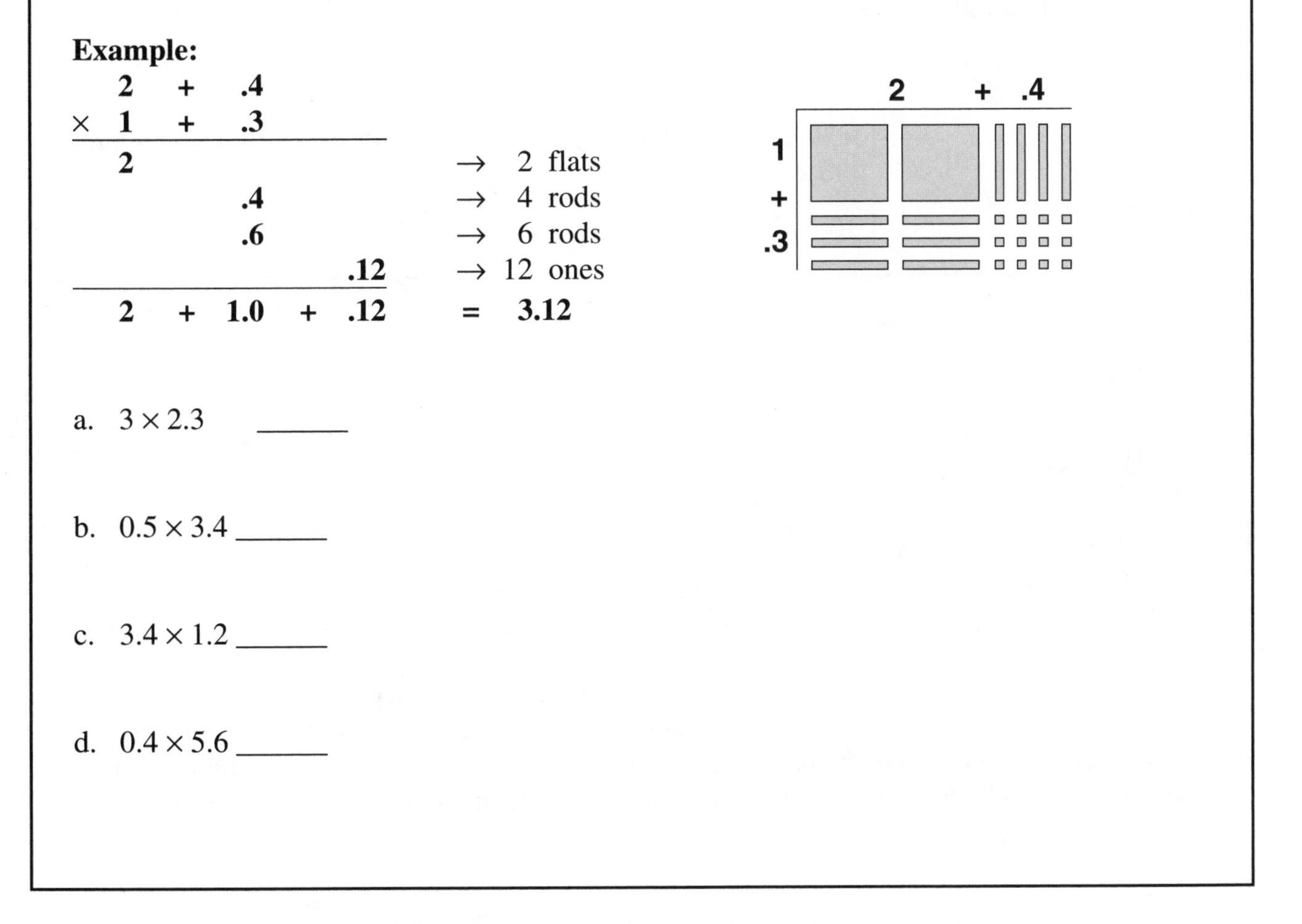

For these problems, a flat = 1, a rod = 0.1, and a small cube = 0.01. Model the multiplication by using base-ten blocks to construct a rectangle in the multiplication and division frame. Then determine the product of the two factors. Record your work as shown in the example.

Example:

	2	**+**	**.4**				
×	**1**	**+**	**.3**				
	2					→	2 flats
			.4			→	4 rods
			.6			→	6 rods
					.12	→	12 ones
	2	**+**	**1.0**	**+**	**.12**	=	**3.12**

a. 3×2.3 ______

b. 0.5×3.4 ______

c. 3.4×1.2 ______

d. 0.4×5.6 ______

EXTENSIONS Construct some rectangular arrays to represent the product of two decimals. Give the arrays to another student, identifying the correct decimal value for each base-ten block. Instruct the student to determine the factors of your number.

Activity 9: Decimal Decisions

PURPOSE Use estimation to develop a rule for the placement of the decimal point in the product of two decimal numbers.

GROUPING Work individually.

GETTING STARTED In the problems below, the values of the two factors in $436 \times 27 = 11772$ have been changed through different placement of the decimal point. Use a rounding strategy to determine a whole-number estimate for the product. Then use the estimate to help place the decimal point in the final product.

Example:

$$\begin{array}{rcr} 4.36 & \approx & 4.00 \\ \underline{\times 27} & \approx & \underline{\times 30} \\ & & 120 \end{array}$$

$4 \times 30 = 120$, so $4.36 \times 27 \approx 120$. Since the exact digits in the product (4.36×27) are 11772,

$4.36 \times 27 = 117.72$.

1. $\begin{array}{rcl} 43.6 & \approx & \_\_\_\_\_\_ \\ \underline{\times 27} & \approx \times & \_\_\_\_\_\_ \\ 11772 & & \_\_\_\_\_\_ \end{array}$

2. $\begin{array}{rcl} 43.6 & \approx & \_\_\_\_\_\_ \\ \underline{\times 2.7} & \approx \times & \_\_\_\_\_\_ \\ 11772 & & \_\_\_\_\_\_ \end{array}$

3. $\begin{array}{rcl} 4.36 & \approx & \_\_\_\_\_\_ \\ \underline{\times 2.7} & \approx \times & \_\_\_\_\_\_ \\ 11772 & & \_\_\_\_\_\_ \end{array}$

HINT: $0.436 \approx \frac{1}{2}$

4. $\begin{array}{rcl} 0.436 & \approx & \frac{1}{2} \\ \underline{\times 27} & \approx \times & \_\_\_\_\_\_ \\ 11772 & & \_\_\_\_\_\_ \end{array}$

5. $\begin{array}{rcl} 436 & \approx & \_\_\_\_\_\_ \\ \underline{\times 0.27} & \approx \times & \_\_\_\_\_\_ \\ 11772 & & \_\_\_\_\_\_ \end{array}$

6. $\begin{array}{rcl} 0.436 & \approx & \_\_\_\_\_\_ \\ \underline{\times 0.27} & \approx \times & \_\_\_\_\_\_ \\ 11772 & & \_\_\_\_\_\_ \end{array}$

In each of these problems, you used estimation to determine the placement of the decimal point in the product. Now look at the relationship between **the place values of the last digit in each factor** and **the place value of the last digit in the product.** Use your findings to write a general rule that describes the placement of the decimal point in the product of two decimal numbers.

Use your rule to place the decimal point in each product.

1. $\begin{array}{r} .025 \\ \underline{\times 0.81} \\ 2025 \end{array}$

2. $\begin{array}{r} .037 \\ \underline{\times 0.042} \\ 1554 \end{array}$

3. $\begin{array}{r} .0063 \\ \underline{\times 0.017} \\ 1071 \end{array}$

Activity 10: Target Number Revisited

PURPOSE Develop number and operation sense, estimation skills, and an increased understanding of multiplication and division of decimals.

MATERIALS A calculator for each pair of students

GROUPING Work in pairs.

GETTING STARTED Recall the Target Number game in Activity 13 in Chapter 3. All computations will be done on the calculator, but determining the guesses to enter must be done mentally.

Game 1 Player 1 chooses the target product and Player 2 chooses the factor. Then follow the steps in the flow chart.

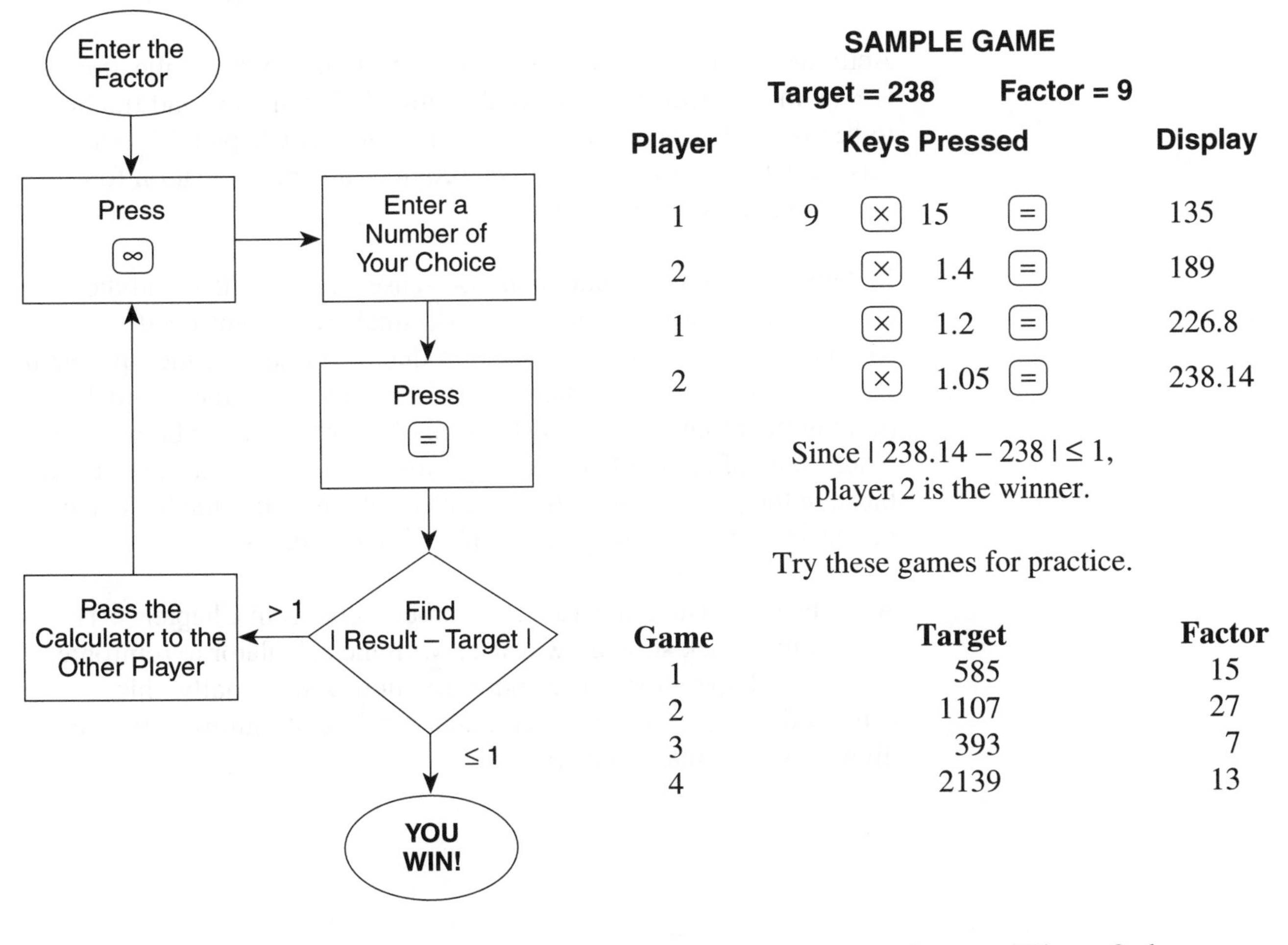

SAMPLE GAME

Target = 238 **Factor = 9**

Player	Keys Pressed	Display
1	9 × 15 =	135
2	× 1.4 =	189
1	× 1.2 =	226.8
2	× 1.05 =	238.14

Since $|238.14 - 238| \leq 1$, player 2 is the winner.

Try these games for practice.

Game	Target	Factor
1	585	15
2	1107	27
3	393	7
4	2139	13

EXTENSIONS **Game 2** Player 1 chooses a target quotient, and Player 2 chooses a product. Follow the steps in the flow chart, but enter the product and use the ÷ key instead of the × key. *The winner is the player who gets within* 0.5 *of the target quotient.*

Chapter Summary

The activities in this chapter stressed the conceptual development of decimals. The activities continued the constructivist design of Chapter 5. In Activity 3, models were used to demonstrate the relationship between fractions and decimals. The activities progressed from the concrete to the representational to the abstract or symbolic level.

The game format used in Activities 5 and 6 gave you the opportunity to physically place rods and cubes on a flat during the addition and subtraction process and then to make the necessary trades in order to represent the answer with the least number of blocks. This physical modeling of the operations and the trading and regrouping processes is essential for understanding the computational algorithms.

Activities 7 and 8 used rectangular arrays to model multiplication and to extend the processes developed with whole numbers and fractions to decimals. The area model will be revisited in Chapter 12. Using this model in a variety of settings provides a firm foundation for understanding the concept of area.

In Activity 8, number and operation sense were revisited and the connections between fractions and decimals were reinforced. Rounding of factors to estimate a product in a multiplication problem was introduced. The estimate was then used to place the decimal point in the product. An investigation of the relationship between the place value of the digits after the decimal point in the two factors and those in the product led to the discovery of a general rule for placing the decimal point in the product of decimal numbers.

Activity 10 revisited the Target Number game from Chapter 3. The Guess and Check strategy was used with the calculator to reinforce number and operation sense and estimation skills. Finally, this activity developed an increased understanding of multiplication and division of decimal numbers.

Chapter 7
Applications of Mathematics

"Understanding the concept of variable is crucial to the study of algebra: a major problem in students' efforts to understand and do algebra results from a narrow interpretation of the term. . . . Students must explore interesting problems, applications, and situations so that they will make the appropriate connection.

"Mathematics arises not only in science, but in other disciplines as well. In social studies, the study of maps is an excellent time to study scaling and its relation to similarity, ratio, and proportion."

—*Curriculum and Evaluation Standards for School Mathematics*

Pure mathematics is a creation of the human mind. However, using variables to represent quantities and develop equations that model real-world situations enables us to use mathematics to solve problems in science, business, social science, economics, etc. Solving the equations may then provide a general solution to many related problems.

In this chapter, variables will be introduced as a means of translating written statements into the symbolic language of mathematics. The concepts of ratio and proportion will be studied in real-world applications that make connections to science, nature, and geography.

Development of the concept of percent is most effective when it is based on the same models and understanding as fractions and decimals. The word percent means part of 100. The idea of part of a whole relates to the concept of a fraction. When the whole is 100, there is also a direct connection to decimals. The activities in this chapter continue the extension of whole-number concepts to percent by applying models that have been used previously with whole numbers, fractions, and decimals.

Activity 1: Equations

PURPOSE Extend the charged-particle model for integers to model and solve linear equations.

MATERIALS Two different-colored chips (or colored squares), 15 of each color, and two different-colored paper cups, five of each color

GROUPING Work individually or in groups of 2–3.

GETTING STARTED In this activity, represents an **unknown** number of protons or electrons (x) and represents an **unknown** number of particles with the opposite charge (^{-}x). Because and have opposite charges, when they are paired together they have an electrical charge of zero. Individual protons and electrons are represented by ● and ○, respectively.

1. Complete each of the following statements by placing the appropriate number of protons or electrons in the last oval. Explain your answers.

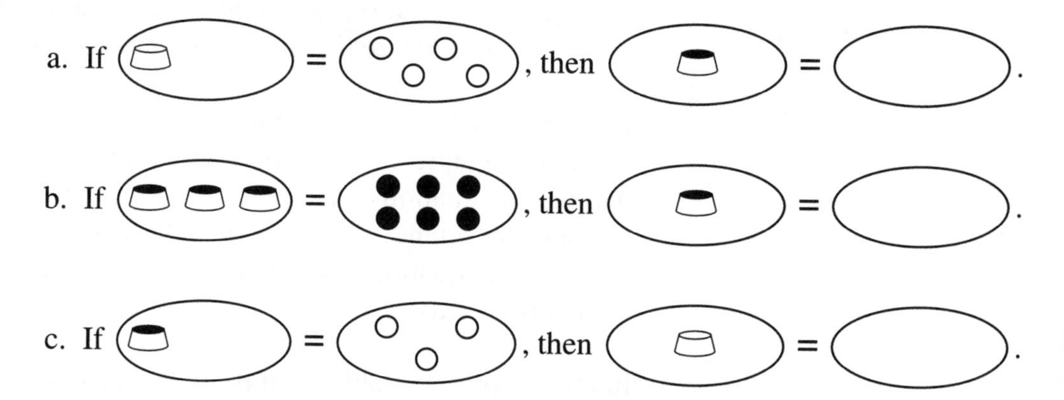

2. What equation is represented by each of the following?

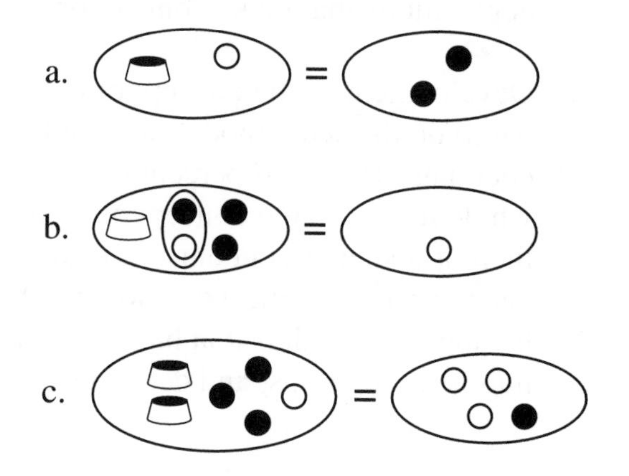

3. Use two different-colored chips and cups to represent each of the following equations, and sketch the result.

   a. $x + {}^-5 = 4$ b. $2x + 3 = {}^-5$ c. ${}^-3x + 5 = {}^-1$

Example 1: Solve the equation $x + 2 = {}^-3$.

Begin by modeling the equation.

If necessary, rename by adding 0.

Perform the operations needed to get just [cup] on one side of the equation.

Solution: $x = {}^-5$

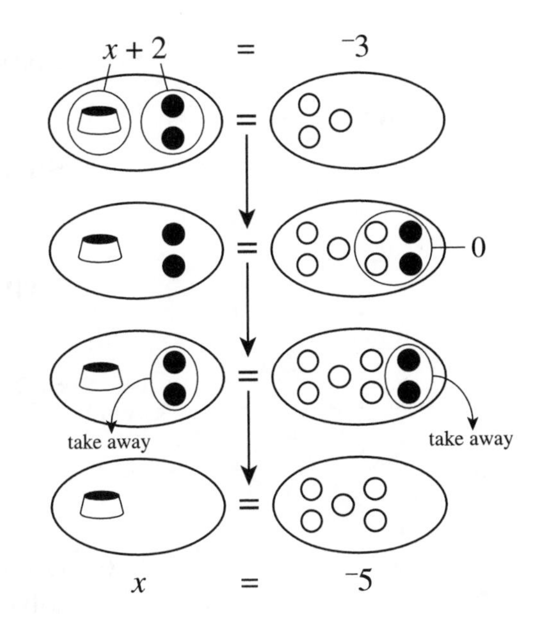

Example 2: Solve the equation $2x - 3 = {}^-5$.

Rewrite the subtraction as addition of the opposite.

Model the equation.

Perform the operations needed to get just [cup] on one side.

Solution: $x = {}^-1$

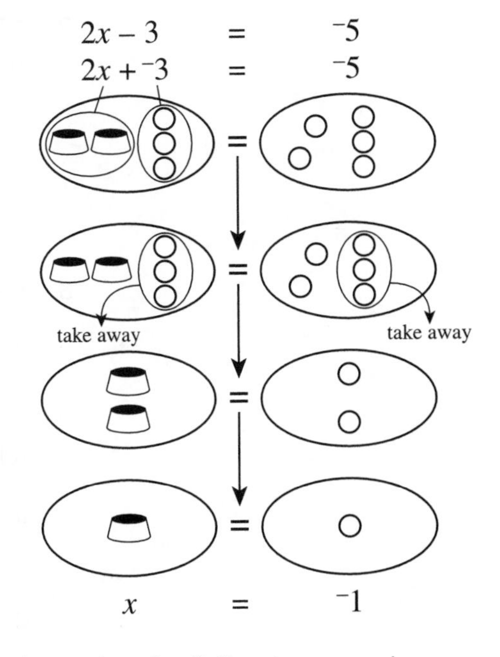

4. Use the cups and chips model to solve the following equations.

   a. $x + 3 = 7$ b. $4 + x = {}^-2$ c. $x - {}^-2 = 4$

   d. $3x + 4 = {}^-5$ e. $2x - {}^-5 = {}^-3$ f. $2x - 3 = x + 1$

5. One student's method for solving the equation $2x - 3 = {}^-5$ is shown below. Explain what she did at each step and why it works.

$$2x - 3 = {}^-5$$

Step 1: $2x + {}^-3 = {}^-5$

Step 2:

Step 3:

Step 4:

Step 5:

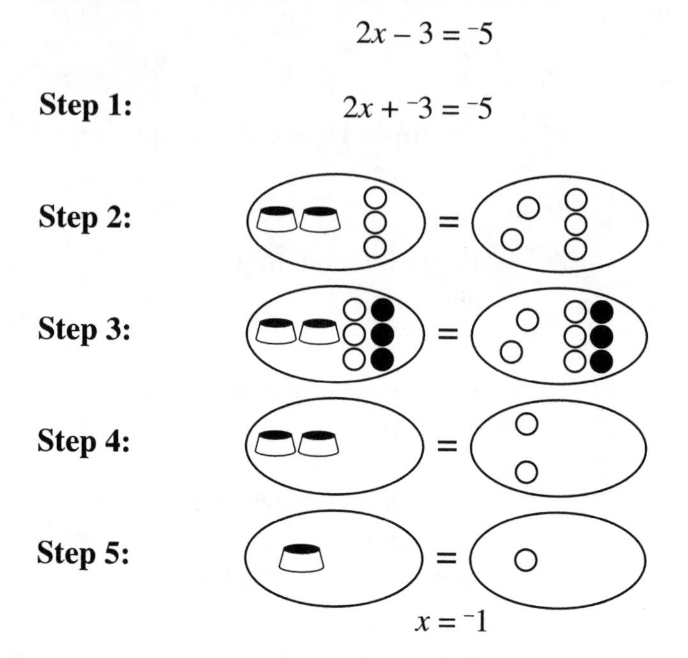

$$x = {}^-1$$

6. Use the cups and chips model and the method in Exercise 5 to solve the following equations.

   a. $x + 4 = {}^-2$ b. $x - 2 = {}^-3$ c. $2x + 5 = 3$

7. Make a sketch to explain how you could use the cups and chips model to solve the equation ${}^-x + 3 = 4$.

8. Use the cups and chips model to solve the following equations.

   a. ${}^-x + 5 = {}^-1$ b. ${}^-4 + {}^-x = 2$ c. ${}^-3x + 2 = {}^-4$

9. Solve the following equations using the cups and chips model when necessary. Show your work to justify your answers.

   a. $3 + x = 8$ b. ${}^-4 = {}^-5 + x$ c. $4 = {}^-5 - x$

   d. $6 - {}^-x = 5$ e. $2x + 1 = {}^-x + {}^-8$ f. $6x + 3 = x - 2$

   g. ${}^-10 + {}^-x = x - 2$ h. $2x = {}^-x - {}^-12$ i. ${}^-4 - {}^-5x = x + 20$

Activity 2: Magic Number Tricks

PURPOSE Develop increased understanding of translating the words in a written problem into algebraic expressions.

MATERIALS Calculator

GROUPING Work individually.

Dr. Wonderful, the Mathematical Magician, astounds crowds with his amazing ability to read a person's mind. Here are the directions that he gives to five people in the crowd. Follow Dr. Wonderful's directions and complete the table below. Choose a different number for each person.

PERSON	1	2	3	4	5
1. Pick any number.	_____	_____	_____	_____	_____
2. Multiply by 3.	_____	_____	_____	_____	_____
3. Add 30.	_____	_____	_____	_____	_____
4. Divide by 3.	_____	_____	_____	_____	_____
5. Subtract your original number.	_____	_____	_____	_____	_____

Dr. Wonderful tells the people to write their answers on a sheet of paper but not to reveal them to anyone else. He closes his eyes, concentrates deeply, and then claims that he knows each person's answer.

Let n be the number. Write an algebraic expression and record the result for each step above.

Step 1: Step 2:

Step 3: Step 4:

Step 5: The result is ________.

One of Dr. Wonderful's other mind-reading tricks involves birthdays. Use a calculator and follow along with the crowd as he gives the directions. Press [=] on your calculator after each step.

1. Enter the month of your birthday.
2. Multiply by 5.
3. Add 20.
4. Multiply by 4.
5. Subtract 7.
6. Multiply by 5.
7. Add the day of your birthday.
8. Subtract the number of days in a non-leap year.

"Oh," "Ah," and "Look at that" can be heard throughout the crowd. What do people see on the display of their calculator?

To the right of each step above, write an algebraic expression that correctly describes Dr. Wonderful's direction. Study the sequence of expressions and then explain how place value helps to explain how this "magic number trick" works.

EXTENSIONS Write some magic tricks of your own. Write the algebraic expression for each step so that you can justify your final result. Try them out on your classmates.

Activity 3: People Proportions

PURPOSE Reinforce the concept of ratio and apply ratios in a real-world setting that makes connections to science.

MATERIALS A calculator and two 150-cm measuring tapes for each group of students

GROUPING Work in groups of 4–5.

GETTING STARTED To gain an understanding of the comparisons of body measurements, measure your hand span—the largest spread between the tip of the thumb and the tip of the middle finger. Wrap your hand around your wrist to compare the hand span with the circumference of the wrist; they should be about equal. Place both hands around your neck. The circumference of your neck is twice your hand span, or twice the circumference of your wrist.

Measure the body parts listed below for each student in your group. All measurements should be in centimeters and rounded to the nearest 0.5 cm. Follow the directions below when making the measurements.

A. Height and navel to floor: Measure without shoes.

B. Wing span: Measure fingertip to fingertip with arms outstretched.

C. Tibia: Rest the foot on the floor and measure from the ankle bone to the top of the tibia on the outside of the kneecap.

D. Radius: Rest the elbow on the desk with the hand up and measure from the tip of the elbow to the wrist bone.

Enter the name of the person and each measurement for that person in Table 1.

TABLE 1

Name of Person	Height	Navel to Floor	Wing Span	Wrist	Radius	Neck	Tibia

Use the measurements from Table 1 to write the fractional form of the ratios for each person in the columns labeled F in Table 2. Do not convert the fractions to decimals.

TABLE 2

Name of Person	Wing Span / Height		Height / Navel to Floor		Tibia / Height		Radius / Height		Wrist / Neck	
	F	D	F	D	F	D	F	D	F	D

1. Examine the ratios for the people in your group. Are any of the ratios nearly the same for all people? ______
   If so, which ones?

2. Now convert all the fractions to decimals, rounded to the nearest thousandth. Compare the decimal equivalents for each of the ratios. Which ratios are approximately the same for all people?

3. Find the average (mean) of the decimal ratios for tibia to height and radius to height.

   a. The length of the tibia is approximately what fraction of the height? ______

   b. The length of the radius is approximately what fraction of the height? ______

4. Find the average of the ratios for height to navel-to-floor as a decimal. ______

   For navel-to-floor to height. ______

5. The average ratio for height to navel-to-floor for all humans is approximately 1.618. Compare this decimal with the average from your group in Exercise 4.

6. Write the reciprocal for the decimal 1.618 rounded to the nearest thousandth. ____

   What did you find?

   Can you find any other decimal with this property?

Paleontologists may find only a few bones of an ancient person, yet be able to reconstruct an entire skeleton. A forensic medical examiner may provide police with the height, weight, age, sex, and other features of a body from as little evidence as a skull or two or three bones. How can the skeleton of a human be reconstructed with so little evidence?

You found that several ratios for body measurements for all the people in your group were approximately the same. Would you expect the same results for people of any age, size, or shape?

Make the same measurements for several people whose age or size differs from the people in your group. Determine if the ratios are about the same as those of your group.

Research the *Golden Ratio.* Compare the *Golden Ratio* to your answer in Exercise 4. Find other body ratios that are close to the *Golden Ratio*. Describe at least one application of the *Golden Ratio* in art, music, nature, architecture, psychology, phyllotaxis, or sea shells.

WHY NOT?

Use a proportion to solve each of the following:

1. Josh Jumper can long jump 5 m with a 15-m run. How far could he jump with a 75-m run? a 150-m run?

2. Wally Walker says he can walk 6 km in an hour. How long would it take him to walk 60 km? 150 km?

3. Tall Tanya was 145 cm tall when she was 12 years old. When she was 15, she was 165 cm tall. How tall will she be when she is 24? 30?

For each problem, explain why solving the proportion results in numbers that do not fit the real-life situation.

Activity 4: A Call to the Border

PURPOSE Apply ratio and proportion, measurement, scale interpretation, estimation, and communication in a problem-solving context.

MATERIALS Almanac or atlas, state map, ruler, string, and scissors

GROUPING Work in groups of four.

GETTING STARTED

The Daily Tabloid

Governor Cuts State Spending
Eliminates National Guard

The governor defends this action by claiming that if he ordered every man, woman, and child to go to the border of the state, all residents could stand side-by-side around the state to defend the border.

Do you think the governor's claim is true?

If the residents of your state lined up along the state's border, do you think people standing next to each other would be able to:

- See each other?
- Carry on a conversation?
- Touch fingertip-to-fingertip?
- Stand shoulder-to-shoulder?
- Stand belly-to-back?
- Hold hands?

1. List the information you will need to verify the governor's claim and your prediction of how close people will be able to stand.

2. Explain how the perimeter of your state is affected by irregularities such as coastline, islands, river boundaries, mountains, etc.

3. Record the following:

   State __________ Population __________ Perimeter of state __________

4. Is the governor's claim correct? Why or why not? Explain your reasoning.

EXTENSIONS In which states might this solution to state spending cuts NOT work? Which states pose special problems in implementing your method for solving the problem? Name some states where residents might be standing closer together or farther apart than in your state.

Activity 5: What Is Percent?

PURPOSE Develop the concept of percent as a part of 100.

MATERIALS A transparent copy of a 100 grid (page A-36)

GROUPING Work individually.

GETTING STARTED Note that the % symbol is a special arrangement of a "1" and two "0s."

0 / 0 100

Percent means *part of* 100.

Example: 40% means 40 equal parts out of 100.

80 correct out of 100 problems $= \frac{80}{100} = 80\%$.

Place a copy of the 100 grid over each of the square. Count the shaded squares to determine what percent of each figure is shaded.

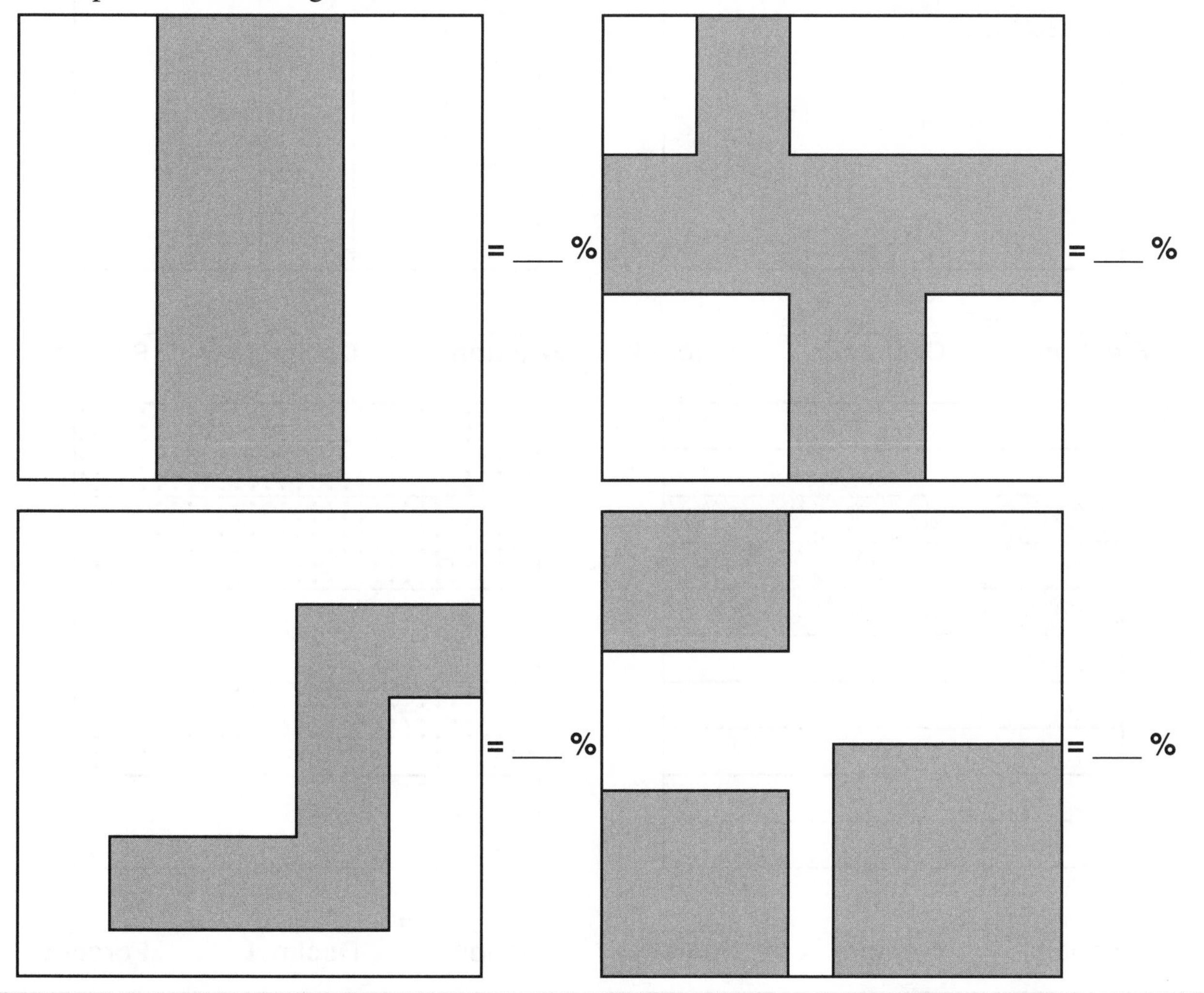

Activity 6: From Fractions to Decimals to Percents

PURPOSE Develop an understanding of the relationships among fractions, decimals, and percents.

MATERIALS A copy of a 100 grid

GROUPING Work individually.

GETTING STARTED Recall that any part of a whole can be expressed as a fraction, ratio, decimal, or percent.

Name the fractional part of each figure that is shaded. Use your 100 grid to determine the equivalent decimal and percent.

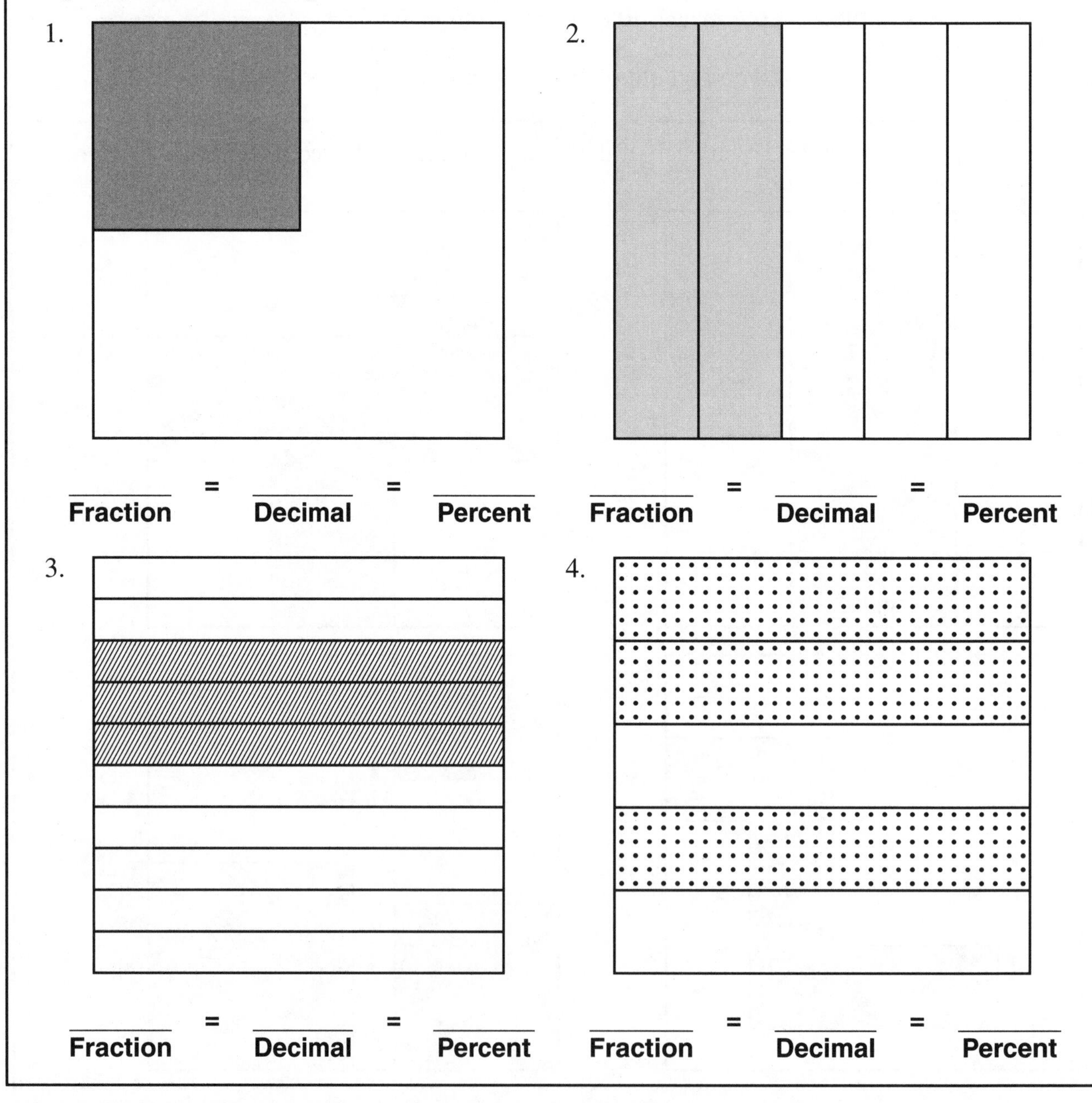

Activity 7: What Does It Mean?

PURPOSE Reinforce the relationships among fractions, ratios, and percents.

MATERIALS A copy of a 100 grid

GROUPING Work individually.

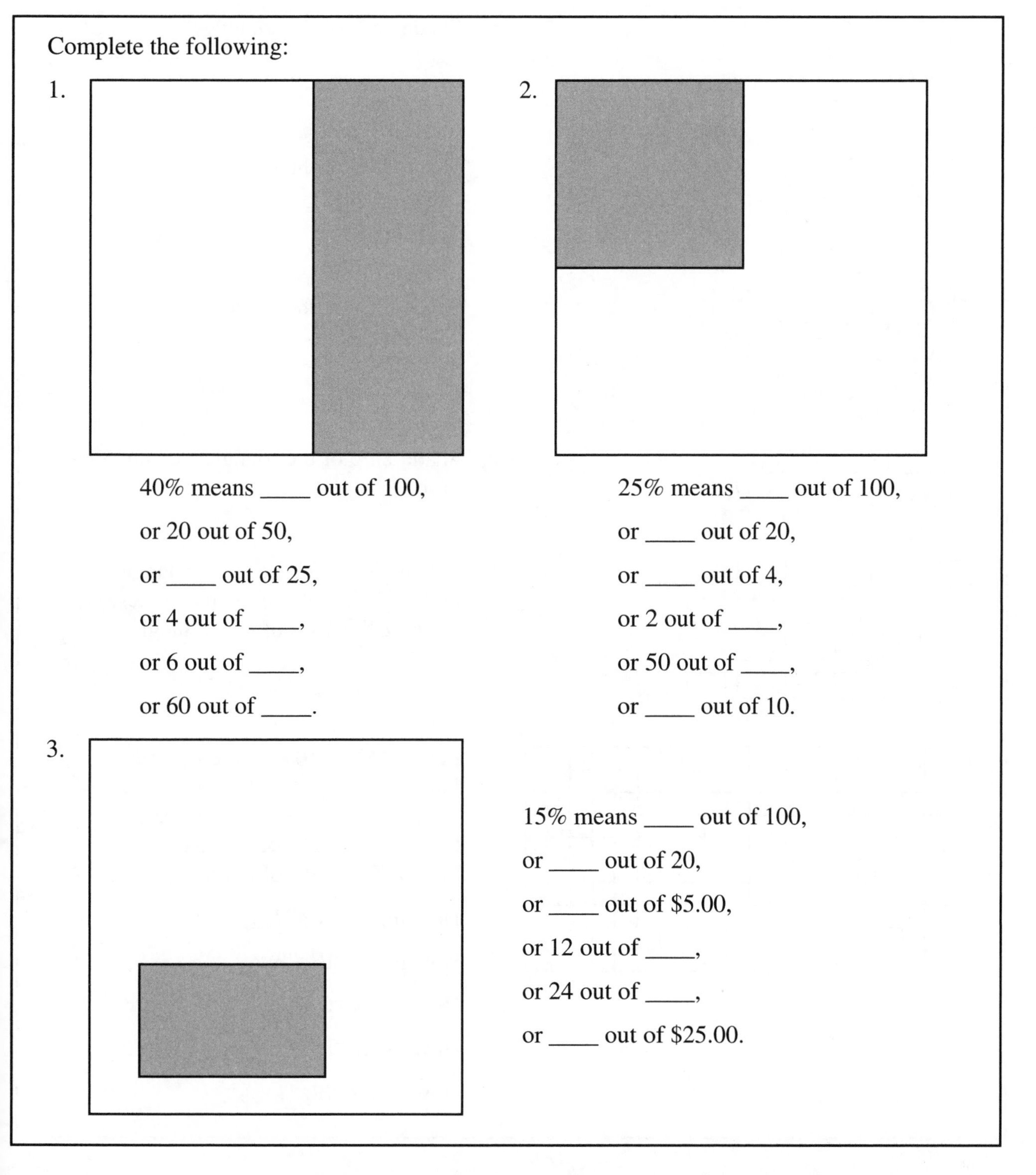

Complete the following:

1.

40% means ____ out of 100,
or 20 out of 50,
or ____ out of 25,
or 4 out of ____,
or 6 out of ____,
or 60 out of ____.

2.

25% means ____ out of 100,
or ____ out of 20,
or ____ out of 4,
or 2 out of ____,
or 50 out of ____,
or ____ out of 10.

3.

15% means ____ out of 100,
or ____ out of 20,
or ____ out of $5.00,
or 12 out of ____,
or 24 out of ____,
or ____ out of $25.00.

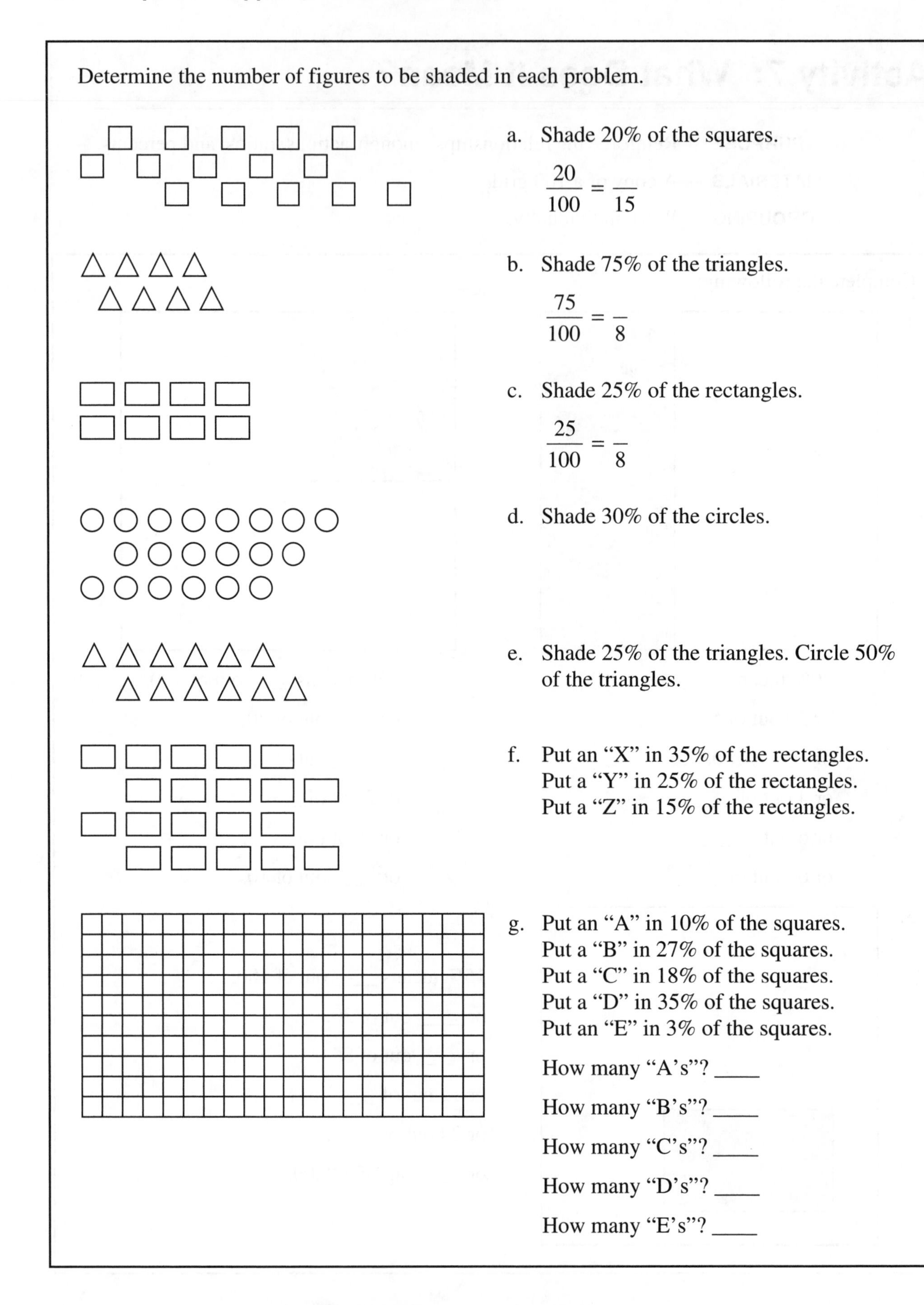

Determine the number of figures to be shaded in each problem.

a. Shade 20% of the squares.

$$\frac{20}{100} = \frac{}{15}$$

b. Shade 75% of the triangles.

$$\frac{75}{100} = \frac{}{8}$$

c. Shade 25% of the rectangles.

$$\frac{25}{100} = \frac{}{8}$$

d. Shade 30% of the circles.

e. Shade 25% of the triangles. Circle 50% of the triangles.

f. Put an "X" in 35% of the rectangles.
Put a "Y" in 25% of the rectangles.
Put a "Z" in 15% of the rectangles.

g. Put an "A" in 10% of the squares.
Put a "B" in 27% of the squares.
Put a "C" in 18% of the squares.
Put a "D" in 35% of the squares.
Put an "E" in 3% of the squares.

How many "A's"? ____

How many "B's"? ____

How many "C's"? ____

How many "D's"? ____

How many "E's"? ____

Activity 8: Finding Percents

PURPOSE Use estimation to find the percent of a number, develop a method for determining the percent of a number, and apply percent to real-world problems.

GROUPING Work individually.

Complete the following:

a. Carmen's score on her last mathematics test was 80%. If the test had 40 problems, how many did she get correct?

Recall that 80% means 80 parts out of 100.

80% of 40 means $\frac{80}{100}$ of 40,

or $\frac{80}{100} \times 40 =$ ______,

or $0.80 \times 40 =$ ______,

or $\frac{4}{5} \times 40 =$ ______.

b. A new ski jacket is on sale at a discount of 25%. How much will you save if the original price was $128.00?

25% of 128 means $\frac{25}{100}$ of 128,

or $\frac{25}{100} \times 128 =$ ______,

or $0.25 \times 128 =$ ______,

or $\frac{1}{4} \times 128 =$ ______.

This sign advertises a fall discount sale.

Is the sale price correct? ______

If not, what should it be? ______ Explain.

Garden Hose
40% Discount

Regular Price $19.95

Sale Price $12.95

MENTAL MATH AND PERCENTS

You can use estimation and mental math to determine a good approximation for the percent of a number.

Example: The total bill at a restaurant is \$44.20. Determine the amount of a 15% tip.

Estimate: 10% is $\frac{1}{10} \times \$44.20 \approx \$4.40$

5% is $\frac{1}{2}$ of 10% 5% is $\frac{1}{2} \times \$4.40 \approx \underline{\$2.20}$

Therefore a 15% tip is approximately \$6.60.

Use this method to determine the amount of a 15% tip on the following restaurant bills.

a. \$36.85 b. \$56.75 c. \$14.39

d. \$87.25 e. \$167.40 f. \$70.90

Example: A popcorn popper is purchased for \$36.95. The state sales tax is 6%. Estimate the amount of tax.

Estimate: 6% is a little more than 5%.

10% is $\frac{1}{10} \times \$36.95 \approx \$3.70$

5% is $\frac{1}{2}$ of 10% 5% is $\frac{1}{2} \times \$3.70 = \$1.85$

Therefore a good estimate for the 6% tax is a little more than \$2.00.

Use this method to determine a good approximation for the 6% sales tax on the following purchases.

a. \$14.89 b. \$27.50

c. \$6.37 d. \$123.67

If the combined state and local sales tax is 9% instead of 6%, explain how you would estimate the amount of tax.

Use your method to determine an estimate for the 9% tax on the above items.

Activity 9: Flex-it

PURPOSE Reinforce the concept of percent and apply the concept of percent increase to measurements of the human body.

MATERIALS A centimeter tape measure and a calculator

GROUPING Work in groups of 4–5.

GETTING STARTED Enter the names of the people in your group in the table below. Predict the order (largest to smallest) for the percent increase for each muscle for each member of the group and enter your guesses in the table. The highest rating is 1.

Make the following measurements in centimeters and round to the nearest 0.5 cm.

A. Extend an arm. Measure the largest circumference of the biceps while it is relaxed.

B. Extend a leg. Measure the largest circumference of the calf while it is relaxed.

C. Flex the arm and leg as much as possible. Measure the largest circumference of the biceps and the calf muscles.

Determine the percent increase for each muscle and record it in the table.

Name	Biceps Extended	Biceps Flexed	Percent Increase	Rank Order of Increase Est.	Rank Order of Increase Act.	Calf Extended	Calf Flexed	Percent Increase	Rank Order of Increase Est.	Rank Order of Increase Act.

1. How does the final order of the percent increases compare with your predictions?

2. Did the most muscular person in your group have the greatest percent increase? ____ Explain.

3. Explain how a person's size and body structure affect the percent increase in the size of the muscles measured in this activity.

Activity 10: Rectangles and Curves

PURPOSE Reinforce the concepts of perimeter and area. Display discrete and continuous data using graphs and develop the idea of a limit.

MATERIALS Centimeter graph paper, scissors, 36 colored squares, and a loop of string 24 cm in circumference

GROUPING Work individually or in pairs.

1. Determine all possible rectangles that can be constructed using the 36 squares. As you determine each rectangle, outline it on a piece of graph paper and label the base (b) and the height (h). Measure the length of the base, the height, and the perimeter (P) for each rectangle and record the measurements in Table 1. The length of the side of a square equals 1 unit.

TABLE 1: Area = 36

Base (b)									
Height (h)									
Perimeter (P)									

2. Cut out each rectangle. Construct a graph showing only the first quadrant. Label the horizontal axis (b) and the vertical axis (h). Place each rectangle on the axis as illustrated in the graph that follows. Each rectangle must be placed so that one vertex is at (0, 0). Make a drawing of the completed graph.

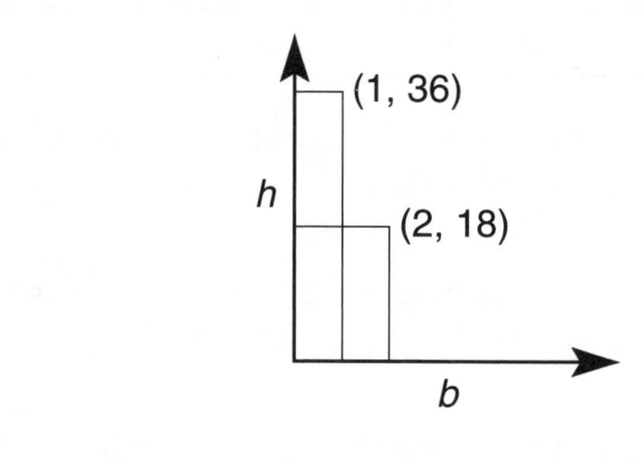

1. Using the data in Table 1, plot the ordered pairs (b, h) that represent the base and height of each rectangle on Graph 1.

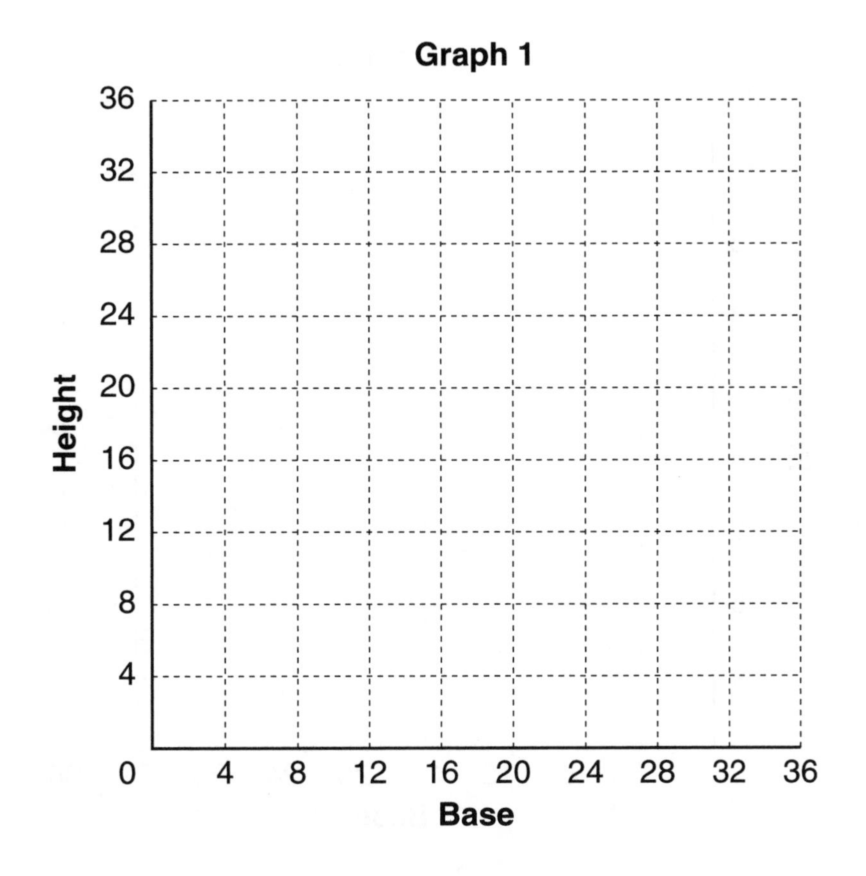

The data displayed in Graph 1 represents *all* rectangles that can be constructed using 36 squares.

2. Are there any other rectangles with an area of 36? Explain.

3. How many are there?

4. List the dimensions of at least three other rectangles whose areas are 36.

5. Are these rectangles represented in Graph 1? If not, describe where on the graph the new points should be placed.

1. On Graph 2, plot and connect the points that represent the ordered pairs (b, h) for *all* possible rectangles whose area is 36. (**HINT:** Use the points on Graph 1.)

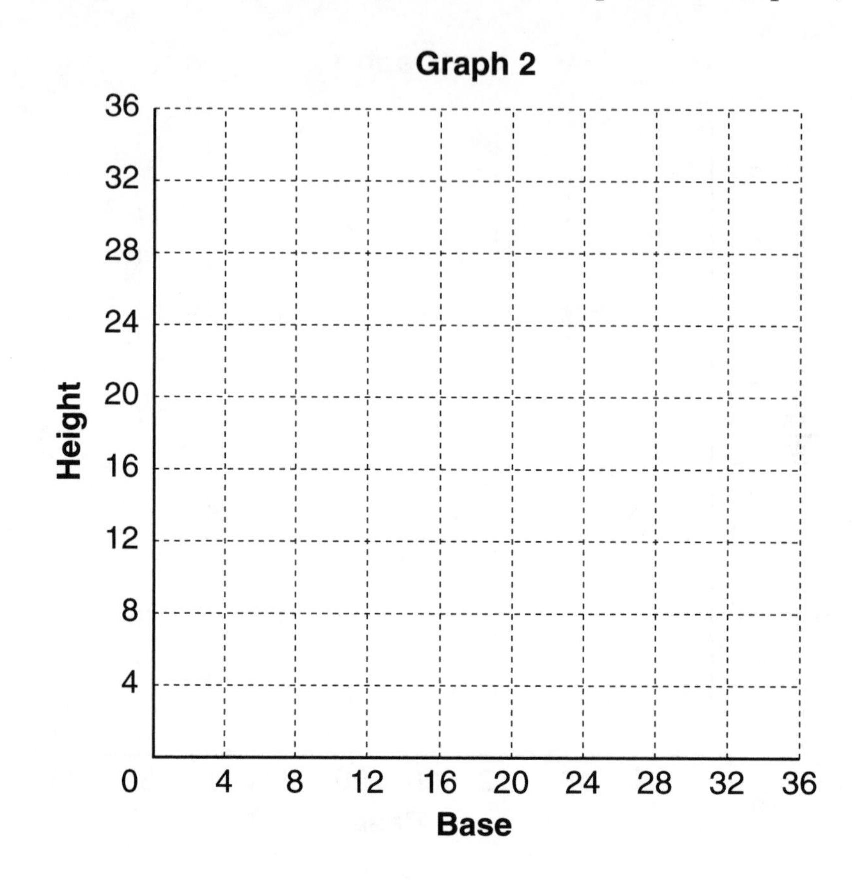

2. Will the graph of this data intersect either axis? If so, where? If not, explain.

3. Can any point on the graph lie below the horizontal axis? Explain.

1. Using the data from Table 1, plot the ordered pairs (b, P) on Graph 3. Connect the points on the graph with a smooth curve.

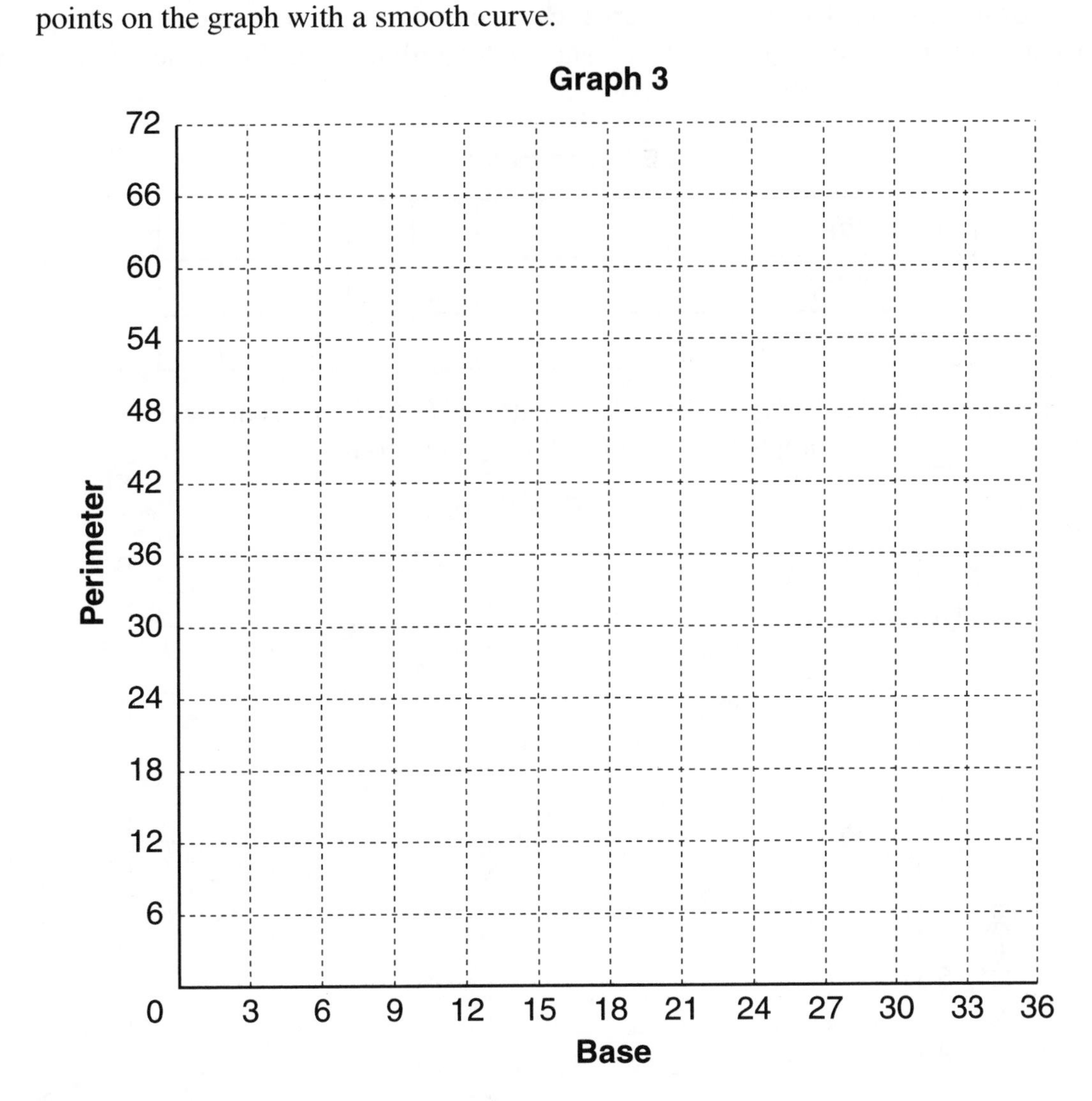

2. Each rectangle represented on the graph has an area of ________.
3. What is the least perimeter of any rectangle?
4. What are the dimensions of the rectangle with the least perimeter?
5. Is there a rectangle with a maximum perimeter? Explain.
6. The area of all rectangles represented on the graph is 36. Explain how it is possible for a rectangle to have an area of 36, yet have a perimeter that is unlimited.
7. Describe a physical model that you could use to illustrate the concept of a finite area being enclosed by an unlimited perimeter.

1. Determine all possible rectangles with integral dimensions that can be enclosed by the loop of string on centimeter graph paper. Outline each rectangle on a piece of graph paper and label the base (b) and the height (h). Record the base (b), the height (h), and the area (A) for each rectangle in Table 2.

TABLE 2: Perimeter = 24

Base (*b*)									
Height (*h*)									
Area (*A*)									

2. Using the data in Table 2, plot the ordered pairs (b, h) that represent the base and height of each rectangle on Graph 4. Draw a line through the points.

Graph 4

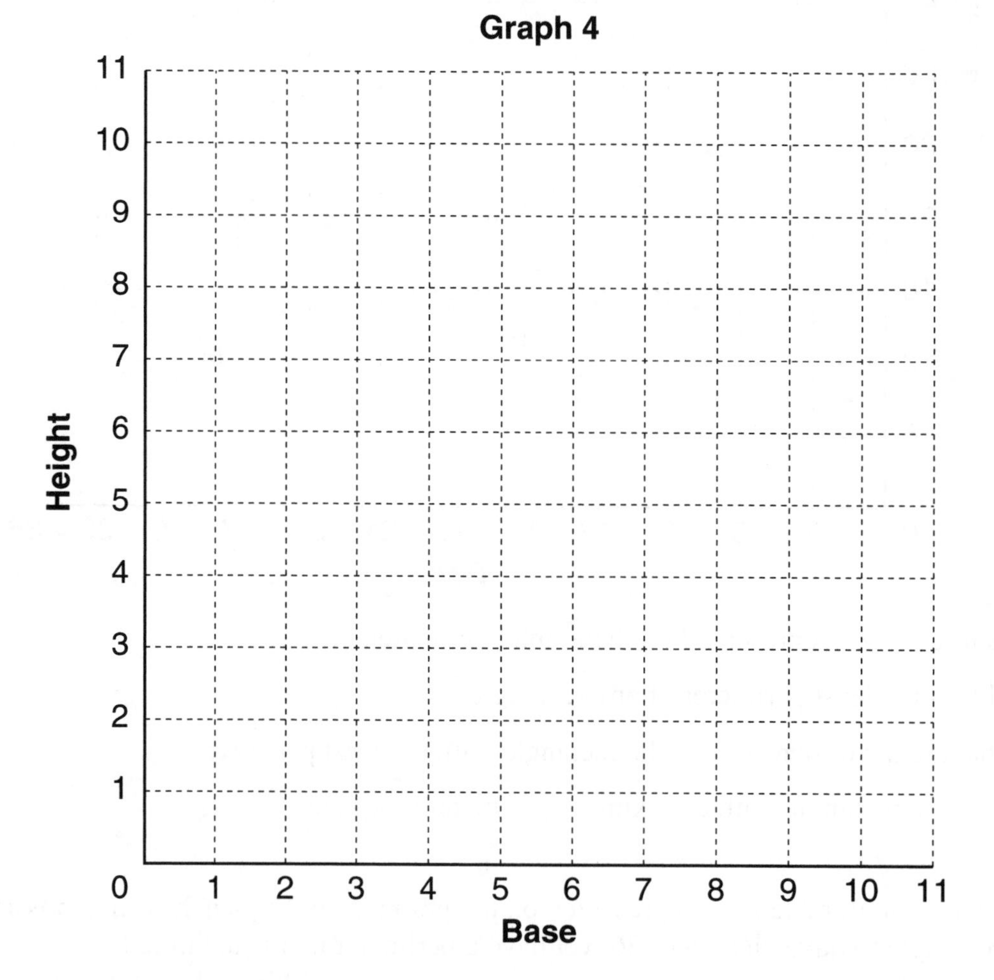

3. Graph 4 displays the data for *all* rectangles with a perimeter of 24 cm. Will the graph of this data intersect either axis? If so, where?

4. If the graph did intersect the horizontal axis, what would be the coordinates of that point? __________ How is the b coordinate of that point related to the perimeter?

1. Using the data from Table 2, plot the ordered pairs (b, A) on Graph 5. Connect the points on the graph with a smooth curve.

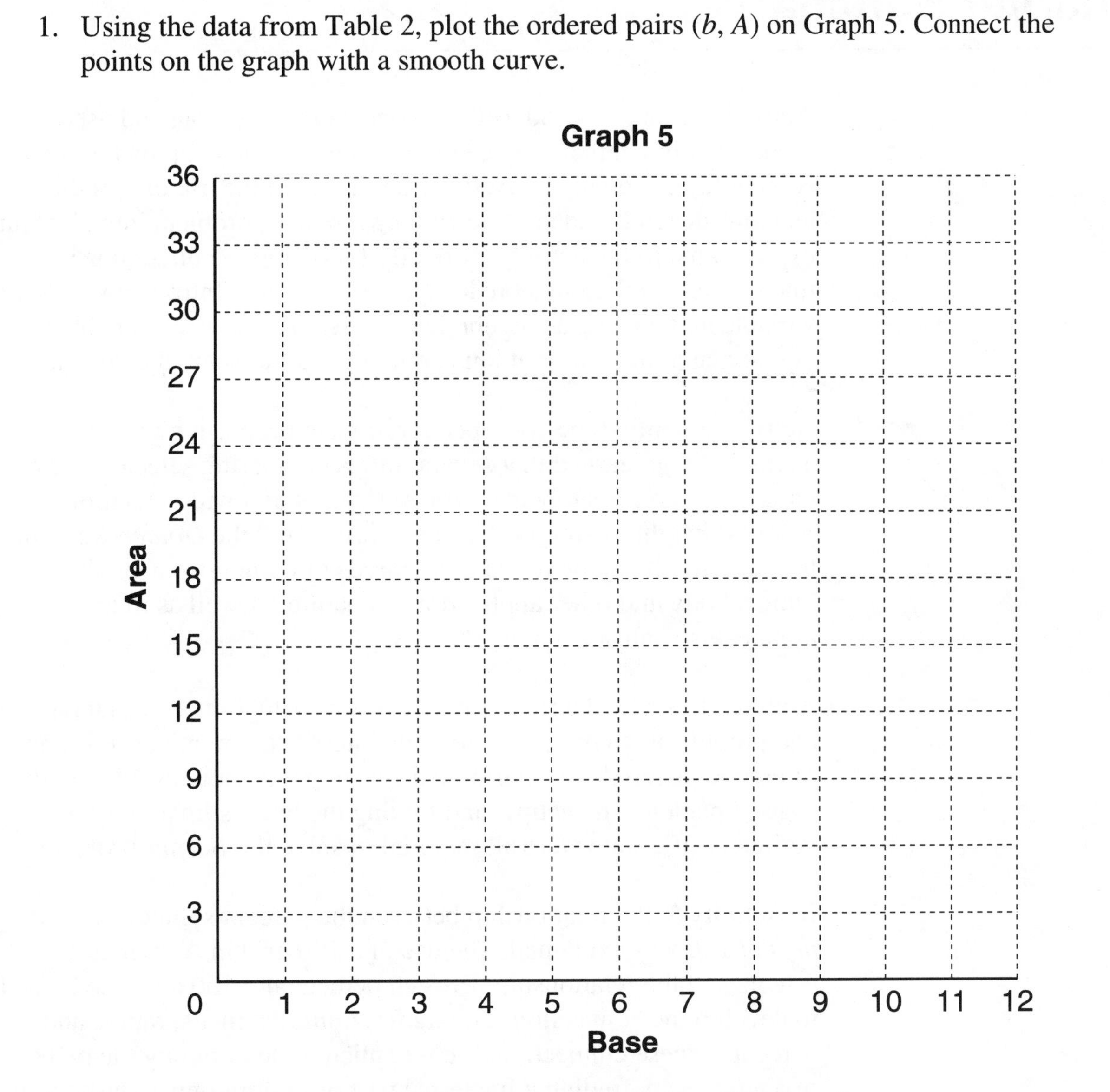

2. If a rectangle has an area of 24, what is the measure of the base?

3. What is the maximum area of any rectangle?

4. What are the dimensions of the rectangle with maximum area?

5. Is there a rectangle with a least area? Explain.

6. As the area of the rectangle approaches zero, the measure of the base approaches a maximum value of ____________. Explain.

Chapter Summary

Activities 1 and 2 introduced the concept of a variable and used the models for integers introduced in Chapter 4 to develop methods for solving linear equations. Activity 2 reinforced the use of variables and introduced the idea of translating a written problem into algebraic expressions. In the activity, correctly following the directions in a magic trick resulted in a predetermined outcome. Introducing a variable and writing an algebraic expression at each step enabled you to generalize the solution and thereby understand the "magic."

Activity 3 applied the concepts of ratio and proportion to human anatomy. You learned that several ratios comparing selected body parts are approximately the same regardless of a wide variation in peoples' height or weight. You also discovered the *Golden Ratio* in this activity. It can be found in hundreds of comparisons on the human body and other applications in nature, as well as in art, architecture, music, and psychology.

Activity 4 involved a rich problem-solving situation in which ratio and proportion were applied to scale interpretation on maps. Using reference materials to determine the state population, estimating the space a person can occupy, and finding the best estimate for the perimeter of a state were all required to solve the *Border* problem.

In Activity 5, the relationship between the percent symbol, %, and *part of 100* was explained. The use of a 100 grid in Activities 5–7 established the relationship between percent and 100 and also helped to develop the connections among fractions, decimals, ratios, and percents. These connections were applied in determining the percent of a number by finding a fractional part or multiplying by a decimal.

Many everyday uses of percent can be estimated mentally using combinations of 10% or one tenth of a number, as demonstrated in Activity 8. As shown in the advertisement in this activity, percent is often misunderstood and misused. Study advertisements, newspaper articles, and graphs that use percent. See how many examples of the misuse of percent you can find.

Chapter 8
Probability

"Probability provides concepts and methods for dealing with uncertainty and for interpreting predictions based on uncertainty. Probabilistic measures are used to make marketing, research, business, entertainment, and defense decisions, and the language of probability is used to communicate these results to others. . . . Students should extend their . . . experiences with simulations and experimental probability to continue to improve their intuition. These experiences provide students with a basis of understanding from which to make informed observations about the likelihood of events, to interpret and judge the validity of statistical claims in view of the underlying probabilistic assumptions, and to build more formal concepts of theoretical probability."

—*Curriculum and Evaluation Standards for School Mathematics*

The activities in this chapter develop the basic ideas and vocabulary of probability and introduce several different models used to determine probabilities. You will learn to use ratios to assign a probability to an event and to use and interpret both experimental and theoretical probabilities. In the process, you will develop an understanding of fair and unfair games and random events.

Most people have an intuitive conceptual understanding of probability even though they have had limited formal educational experiences with it. The goal of this chapter is to extend your intuitive ideas to a sound mathematical understanding of probability.

Activity 1: What Are the Chances?

PURPOSE Use intuitive ideas about chance to introduce the concept of probability.

GROUPING Work individually.

GETTING STARTED We frequently encounter situations in which we cannot tell the outcome in advance. In such situations, we often talk about the chances of an outcome occurring. If we think the chances that something will happen are good, we might say it is likely or it is probable. On the other hand, when we think the chances are poor, we often say the outcome is unlikely or not very probable.

A paper bag contains 8 marbles: 5 green, 2 blue, and 1 yellow. Suppose 1 marble is drawn from the bag. Use the scale below to describe the chances of each of the following events occurring. Explain your decision for each event.

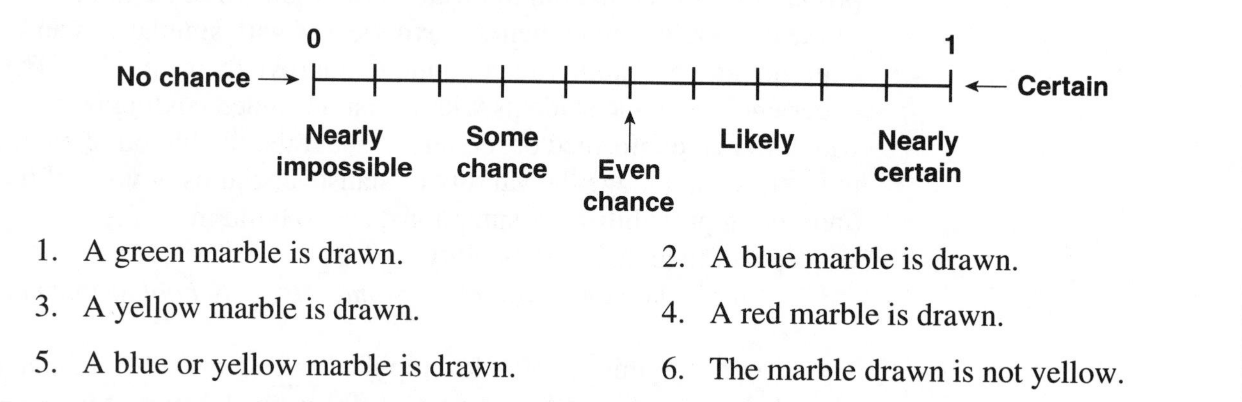

1. A green marble is drawn.
2. A blue marble is drawn.
3. A yellow marble is drawn.
4. A red marble is drawn.
5. A blue or yellow marble is drawn.
6. The marble drawn is not yellow.

Make a spinner face by dividing each circle into three sections and labeling each section with a color (RED, WHITE, or BLUE) so that the given condition is true.

1. The spinner is certain to stop on RED.

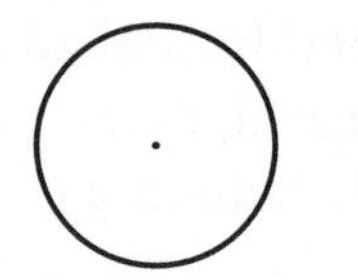

2. The spinner is likely to stop on WHITE.

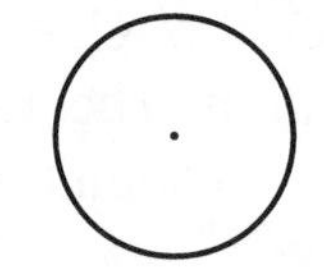

3. The spinner can't stop on BLUE.

4. There is little chance the spinner will stop on RED.

5. The spinner will probably stop on RED or BLUE.

6. The spinner has the same chance of stopping on RED, WHITE, or BLUE.

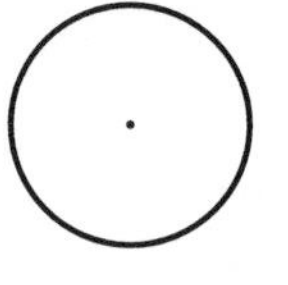

Activity 2: The Spinner Game

PURPOSE Develop and extend the concept of probability and introduce the idea of a fair game.

MATERIALS A #1 paper clip (approximately $1\frac{5}{16}$ in. long) for each student

GROUPING Work in pairs.

GETTING STARTED Make a spinner by bending a paper clip into the shape shown below. The long straight part will be the pointer. Place your pencil through the loop of the paper clip and put the point of the pencil on the center of the spinner. Spin the spinner by flicking the paper clip with your finger.

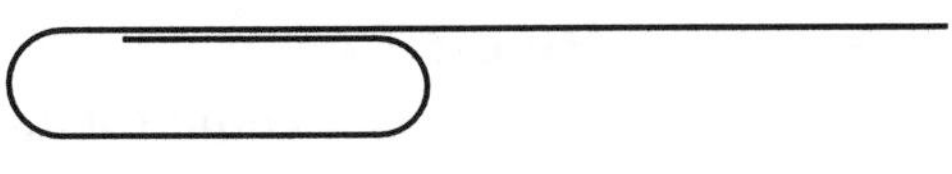

Rules for the Spinner Game:

- This is a game for two players. One player spins Spinner A and the other spins Spinner B.
- Both players spin their spinner at the same time. The player who spins the highest number is the winner.

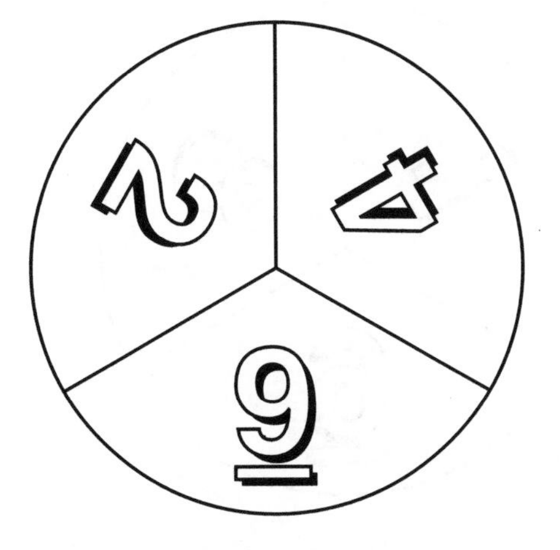

Spinner A

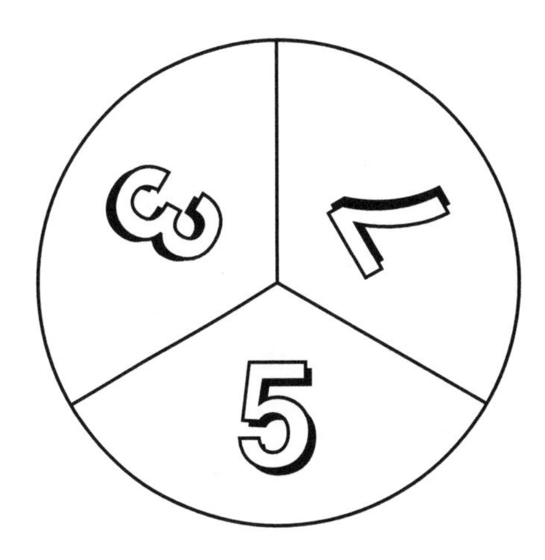

Spinner B

Play the game 20 times. Record the winner of each game in the table. You may not change spinners once the game has begun.

Do you think this is a fair game? Explain.

Winning Spinner		
Spinner	**Tally**	**Frequency**
A		
B		
Total	X	

Combine your data for the Spinner Game with the data from the other teams in your class to find a class total for the number of wins for Spinner A and Spinner B. Record the data in the table.

Does this data change your opinion about whether the game is fair? Explain.

Winning Spinner	
Spinner	**Frequency**
A	
B	
Total	

EXTENSIONS Play the following spinner game.

Rules:

- This is a game for two players.
- The first player chooses a spinner from Spinners A, B (preceding page), or C (below).
- The second player chooses a spinner from the two remaining spinners.
- Both players spin at the same time. The player who spins the highest number is the winner.

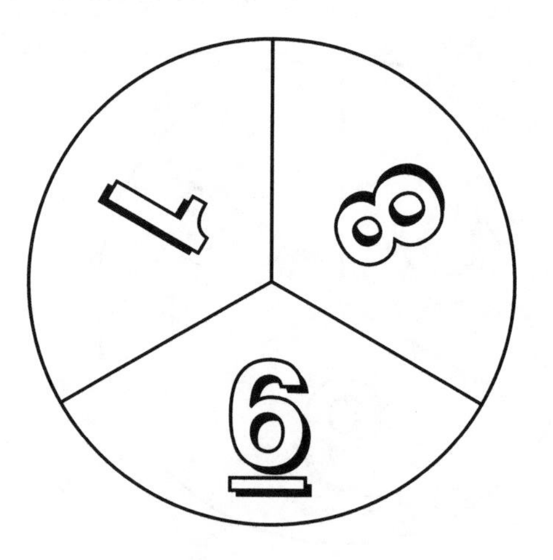

Spinner C

Play the game using different combinations of spinners. Then answer the following questions.

1. How does allowing the first player to choose a spinner affect the fairness of this game?

2. Is there a strategy for choosing the spinners that would give one player an advantage over the other? Explain.

Activity 3: Numbers That Predict

PURPOSE Introduce experimental and theoretical probability and mutually exclusive, complementary, equally likely, certain, and impossible events.

GROUPING Work individually or in pairs.

GETTING STARTED A meteorologist says the chance of snow today is 20%. The fine print in the sweepstakes announcement indicates that your odds of winning are 1 in 2.8 million. In a group of 25 nineteen-year-old women, about 22 will never have been married. These are just three examples of situations in which numbers are used to predict.

EXPERIMENTAL PROBABILITY

1. Spin Spinner A from Activity 2 thirty times and record the results in the table.

Outcome	Tally	Frequency
2		
4		
9		
Total	X	

A *probability* is a ratio that predicts the chance or likelihood of something happening.

$$\text{Experimental probability} = \frac{\text{number of times the event occurs}}{\text{total number of trials}}$$

2. Use the data in the table to calculate the following experimental probabilities. [P(2) = the probability the spinner stops in the region labeled 2.]

   P(2) = _____ P(4) = _____ P(9) = _____

   P(square number) = _____ P(odd number) = _____ P(even number) = _____

3. What is P(1)? Why?

4. What is P(a number less than 10)? Why?

5. a. What does it mean for an event to have a probability of 0?

   b. Of 1?

6. a. What do you observe about P(even number) + P(odd number)?

   b. If the spinner does not stop on an even number, then it must stop on an odd number. For this reason, the events "the spinner stops on an even number" and "the spinner stops on an odd number" are called *complementary events*. What can you conclude about the probabilities of complementary events?

THEORETICAL PROBABILITY

When you spin Spinner A, there are three possible outcomes. Since each region on the spinner has the same shape and size, over many trials, each of the possible outcomes should occur about the same number of times. That is, the outcomes are *equally likely*.

If the outcomes of an experiment are equally likely, the theoretical probability of an event may be calculated without conducting an experiment.

$$\text{Theoretical probability} = \frac{\text{number of outcomes making up the event}}{\text{total number of possible outcomes}}$$

1. Calculate the following theoretical probabilities for spinning Spinner A.

   P(2) = _____ P(4) = _____ P(9) = _____

   P(square number) = _____ P(odd number) = _____ P(even number) = _____

2. Compare the probabilities in Exercise 1 above to the experimental probabilities calculated in Exercise 2 on the preceding page. Explain any similarities or differences.
3. a. What is P(2 or 4)?

   b. The events "the spinner stops in the region labeled 2" and "the spinner stops in the region labeled 4" cannot happen at the same time. Events that cannot occur simultaneously are said to be *mutually exclusive*. What is P(2) + P(4)?

   c. What is P(odd number or square number)?

   d. What is P(odd number) + P(square number)?

   e. Are the events "the spinner stops on an odd number" and "the spinner stops on a square number" mutually exclusive? Explain.

   f. Based on your answers to Parts (a)–(c), if A and B are mutually exclusive events, what is P(A or B)?
4. What is the difference between mutually exclusive events and complementary events?

---

1. Design a spinner that has two regions, one labeled Red and the other Blue, and such that P(Red) = $\frac{1}{4}$.
2. Spin the spinner 20 times and record the number of times it stops on Red. Use the results to calculate the following experimental probabilities.

   P(Red) = _____ P(Blue) = _____
3. Did the probabilities come out exactly as you expected? Explain.
4. Combine the results of your 20 spins with those of four classmates. Calculate the experimental probabilities for the 100 spins. What do you observe?

Activity 4: It's in the Bag

PURPOSE Determine theoretical and experimental probabilities and use probabilities to make predictions.

MATERIALS A paper bag and 8 green-, 8 yellow-, and 8 blue-colored squares

GROUPING Work in pairs.

1. A paper bag contains 5 green squares, 2 blue squares, and 1 yellow square. If 1 square is drawn from the bag, what is the theoretical probability of each event below?
   a. A green square is drawn.
   b. A blue square is drawn.
   c. A yellow square is drawn.
   d. A red square is drawn.
   e. A blue or yellow square is drawn.
   f. The square drawn is not yellow.
2. a. Plot the probabilities from Exercise 1 on the following number line.

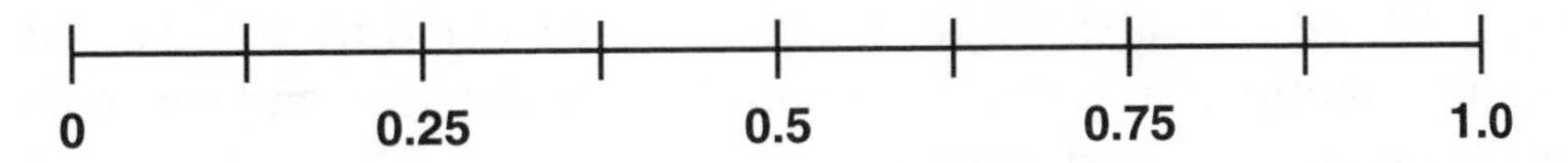

   b. For each event, select the phrase that most accurately describes the chance of it occurring: *no chance*, *nearly impossible*, *some chance*, *an even chance*, *likely*, *nearly certain*, or *certain*.
3. How many blue squares must be added to the bag in Exercise 1 so that P(blue) = 0.5? Explain how you determined your answer.
4. Simulate the experiment in Exercise 1. Put 5 green squares, 2 blue squares, and 1 yellow square in a paper bag. Shake the bag and draw 1 square. Record the color on a tally sheet and put the square back in the bag. Repeat the experiment until you have completed 40 trials.
5. Use the results from the simulation in Exercise 4 to find the following experimental probabilities.
   a. P(green)
   b. P(blue)
   c. P(yellow)
   d. P(red)
   e. P(blue or yellow)
   f. P(not yellow)
6. Compare the experimental probabilities to the theoretical probabilities in Exercise 1.
7. a. Make up a bag containing 9 squares of three different colors. Exchange bags with your partner.
   b. Repeat the sampling procedure in Exercise 4 until you have completed 45 trials.
   c. Predict how many squares of each color are in the bag. Then check your prediction by looking in the bag.
   d. How could you use experimental probabilities to help make your predictions?

Activity 5: The Mystery Cube

PURPOSE Reinforce the concept of expected values and the process of making predictions based on a sample.

MATERIALS 3 blank cubes per student (paper squares and a sack may be used in place of the cubes) and an inflatable model of the Earth (a globe)

GROUPING Work in pairs.

Write one of the digits 1, 2, 3, 4, 5, or 6 on each face of a cube. It is not necessary to use all six digits; you may write the same digit on different faces. **Do not show the cube to your partner.** (If paper squares and a sack are used, write one digit on each of six paper squares and place them in the sack.)

Roll the cube **without letting your partner see it** and tell your partner the digit on the top face. Continue rolling the cube and announcing the outcome until your partner decides to guess how many faces of the cube are labeled with each digit. Be sure to keep track of the number of times you roll the cube. (If paper squares are used, draw a square from the sack and read the number written on it. Put the square back in the sack and shake the sack thoroughly before making the next draw.)

If your partner correctly identifies how many faces are labeled with each digit, record the number of rolls in the table below. If the labels are not guessed correctly, tell your partner the number of digits that are correct and then continue rolling the cube until your partner can correctly identify how many faces are labeled with each digit.

Trial	1	2	3	4	5	6
Number of Rolls						

When your partner has identified the labels correctly, switch roles. Your partner will write digits on the faces of a cube and roll it while you try to guess how the faces are labeled. Repeat this procedure until all 6 cubes have been used.

Example:

Roll:	1	2	3	4	5	6
Outcome:	5	1	5	5	1	5

Guess: There is a 1 on two faces and a 5 on four faces.
Response: You have one of the digits correct.
(The number of faces labeled with a 1 is correct.)

Roll:	7	8	9	10
Outcome:	2	5	1	5

Guess: There is a 1 on two faces, a 2 on one face, and a 5 on three faces.
Response: That is correct.

Labels on the Cube

	1		
2	5	5	5
	1		

Enter 10 rolls in the table and switch roles.

1. What was the average number of rolls needed to determine how many faces of the cube were labeled with each digit?

2. Which labeling was the most difficult to guess? Why?

3. Explain how you used (or could have used) probabilities to make your predictions.

The eight sides of an octahedral die are congruent equilateral triangles. Each side is painted either yellow, green, or blue. The results of rolling the die 240 times are summarized in the table below. Predict the number of yellow sides, the number of green sides, and the number of blue sides on the die. Explain how you arrived at your predictions.

COLOR	FREQUENCY
Green	58
Blue	27
Yellow	155
Total	**240**

EXTENSIONS Appoint one person in the class to be the recorder. Toss the inflated globe from person to person in the classroom. When a person catches the globe, she/he should note whether her/his right index finger is touching land or water, tell the recorder, and then toss the globe to another person. The recorder tallies how many times the globe was caught with the index finger on land and on water after 10 catches, 25 catches, and 50 catches.

1. Estimate the percent of the Earth's surface that is covered by water based on
   a. 10 catches.
   b. 25 catches.
   c. 50 catches.

2. Which estimate in Exercise 1 do you think is the most accurate? Why?

3. Look up in an almanac the percent of the Earth's surface that is covered by water. Compare this figure to your estimates and explain the reasons for any similarities or differences.

Activity 6: Scissors-Paper-Rock

PURPOSE Introduce the use of a matrix and a tree diagram to calculate theoretical probabilities, and reinforce fair games.

GROUPING Work in pairs.

GETTING STARTED The scissors-paper-rock game has been popular for many years. The two-player game is played as follows:

- Each player makes a fist.
- On the count of three, each player shows either scissors by showing two fingers, paper by showing four fingers, or rock by showing a fist.
- If scissors and paper are shown, the player showing scissors wins, since scissors cut paper.
- If scissors and rock are shown, the player showing rock wins, since a rock breaks the scissors.
- If paper and rock are shown, the player showing paper wins, since paper wraps a rock.

1. Play the game 20 times and record the results in the table.
2. Do you think this is a fair game? Explain.
3. Use the data in the table to calculate the following probabilities.

   P(you win) = ______

   P(your partner wins) = ______

Winner	Tally	Frequency
You		
Your Partner		

You can determine if the game is fair without conducting an experiment.

1. Complete the *matrix* at the right.
2. If the players choose the sign they show randomly, each of the nine outcomes in the matrix are *equally likely*. Find the following probabilities in this case.

   P(A wins) = ______

   P(B wins) = ______

   P(Tie) = ______

A = A wins
B = B wins
T = Tie

		Player B		
		Scissor	Paper	Rock
Player A	Scissor	T		
	Paper			A
	Rock			

3. Use the probabilities in Exercise 2 to determine whether this is a fair game. Explain your decision.

By thinking of the game as a multistage experiment, you can analyze it using a *tree diagram*.

1. Complete the tree diagram at the right.
2. Since the choices made by Player A and Player B are *independent* of one another, the probability of the outcome along any path is equal to the product of the probabilities along the path. Find the following probabilities.

   P(PS) = _____

   P(SP or SR) = _____

   P(A wins) = _____

   P(B wins) = _____

   P(at least one player shows scissors) = _____

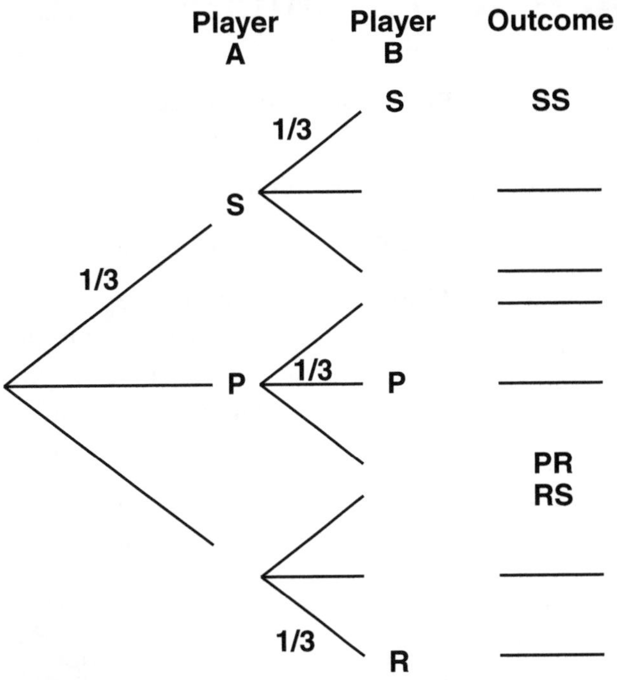

Player A | Player B | Outcome

S | 1/3 S | SS

S | ____ | ____

S | ____ | ____

1/3 P | 1/3 P | ____

P | ____ | PR

R | ____ | RS

R | ____ | ____

R | 1/3 R | ____

THE SPINNER GAME REVISITED

1. Use your individual data from Activity 2 to calculate the following experimental probabilities.

   P(Spinner A wins) = _____ P(Spinner B wins) = _____

2. Use your class totals to calculate the following experimental probabilities.

   P(Spinner A wins) = _____ P(Spinner B wins) = _____

3. a. Construct a matrix for the Spinner Game in Activity 2.

   b. Are the outcomes in the matrix equally likely? Explain.

4. Use the data in the matrix to calculate the following theoretical probabilities.

   P(Spinner A wins) = _____ P(Spinner B wins) = _____

5. a. Compare the probabilities in Exercise 4 to those in Exercise 1.

   b. Compare the probabilities in Exercise 4 to those in Exercise 2.

6. Is the Spinner Game a fair game? Explain.

EXTENSIONS Do the Extension in Activity 2, but instead of playing the game, analyze it using matrices.

Activity 7: Catch the Leprechaun

PURPOSE Introduce the use of an area model to determine the theoretical probabilities for a multistage experiment.

MATERIALS One die

GROUPING Work individually or in pairs.

GETTING STARTED Help! The leprechaun is escaping through the maze with his pot of gold! The exits to the maze are at the letters "A" through "G." If you stand at any exit, you can catch the leprechaun if he comes out that exit or an adjacent exit. For example, if you stand at exit C, you can catch the Leprechaun if he comes out exits B, C, or D.

1. Where do you think you should stand in order to have the greatest chance of catching the leprechaun? Why?

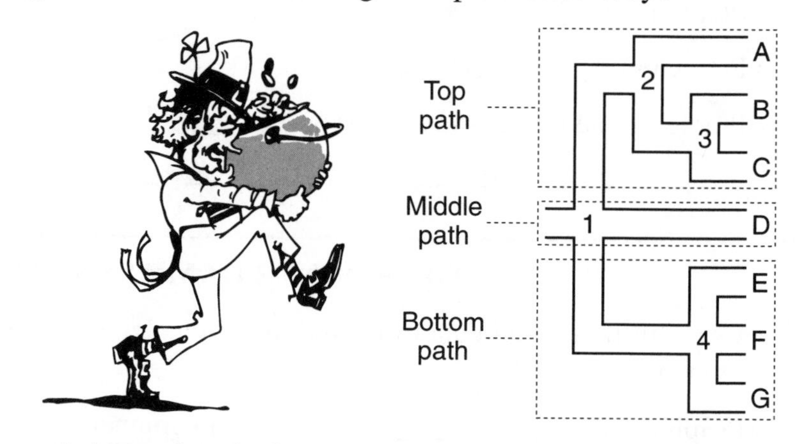

The probability that the leprechaun comes out each exit can be determined by using an area model. The grid on the left below represents the choices the leprechaun can make at junction 1 in the maze. Because each choice is equally likely, each is represented by an equal area. If the leprechaun chooses the bottom path, exits E, F, and G all have the same chance of being chosen. This is shown by dividing the area for the bottom path into 3 parts with equal areas and labeling the squares in each part with the letter of one of the exits, as in the grid at the right below.

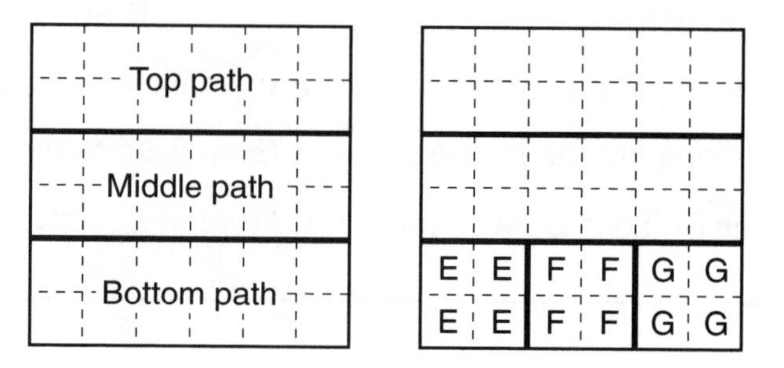

2. Continue this process to divide and label the squares in the regions for the top and middle paths.

3. a. How many squares are there in the grid?

   b. For each exit, how many squares are labeled with the corresponding letter?

   c. Use the results from Parts (a) and (b) to determine the probability that the leprechaun will come out through each exit.

4. Do the probabilities in Exercise 3 give you an idea of where you should stand? Explain.

The probability that the leprechaun comes out any particular exit can also be determined experimentally.

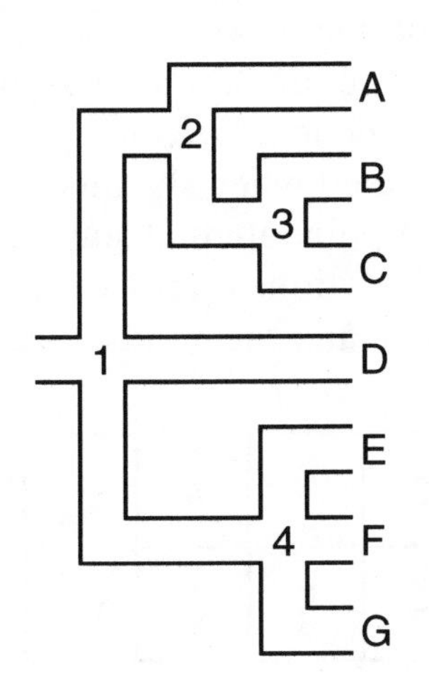

To determine the leprechaun's route through the maze, follow his path up to a numbered junction. To find the direction he turns, roll the die. **At junctions 1 and 4**, turn left if the outcome is a 1 or 2; continue straight ahead if it is a 3 or 4; turn right if it is a 5 or 6. **At junctions 2 and 3**, turn left if the outcome is even; turn right if it is odd. Then advance to the next number.

Continue advancing in this way until the leprechaun comes out through one of the lettered exits.

5. Repeat the experiment 36 times. Record your results in the following table.

Exit	A	B	C	D	E	F	G
Tally							
Frequency							

6. a. Collect the results from four other pairs, record them in the table below, and find the totals.

Exit	A	B	C	D	E	F	G
Yours							
1							
2							
3							
4							
Totals							

   b. Use the totals in the table to determine the probability the leprechaun will come out through each exit.

7. Compare the experimental probabilities calculated in Exercise 6 to the theoretical probabilities calculated in Exercise 3. Explain any similarities or differences.

Activity 8: When You're Hot, You Get Hotter!

PURPOSE Use a simulation to determine experimental probabilities for a multistage experiment and compare the results with theoretical probabilities determined using tree diagrams.

MATERIALS A paper bag and 5 green- and 4 red-colored squares

GROUPING Work in pairs.

GETTING STARTED Annie has an awesome 3-point shot in basketball. Overall, she makes 60% of her attempts, but in a game, every time she makes a 3-point shot she becomes more confident. As a result, the probability that she will make a 3-point shot attempted later in the game increases slightly. Amazingly, her confidence is not affected when she misses a shot, so the probability of her making a later 3-point attempt stays the same after a miss. The table below shows the relationship between the probability that she will make a 3-point shot and the number she has already made in a game.

Number of 3-point shots already made	0	1	2	3	4	...
Probability of making a 3-point shot	$\frac{3}{5}$	$\frac{4}{6}$	$\frac{5}{7}$	$\frac{6}{8}$	$\frac{7}{9}$	...

Suppose Annie attempts **three** 3-point shots in a game. To estimate the probability that she will make a given number of her shots (0, 1, 2, or 3), try the following experiment. In the experiment, drawing a green square from the paper bag means she makes the shot and drawing a red square means she misses.

- At the beginning of the game, the probability that Annie will make the first 3-point shot she attempts is $\frac{3}{5}$. Put 3 green squares and 2 red squares in the bag to represent this situation.
- To simulate each shot, shake the bag and draw 1 square. Record the color and replace the square. If the square was green, add another green square to the bag. If it was red, do not add any additional squares to the bag.

1. Conduct 25 trials of the experiment and record the results in a table like the one below. Since each trial represents a different game, before beginning a trial, make sure the bag contains only 3 green squares and 2 red squares.

Trial	Outcome	Shots Made
Example	GRG	2
1		
2		

2. Record the number of trials in which Annie made each number of shots in the column for Pair 1 in the following table. Complete the table by combining your results with those of three other pairs.

Shots Made	Pair				Total
	1	2	3	4	
0					
1					
2					
3					

3. Use the combined results to find the following experimental probabilities. Express your answers as decimals.

   a. P(she misses all 3 shots)

   b. P(she makes 1 shot)

   c. P(she makes 2 shots)

   d. P(she makes all 3 shots)

   e. P(she makes at least 2 shots)

   f. P(she misses at least 1 shot)

4. The results of Annie's game can also be analyzed using a tree diagram. Complete the tree diagram below.

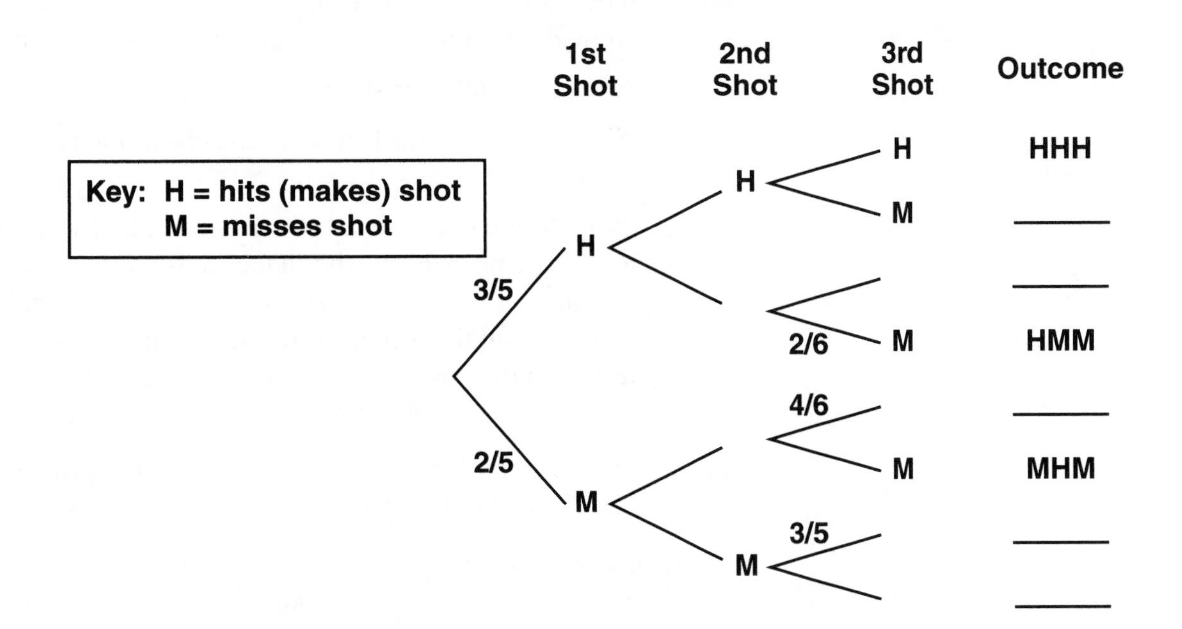

Consider the path leading to the outcome **HMH**. Based on the probabilities along the path, we would expect Annie to do the following:

- **H**it her first shot on $\frac{3}{5}$ of her attempts.
- **M**iss her second shot on $\frac{2}{6}$ of the attempts on which she made her first shot $\left(\frac{2}{6} \text{ of } \frac{3}{5}\right)$.
- **H**it her third shot on $\frac{4}{6}$ of the attempts on which she hit her first shot and missed her second $\left(\frac{4}{6} \text{ of } \frac{2}{6} \text{ of } \frac{3}{5}\right)$.

This shows that the probability of each outcome is the product of the probabilities along the path leading to the outcome.

5. Find the probability of each outcome by multiplying the probabilities along the path leading to it. Express each product as a decimal and write it next to the outcome.
6. Use the results from Exercise 5 to find the following:
   a. P(she misses all 3 shots)
   b. P(she makes 1 shot)
   c. P(she makes 2 shots)
   d. P(she makes all 3 shots)
   e. P(she makes at least 2 shots)
   f. P(she misses at least 1 shot)
7. Compare your answers in Exercise 6 to the experimental probabilities determined in Exercise 3.
8. Suppose Annie attempts **four** 3-point shots in a game. Find the theoretical probability of each of the following events.
   a. She makes all four shots.
   b. She makes at least one shot.
   c. She misses her third shot but makes all the rest.

EXTENSIONS

1. Suppose Annie becomes a little less confident when she misses a shot, so the probability she makes a 3-point shot attempted later in the game decreases slightly. Conduct 25 trials of the experiment, but this time, to simulate a shot do this: Draw a square from the bag, record the color, replace the square, and add another square of the same color to the bag.
2. Use the results of the 25 trials to estimate the probabilities that Annie will make 0 shots, 1 shot, 2 shots, and 3 shots.
3. Construct a tree diagram for this situation and use it to find the theoretical probabilities that Annie will make 0 shots, 1 shot, 2 shots, and 3 shots.

Activity 9: What's the Distribution?

PURPOSE Determine experimental probabilities and introduce probability distributions.

MATERIALS Six pennies, five dice, and one paper cup per student; colored pencils

GROUPING Work in pairs.

1. If you place all six coins in the paper cup, shake it thoroughly, and spill the coins on a table, do you think you will get one combination of heads and tails more often than the others? If so, what will it be and why?

2. Do you think you will get some combinations of heads and tails less often than others? Explain.

3. Perform the experiment of spilling the coins 35 times. After each spill, record the number of heads by coloring in a square in the row corresponding to the number of heads in the grid below.

Number of Heads	
0	
1	
2	
3	
4	
5	
6	

4. Compare the results with your predictions in Exercises 1 and 2.

5. Use a different colored pencil to add the results of your partner's 35 trials to the grid. Use the data for the 70 trials to calculate the following experimental probabilities.

   P(0 heads) = _____ P(1 head) = _____ P(2 heads) = _____

   P(3 heads) = _____ P(4 heads) = _____ P(5 heads) = _____

   P(6 heads)

6. Compare your combined results with that of another team. What similarities and differences do you notice in

   a. the distribution of the outcomes in Exercise 3?

   b. the probabilities in Exercise 5?

1. If you do the experiment again using five coins instead of six, how will the distribution of heads change? Why?

2. Use five coins. Perform the experiment 35 times. Record the results in the grid below.

Number of Heads																														
0																														
1																														
2																														
3																														
4																														
5																														

3. Were your predictions in Exercise 1 correct? Explain.

4. Use the data for the 35 trials to calculate the following experimental probabilities.

P(0 heads) = _____ P(1 head) = _____ P(2 heads) = _____

P(3 heads) = _____ P(4 heads) = _____ P(5 heads) = _____

BIASED COINS

How would the distribution of heads in the experiment of spilling five coins from a cup change if the probability of a coin showing heads was $\frac{1}{3}$ instead of $\frac{1}{2}$? To find out, try the following experiment.

1. Use five dice in place of the five coins. Spill the dice from the cup. A die that lands with a 1 or 2 facing up is counted as a head. One that lands with a 3, 4, 5, or 6 facing up is counted as a tail. Repeat the experiment 35 times and record the number of heads in a grid as above.

2. Describe the similarities and differences between the two distributions and explain the reasons for them.

EXTENSIONS Repeat the biased coin experiment for coins that have a probability of $\frac{5}{6}$ of showing heads.

Activity 10: Pascal's Probabilities

PURPOSE Use tree diagrams and Pascal's Triangle to find theoretical probabilities in situations involving independent events.

GROUPING Work individually or in pairs.

Suppose a family has 3 children. What is the probability that all 3 children are girls? 2 are girls? Only 1 is a girl? None are girls? A tree diagram can be used to determine the theoretical probabilities of each number of girls.

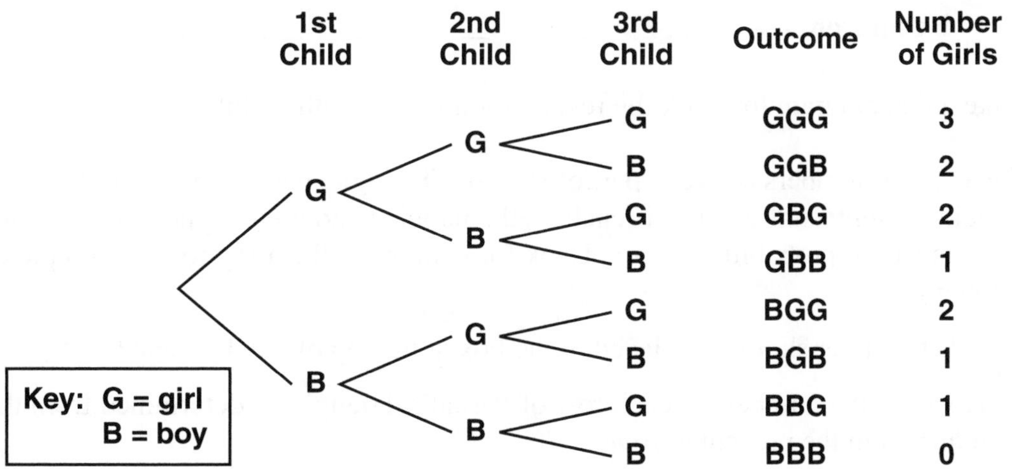

1st Child	2nd Child	3rd Child	Outcome	Number of Girls
G	G	G	GGG	3
		B	GGB	2
	B	G	GBG	2
		B	GBB	1
B	G	G	BGG	2
		B	BGB	1
	B	G	BBG	1
		B	BBB	0

Key: G = girl
B = boy

1. Assuming that when a child is born the probability that it is a girl is 0.5, are the outcomes shown above equally likely? Explain.

2. Complete the following to summarize the results from the tree diagram.

Number of outcomes with

0 girls		1 girl		2 girls		3 girls		Total outcomes
↓		↓		↓		↓		
_____	+	_____	+	_____	+	_____	=	________

3. Use the results above to determine the following theoretical probabilities for the 3-child family.

P(0 girls) = _____ P(1 girl) = _____

P(2 girls) = _____ P(3 girls) = _____

Tree diagrams can also be used to find the outcomes for families with other numbers of children, but such diagrams quickly become very large and complex. The results for families with 1, 2, 3, 4, and 5 children are shown in the diagram below. The results are listed in order from the number of outcomes with 0 girls to the number with all girls.

							☐						
1 Child →						1		1					
2 Children →					1		2		1				
3 Children →				1		3		3		1			
4 Children →			1		4		6		4		1		
5 Children →		1		5		10		10		5		1	
6 Children →	___		___		___		___		___		___		___

1. Make a tree diagram to check the results for a family with 4 children.

2. The array of numbers above is part of Pascal's Triangle, named in honor of Blaise Pascal, a seventeenth-century French mathematician and philosopher who was one of the founders of probability theory. Look for patterns in the array to help complete the following:

   a. What number should be placed in the box at the top of the Triangle?

   b. How can the numbers in each row of Pascal's Triangle be determined from the numbers in the preceding row?

   c. Complete the seventh row of Pascal's Triangle.

3. a. Use the numbers in the sixth row of Pascal's Triangle to calculate the theoretical probability of each number of girls in a family with 5 children.

   b. How do the theoretical probabilities compare to the experimental probabilities for the numbers of heads when 5 coins are spilled from a cup in Activity 9? Explain any similarities or differences.

4. Use the numbers in the seventh row to help answer the following questions. What is the probability that in a family with 6 children

   a. there are exactly 3 girls?

   b. there are at least 3 girls?

   c. there is exactly 1 boy?

   d. there are more than 2 boys?

5. Use Pascal's Triangle to calculate the theoretical probability for each number of girls in a family with 9 children. Explain your procedure.

Activity 11: Simulate It

PURPOSE Analyze probability situations using Monte Carlo simulations.

MATERIALS One die

GROUPING Work individually or in pairs.

DESIGNING A SIMULATION

Robin Hood and Maid Marian are having an archery contest. They alternate turns shooting at a target. The first person to hit it wins. The probability that Robin hits the target is $\frac{1}{3}$ and the probability that Marian hits it is $\frac{2}{5}$. Since Marian is the better archer, Robin shoots first. What is the probability that Marian will win the contest? On average, how many shots will be necessary to determine a winner? To find the answers, you can simulate the problem.

Step 1: Select a model.

- To simulate Robin's shot, roll a die. If the result is a 1 or 2, he hits the target. Otherwise he misses.
- For Marian's shot, roll a die. If the outcome is a 1 or 2, she hits the target. If it is a 3, 4, or 5, she misses. If it is a 6, roll the die again.

Step 2: Conduct a trial and record the result.

A trial consists of rolling a die to alternately simulate a shot for Robin and then one for Marian until someone hits the target.

Example:	**Shot**	**Archer**	**Die Roll**	**Result**
	1	Robin	4	Miss
	2	Marian	6	Roll Again
			5	Miss
	3	Robin	1	Hit—Robin Wins!

Step 3: Repeat Step 2 until the desired number of trials is completed.

Complete 30 trials and record the results in a table like the one below.

Trial	Winner	Number of Shots	Trial	Winner	Number of Shots
Example	Robin	3			

Step 4: Interpret the results.

Based on the results of the 30 trials, what is the probability that Robin wins the contest?

What is the average number of shots for the 30 trials?

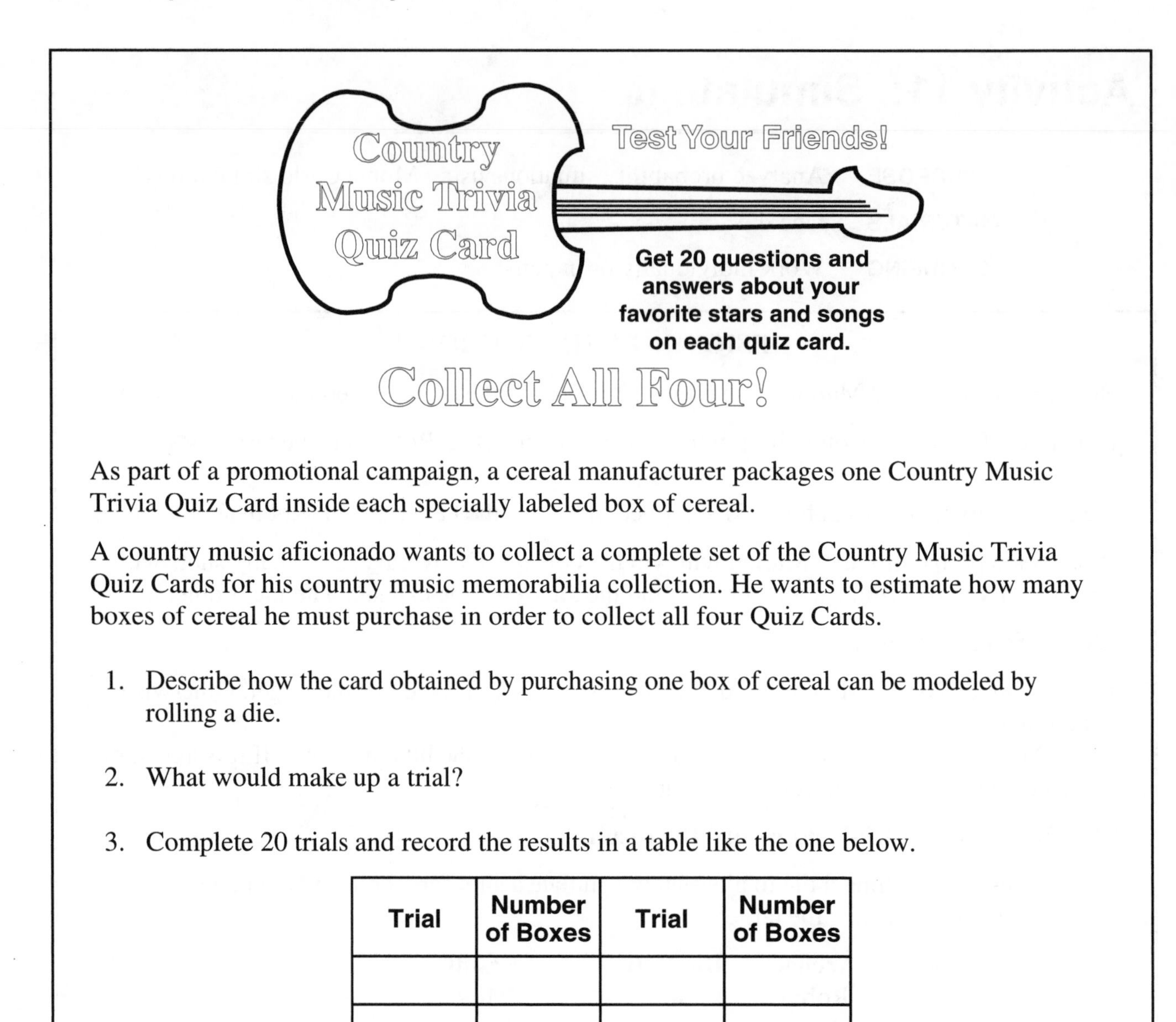

As part of a promotional campaign, a cereal manufacturer packages one Country Music Trivia Quiz Card inside each specially labeled box of cereal.

A country music aficionado wants to collect a complete set of the Country Music Trivia Quiz Cards for his country music memorabilia collection. He wants to estimate how many boxes of cereal he must purchase in order to collect all four Quiz Cards.

1. Describe how the card obtained by purchasing one box of cereal can be modeled by rolling a die.

2. What would make up a trial?

3. Complete 20 trials and record the results in a table like the one below.

Trial	Number of Boxes	Trial	Number of Boxes

4. On average, how many boxes of cereal will the aficionado have to purchase in order to collect a complete set of four Country Music Trivia Quiz Cards?

5. Describe a model that could be used if one card was twice as likely to be in a box of cereal as the others.

6. In this case, what would make up a trial?

EXTENSIONS Suppose Robin and Marian change the rules of their tournament so that the winner is the person who wins two out of three contests. Draw a tree diagram to show the possible outcomes of the tournament. Use the results from the archery simulation to assign probabilities to the branches and determine the probability that Marian wins the tournament.

Activity 12: What's Your Strategy?

PURPOSE Apply ideas about probability, fair games, and expected values to develop strategies for playing the *Sum or Product* game.

MATERIALS Two index cards (one with "SUM" written on it, the other with "PRODUCT" written on it), 9 red- and 9 blue-colored squares per person, and a pair of dice

GROUPING Work in groups of 2–6.

GETTING STARTED The rules of the *Sum or Product* game are as follows:

- Each person chooses a different number from 1 through 6 and a die is rolled. The person whose number is rolled is the Banker. Everyone else is a Player.
- The Banker places either the SUM or the PRODUCT card face down on the table, and each Player pays the Banker either 2 red or 2 blue squares to start each round.
- The Banker turns the card face up, rolls the dice, and determines either the sum or product as indicated on the card.
- If the outcome is **odd**, the Banker pays *1 blue square* to each Player who paid blue squares and *4 red squares* to each Player who paid red squares.
- If the outcome is **even**, the Banker pays *3 blue squares* to each Player who paid blue squares and *0 red squares* to Players who paid red squares.
- Play three rounds of the game and then select a new Banker.

Play at least six rounds of the game and then answer the following questions.

1. As the Banker, does it make any difference if you choose the SUM or PRODUCT card? Explain.
2. As a Player, what strategy would you use to maximize the amount you are paid by the Banker? Explain how you decided on that strategy.
3. Discuss the fairness of choosing the SUM card versus the fairness of choosing the PRODUCT card.

Chapter Summary

Probabilities are used to help interpret and understand situations involving uncertainty. There are two kinds of probabilities: experimental and theoretical. *Experimental probabilities* are determined by observing the outcome of a situation or experiment over a large number of trials. For example, meteorologists collect weather data over a long period of time. When they say that there is a 20% chance, or probability, that it will snow on a given day, they mean that, in the past, it snowed on 20% of the days with similar weather conditions.

Theoretical probabilities are based on mathematical analysis of the possible outcomes of an experiment rather than on observation of the outcomes. A fundamental concept of probability is that over a large number of trials, the experimental probability of an event will approach the theoretical probability. This idea was emphasized throughout the activities by first having you compare your individual results from an experiment with the theoretical results and then combining your results with other people's to obtain the results from a larger number of trials and repeating the comparison. In general, as the number of trials increases, the experimental probability more closely approximates the theoretical probability.

The concept of probability was introduced in Activity 1 by building on your intuitive ideas about the chances of something happening.

Activity 2 introduced the idea of a *fair game*. The intuitive notion is that a game is fair if the chances of winning are equal to the chances of losing. The notion of a fair game was explored further in Activity 6, where games were analyzed using theoretical probabilities. Activities 6 and 7 also introduced several tools that are useful in determining theoretical probabilities: matrices, probability trees, and area models.

Many of the basic concepts of probability were introduced in Activity 3 and followed up in Activity 4. You learned that a probability is a ratio that expresses the likelihood of something happening. You also explored the distinction between an experimental probability and a theoretical probability. Some of the basic ideas introduced in Activity 3 are that

$$\text{Experimental probability} = \frac{\text{number of times an event occurs}}{\text{total number of trials}}$$

and that if the outcomes of an experiment are equally likely, then

$$\text{Theoretical probability} = \frac{\text{number of outcomes making up the event}}{\text{total number of possible outcomes}}.$$

If an event **cannot happen,** its probability is 0, and if an event **is certain to happen,** its probability is 1. If A and B are *complementary events*, then P(A) = 1 – P(B). And finally, if C and D are *mutually exclusive events*, that is, events that cannot occur simultaneously, then P(C or D) = P(C) + P(D).

Activities 4 and 5 explored the concepts of *expected value* and *random sampling*. If you know the probability that an event will occur, then you can estimate how many times it will occur in a given number of trials—the expected value—by multiplying the number of trials by the probability. Thus, if we know that the probability that a nineteen-year-old woman has never been married is 0.887, we would estimate that in a group of 25 nineteen-year-old women, about 22 (0.887×25) will never have been married.

In Activity 4, the probabilities that a bag contains particular colors of squares were estimated by *random sampling*, that is, by drawing squares from the bag and recording the outcomes. The probabilities were then used to predict how many squares of each color were in the bag. Random sampling was also used in Activity 5 to determine how many faces of a die have particular numbers on them. The idea of predicting the characteristics of a population (in these cases, the squares in a bag and the six faces of a die) based on a random sample will be explored again in Chapter 9.

Another topic related to sampling is the distribution of outcomes. In Activity 9, you investigated how the distribution of the outcomes of an experiment is affected by the probabilities of the outcomes. The distribution is often a bell-shaped curve centered on the outcome or outcomes with the highest probabilities. The distribution curve can be skewed to the left or right depending on which outcome has the greatest probability.

In Activity 8, a tree diagram was used to determine the probabilities of the outcomes of a multistage experiment. In the activity, you learned that the probability of each outcome is the product of the probabilities along the path leading to it. Activity 10 showed how the probabilities of certain events can be determined by using Pascal's Triangle.

Many real-world problems involving probability are too complex to analyze theoretically and too difficult, time-consuming, or expensive to observe through actual trials. In these cases, the solution can often be obtained by simulating the problem. The techniques used to design and conduct a *simulation* were developed in Activity 11. Because a large number of trials must be conducted in order to obtain accurate results by simulating an experiment, simulations are usually performed by a computer. Finally, Activity 12 provided an opportunity to apply many of the concepts developed in previous activities to the problem of devising strategies for playing a game.

Chapter 9
Statistics—An Introduction

"In this age of information and technology, an ever-increasing need exists to understand how information is processed and translated into usable knowledge. Because of society's expanding use of data for prediction and decision making, it is important that students develop an understanding of the concepts and processes used in analyzing data. A knowledge of statistics is necessary if students are to become intelligent consumers who can make critical and informed decisions."
—*Curriculum and Evaluation Standards for School Mathematics*

Statistics may be defined as the science of collecting, organizing, and interpreting data. In this chapter, you will learn to collect data by sampling and to organize data using a variety of techniques—line plots, frequency tables, bar graphs, stem-and-leaf plots, and box-and-whisker plots. You will also learn how to describe data using measures of central tendency and to interpret data presented in graphs.

The activities in this chapter are designed to develop your understanding of these concepts and the processes of statistics. The emphasis in the activities is on the visual presentation of data and informal methods of data analysis rather than on formal statistical methods. In the activities, you will learn to use statistics to communicate information relating to sets of data effectively. You will also learn to interpret statistical displays and to make critical and informed decisions based on them.

Activity 1: Graphing *m&m's*®

PURPOSE Use a variety of graphs to display data and explore relationships among the data.

MATERIALS A one-pound bag of *m&m's*® Plain Chocolate Candies*, a one-tablespoon measure, colored pencils, a balance scale and weights, 6 line plot charts (one for each color of *m&m's*®, page A-37), and a calculator

GROUPING Work individually and as a whole class.

GETTING STARTED *m&m's*® Plain Chocolate Candies come in six colors: brown, green, orange, red, blue, and yellow.

Before you take a sample from the bag of *m&m's*®, answer the following:

1. a. Which color of *m&m's*® do you think will occur most in the bag? Why?
   b. in your sample? Why?
2. a. Which color of *m&m's*® do you think will occur least in the bag? Why?
   b. in your sample? Why?

REAL GRAPHS

1. Take a sample of *m&m's*® by dipping the measuring spoon into the bag of candy and removing a spoonful. *CAUTION: Do not eat any of the m&m's®!*

Statistical data are often displayed graphically. Using a graph rather than simply presenting the data as a set of numbers makes it easier to study relationships in the data.

2. Arrange the *m&m's*® in your sample on Graph 1 on the next page. This type of graph is often called a *real graph* because the statistical data are displayed using the actual objects whose frequencies are being compared.

3. Record the number of *m&m's*® of each color and the total number in your sample.

   Brown: ____ Orange: ____ Blue: ____ Green: ____ Red: ____ Yellow: ____ Total: ____

4. What color occurred most often and what color occurred least often in your sample? How do these colors compare with your predictions?

**m&m's*® Plain Chocolate Candies is a registered trademark of Mars, Inc.

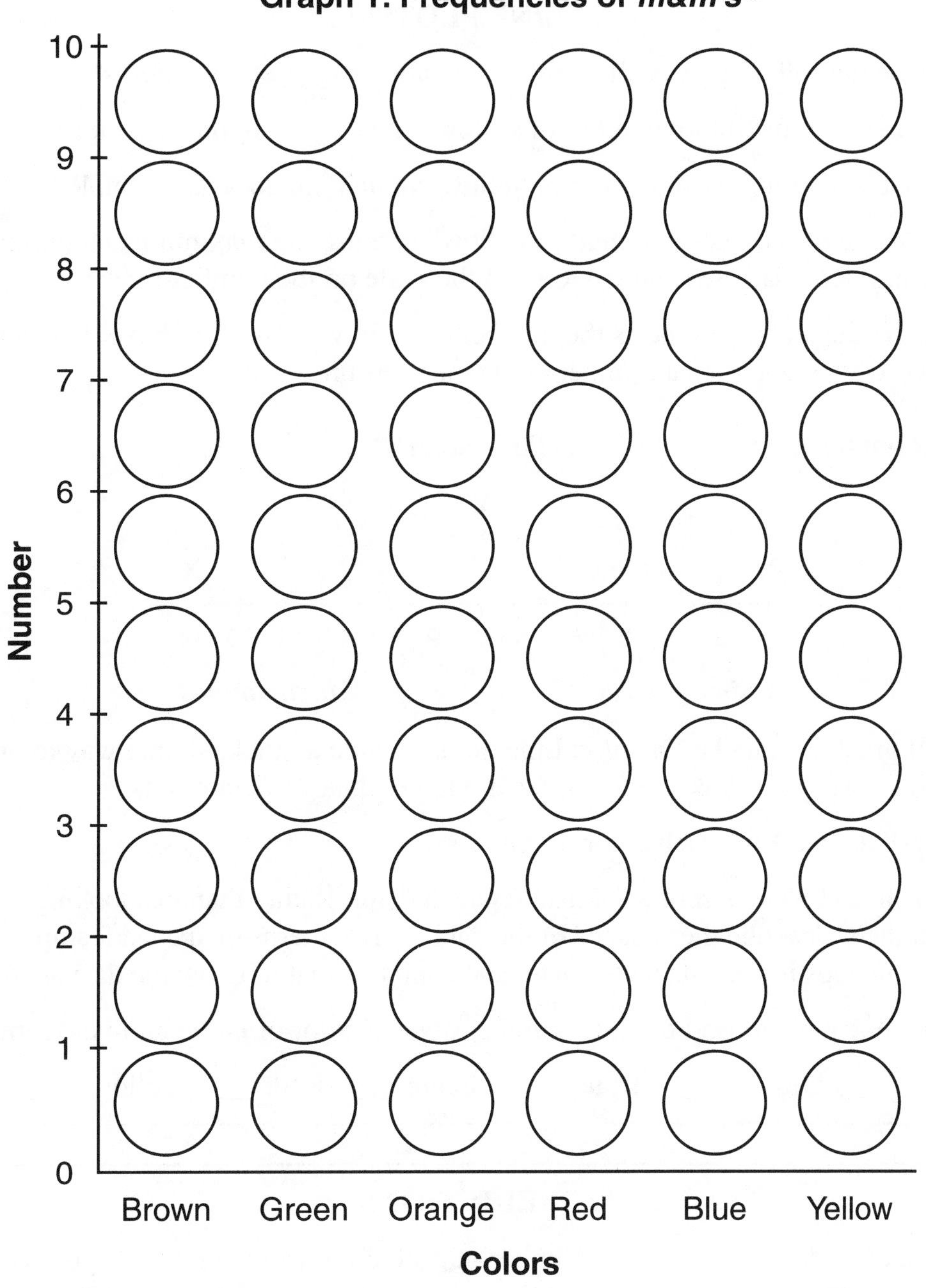

PICTOGRAPHS

1. As you remove each candy from the graph, color its circle the appropriate color. This type of graph is called a *pictograph* because the data are displayed using parallel columns (or rows) of pictures in which each picture represents one or more of the objects being compared.

Now you may eat the m&m's® in your sample!

2. Compare your pictograph with your classmates' pictographs. Describe any similarities and differences and explain why these may have occurred.

LINE PLOTS

1. Complete the following to collect the class data for the yellow *m&m's*®.
   a. What is the maximum number of yellow *m&m's*® in anyone's sample?
   b. What is the minimum number of yellow *m&m's*® in anyone's sample?
   c. Title one of the line-plot charts "Yellow" and use the maximum and minimum values from Parts (a) and (b) to label the scale on the number line.
   d. Each time a person reports the number of yellow *m&m's*® in his or her sample, record an X above that number on the number line.

Example: **Yellow *m&m's*®**

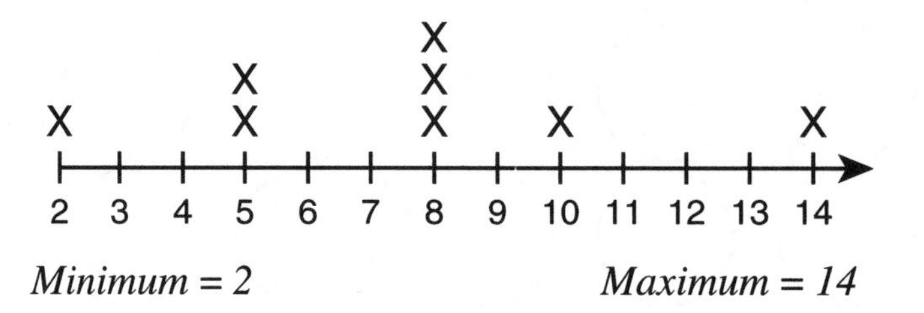

This type of graph is called a *line plot*. Line plots provide a quick, simple way to organize numerical data. They work best when there are fewer than 25 data points.

2. Repeat Exercise 1 for each color of *m&m's*®.
3. Use the line plots to describe the data for each color. Rather than just looking at individual numbers, describe the "shape" of the data—any patterns or special features such as clusters or gaps in the data and isolated data points—that tell how the data are distributed.
4. Use the line plots to find the total number of each color of *m&m's*® in the samples.

   Brown: ____ Orange: ____ Blue: ____ Green: ____ Red: ____ Yellow: ____

PREDICTIONS

1. Use the class data to predict the number of each color of *m&m's*® that you would expect to find in a one-pound bag.

   Brown: ____ Orange: ____ Blue: ____ Green: ____ Red: ____ Yellow: ____
2. Describe the procedure you used to make your predictions.
3. Help your classmates count the *m&m's*® remaining in the bag. Add these counts to the numbers you already have. What was the total number of each color of *m&m's*® in the bag?

   Brown: ____ Orange: ____ Blue: ____ Green: ____ Red: ____ Yellow: ____
4. How do these totals compare with the predictions you made in Exercise 1?

PICTOGRAPHS REVISITED

Construct a pictograph for the number of each color of *m&m's*® in the bag on Graph 2 below. **HINT:** Let each circle represent more than one *m&m's*®.

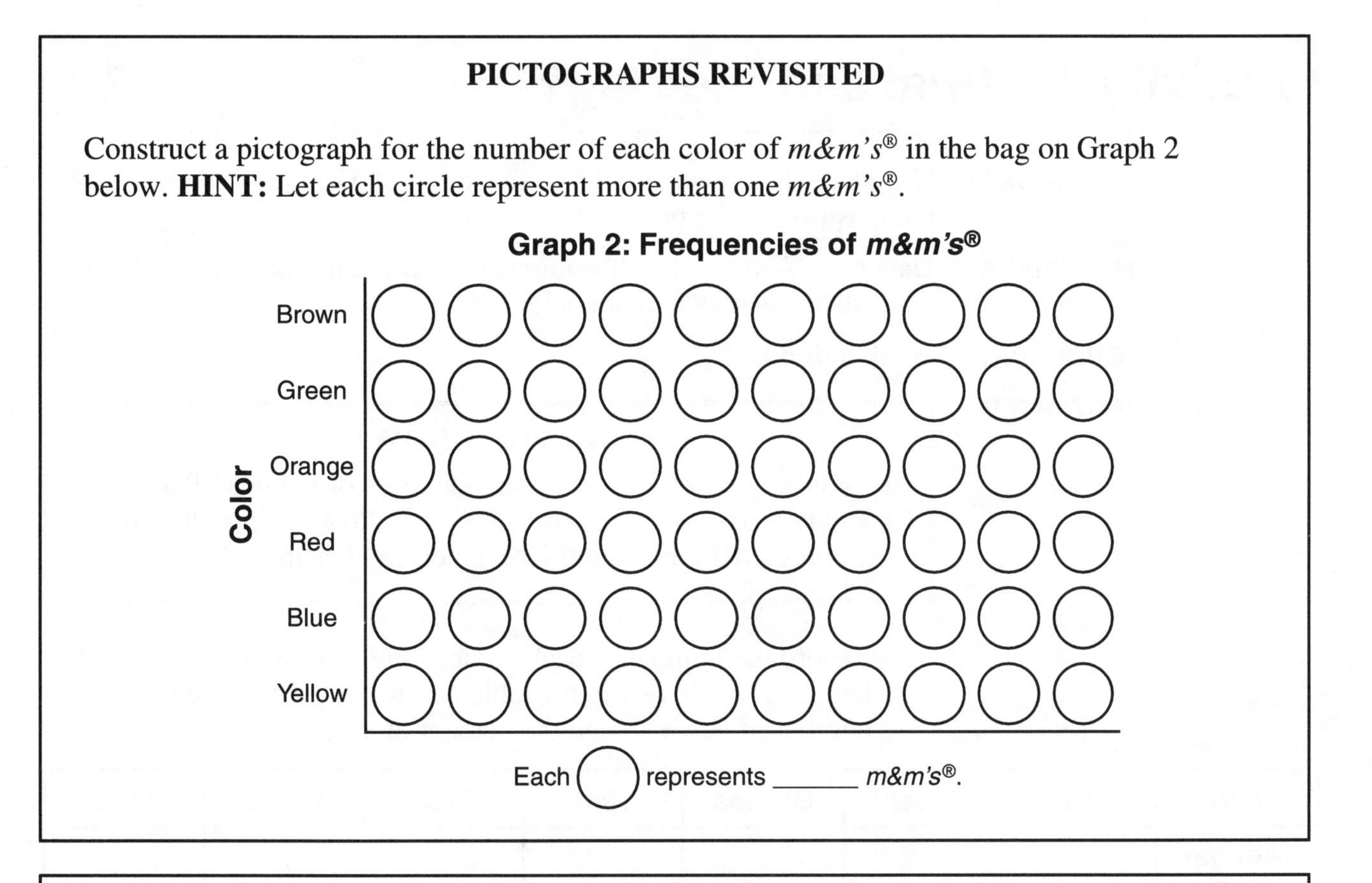

BAR GRAPHS

1. Use Graph 3 below to construct a bar graph for the number of each color of *m&m's*® in the bag. Label the scale on the horizontal axis.

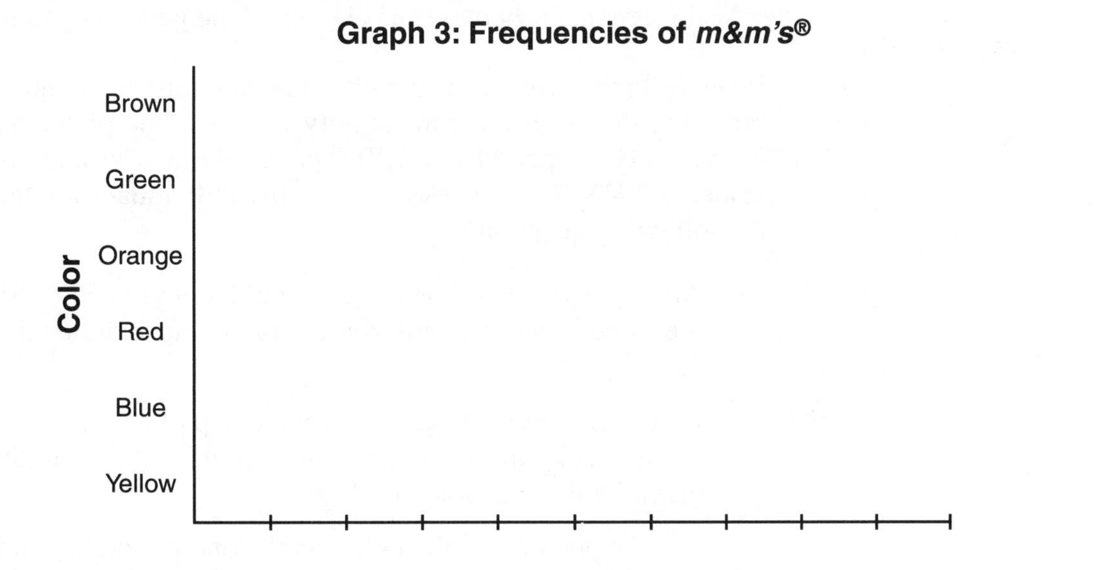

2. a. Which graph, the pictograph or the bar graph, was easier to construct? Why?

   b. Which graph is easier to read? Why?

Activity 2: What's in the Bag?

PURPOSE Use percents to compare the composition of a sample and the composition of the population.

MATERIALS Data from Activity 1 on the number of each color of *m&m's*® Plain Chocolate Candies in a one-pound bag

GROUPING Work individually.

GETTING STARTED

> **DID YOU KNOW?**
>
> According to Mars, Inc., the manufacturers of *m&m's*® Plain Chocolate Candies, there are 30% brown, 20% yellow, 20% red, 10% orange, 10% blue, and 10% green candies in each bag.

1. Record the number of each color of *m&m's*® in the one-pound bag from Activity 1 in the table below. Determine the total number of *m&m's*® and the percent of each color.

Color	Brown	Green	Orange	Red	Blue	Yellow	Total
Number							
Percent							

2. Use the data in the table to support or dispute the claim made by the manufacturer. How could the company explain any discrepancies between its claim and the percents you found?

EXTENSIONS Investigate the effect of using different sampling techniques by repeating the experiment in Activity 1 using a one-pound bag of individually wrapped FUN SIZE® packs of *m&m's*® and giving each student a FUN SIZE® pack for his or her individual sample. Answer the following questions.

1. Are the percents of the different colors in your FUN SIZE® pack the same as the percents you found in your individual sample in Activity 1?
2. In which individual sample (your scoop of *m&m's*® or the FUN SIZE® pack) are the percents of the colors closest to the percents given by the manufacturer?
3. Are the percents of the colors in the one-pound bag of FUN SIZE® packs the same as the percents you found in the one-pound bag of *m&m's*® in Activity 1? Are they the same as the percents given by the manufacturer?

*FUN SIZE® is a registered trademark of Mars, Inc.

Activity 3: Grouped Data

PURPOSE Display data using grouped frequency tables and stem-and-leaf plots.

GROUPING Work individually or in pairs.

GETTING STARTED In many situations, there may be so much data, or the data may be so spread out, that it becomes difficult to construct a line plot or a frequency table for individual items. In such cases, it may be more convenient to group the data.

President	Political Party[1]	Age at Inauguration	President	Political Party[1]	Age at Inauguration
Washington	F	57	Cleveland	D	47
J. Adams	F	61	B. Harrison	R	55
Jefferson	DR	57	Cleveland	D	55
Madison	DR	57	McKinley	R	54
Monroe	DR	58	T. Roosevelt	R	42
J. Q. Adams	DR	57	Taft	R	51
Jackson	D	61	Wilson	D	56
Van Buren	D	54	Harding	R	55
W. H. Harrison	W	68	Coolidge	R	51
Tyler	W	51	Hoover	R	54
Polk	D	49	F. D. Roosevelt	D	51
Taylor	W	64	Truman	D	60
Fillmore	W	50	Eisenhower	R	62
Pierce	D	48	Kennedy	D	43
Buchanan	D	65	L. B. Johnson	D	55
Lincoln	R	52	Nixon	R	56
A. Johnson	U	56	Ford	R	61
Grant	R	46	Carter	D	52
Hayes	R	54	Reagan	R	69
Garfield	R	49	Bush	R	64
Arthur	R	51	Clinton	D	46

[1]F = Federalist, DR = Democratic-Republican, D = Democrat, W = Whig, R = Republican, U = Unionist

SOURCE: *The World Almanac and Book of Facts 1992*

GROUPED FREQUENCY TABLES

Complete the *grouped frequency table* for the presidents' ages at inauguration.

How many different ages can fall within an interval in the table?

Can you determine the range, median, and mode of the ages from the information in the table? Explain why or why not for each statistic.

Age at Inauguration		
Interval	**Tally**	**Frequency**
40–44		
45–49		
50–54		
55–59		
60–64		
65–69		

STEM-AND-LEAF PLOTS

Stem-and-leaf plots provide a natural way to group data in intervals. To construct a stem-and-leaf plot, first find the smallest and the largest data points.

What is the youngest age at inauguration for the presidents? ______ the oldest age? ______

Next, decide on the stems. Since the presidents were inaugurated in their 40s, 50s, and 60s, use the tens digits of the ages as the stems. Write the stem digits in a column from least to greatest on the left side of a vertical line, as shown. If, as in this case, the data is to be grouped in intervals of 5, a two-line stem may be used. The first line for each stem will contain leaves ranging from 0 to 4. The second line for each stem, indicated by the dot, will contain leaves ranging from 5 though 9.

Stem	Leaf
4	
•	
5	
•	
6	
•	

For each age at inauguration, record a leaf by writing the units digit of the age on the right side of the vertical line in the row that contains its stem. The leaves for Washington and J. Adams have been done for you.

Stem	Leaf
4	
•	
5	
•	7
6	1
•	

After all the leaves have been recorded, rearrange the leaves in increasing order.

Use the stem-and-leaf plot to find the range, median, and mode of the ages at inauguration.

How are a stem-and-leaf plot and a grouped frequency table alike? How do they differ?

Stem	Leaf
4	
•	
5	
•	
6	
•	

On the grid, list the stems for the age at inauguration in the center column. Construct a stem-and-leaf plot for the age at inauguration of the Democratic (D) presidents to the left of the stems and one for the Republican (R) presidents to the right of the stems.

How do the ages at inauguration of Democratic presidents compare to those of Republican presidents?

Democratic		Republican

Activity 4: What's the Average?

PURPOSE Investigate mean, median, and mode and examine how each average is affected by extremes in the data.

MATERIALS 20 strips of 1-cm graph paper and a pair of scissors for each group

GROUPING Work individually or in groups of 3–4.

GETTING STARTED When we describe a set of data, it is often convenient to use a single number, often called the *average*, to indicate where the data are centered or concentrated. The mean, the median, and the mode are three commonly used *averages*.

Write the name of each of the following states on a strip of graph paper. Use one strip of paper for each state and one square for each letter in the name. Cut off the unused squares on the end of each strip.

Arizona, Hawaii, Ohio, Maine, Oregon, Idaho,
Texas, Louisiana, Kentucky

Arrange the names from shortest to longest, as shown in the example.

Count the number of letters in the name of each state. On a separate strip of graph paper, **write the numbers in order from least to greatest.** Write one number in each square and do not leave any blank squares between numbers. Cut off the unused squares on the end of the strip.

Example:

N	E	V	A	D	A			
M	O	N	T	A	N	A		
V	I	R	G	I	N	I	A	
W	I	S	C	O	N	S	I	N
M	I	N	N	E	S	O	T	A

6	7	8	9	9

THE MODE

Look at the numbers on the strip. What number of letters occurs most often in the names of the states?

This is the **mode** of the lengths of the names of the states.

In the example, the mode is 9.

N	E	V	A	D	A			
M	O	N	T	A	N	A		
V	I	R	G	I	N	I	A	
W	I	S	C	O	N	S	I	N
M	I	N	N	E	S	O	T	A

Mode (9 letters): WISCONSIN, MINNESOTA

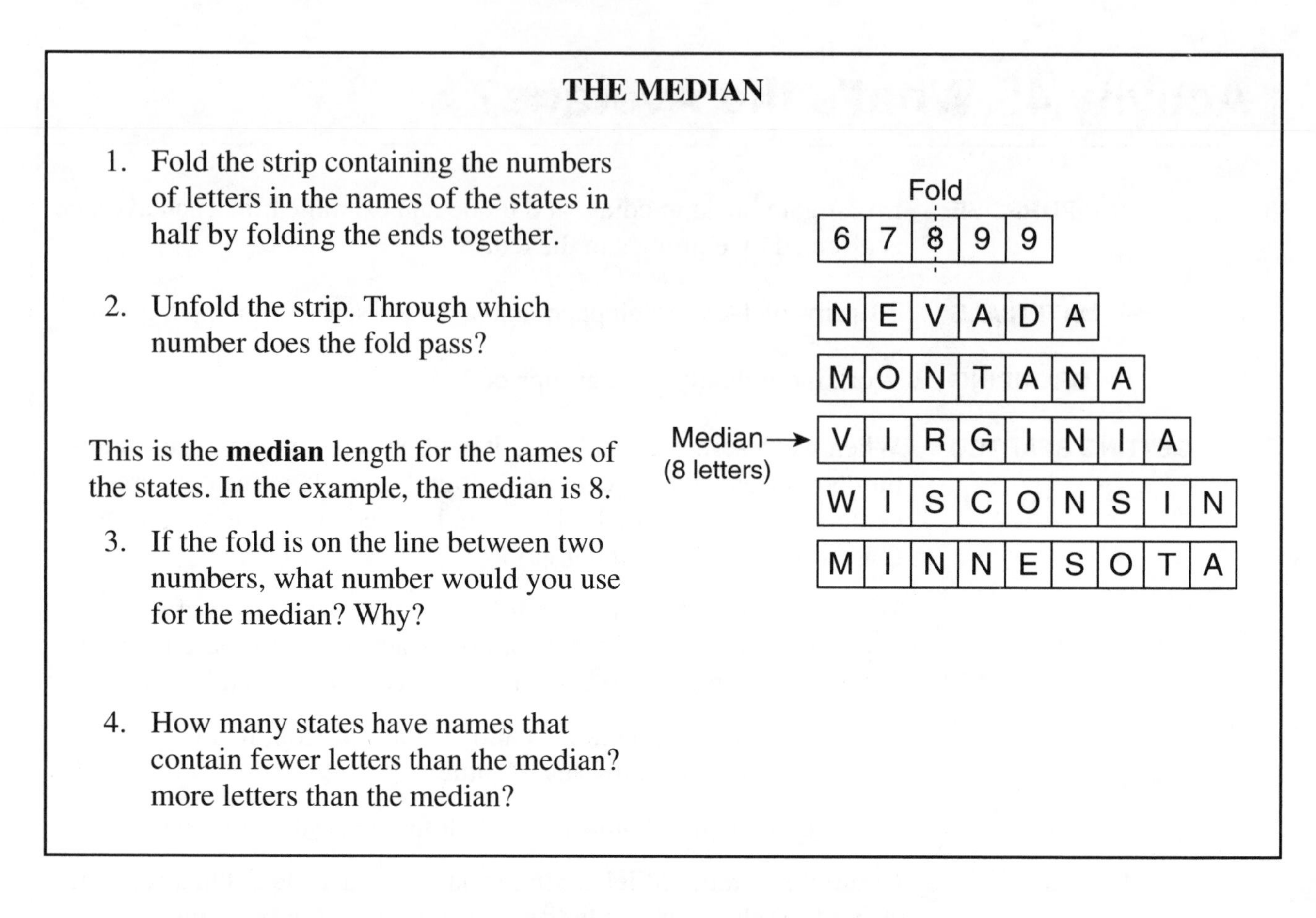

THE MEDIAN

1. Fold the strip containing the numbers of letters in the names of the states in half by folding the ends together.

2. Unfold the strip. Through which number does the fold pass?

This is the **median** length for the names of the states. In the example, the median is 8.

3. If the fold is on the line between two numbers, what number would you use for the median? Why?

4. How many states have names that contain fewer letters than the median? more letters than the median?

THE MEAN

To find the **mean** of the lengths of the names, cut off letters from the longer names and move them to fill in the shorter ones. Continue cutting off and moving letters until all the rows have as close to the same number of letters as possible.

N	E	V	A	D	A					N	E	V	A	D	A	N	
M	O	N	T	A	N	A				M	O	N	T	A	N	A	A
V	I	R	G	I	N	I	A		→	V	I	R	G	I	N	I	A
W	I	S	C	O	N	S	I	N		W	I	S	C	O	N	S	I
M	I	N	N	E	S	O	T	A		M	I	N	N	E	S	O	T

The mean in this example is a little less than 8 because all the rows except one contain eight letters.

What is the mean of the lengths of the names of the nine states?

1. Write each letter of Massachusetts in a square on a strip of graph paper. Add this to the data for the other nine states. Then repeat the steps to find the median, mode, and mean of the lengths of the names of the ten states.

   The median is ______. The mode is ______. The mean is ______.

2. Compare these *averages* with those for the original nine states. Describe how the addition of Massachusetts affected each *average* and explain the differences.

3. Remove the data for Massachusetts. Write the names Maryland, Michigan, and Oklahoma on strips of graph paper. Add them to the data for the original nine states. Then repeat the steps to find the median, mode, and mean.

   The median is ______. The mode is ______. The mean is ______.

4. Compare these *averages* with those for the original nine states. Describe how these additions affected each *average* and explain the differences.

5. To find the mean of the lengths of the names of N states, the letters making up the names of the states must be separated into N sets with the same (or nearly the same) number of letters in each set. Explain how you could find the number of letters in each set without writing each letter on a square.

WHICH WOULD YOU USE?

Sam Slugger's contract with the Columbus Mudcats baseball team says his annual salary will be $1,000,000 times the ***average*** of his batting averages for the preceding five seasons. Sam's batting averages for the past five seasons were .145, .130, .160, .130, and .495.

1. If you were Sam, which *average*—mean, median, or mode—would you want to use to compute your salary? Why?
2. If you were the owner of the Mudcats, which *average* would you want to use? Why?
3. Sam's contract went to arbitration. You are the arbitrator. Which *average* would you use to determine Sam's salary? How would you justify your decision?

IDENTIFY THE AVERAGE

Which *average*—mean, median, or mode—do you think was used in each of the following statements? Explain your choice in each case.

The *average* lady's shoe size is $7\frac{1}{2}$.

The *average* size of a household in the United States is 2.67 people.

The *average* annual family income in the United States is $28,236.

Activity 5: Finger-Snapping Time

PURPOSE Display and compare data using box-and-whisker plots.

MATERIALS Scissors, three strips of 1-cm graph paper, and a clock or watch to measure elapsed time in seconds

GROUPING Work in pairs.

GETTING STARTED Snap your fingers as fast as you can for 15 seconds. Have your partner time you while you snap your fingers and count the number of snaps. Then do the same thing for your partner. Record the data in the table below.

Ask 13 other people how many times they snapped their fingers in 15 seconds and record the information in the table.

Finger Snaps

Person	Finger Snaps in 15 sec	Person	Finger Snaps in 15 sec	Person	Finger Snaps in 15 sec
You		4		9	
Partner		5		10	
1		6		11	
2		7		12	
3		8		13	

Box-and-whisker plots provide a useful method for summarizing and comparing data such as the number of times people can snap their fingers.

- **The first step in constructing a box-and-whisker plot is to order the data values from least to greatest.**

1. Write the finger-snapping data in order from least to greatest on a strip of graph paper. Write one number in each square. Do not leave any blank squares between numbers. Cut off the unused squares on the end of the strip.

Example:

20	35	37	39	41	45	45	47	59

LE = 20; UE = 59

- **The second step is to find the extremes of the data.** The smallest data point is called the *lower extreme* (LE), and the largest data point is called the *upper extreme* (UE).

2. Find and record the lower extreme and the upper extreme of the finger-snapping data.

LE = ________ UE = ________

20 30 40 50 60

LE UE

3. Use the extremes to select an appropriate scale and label the number line below.

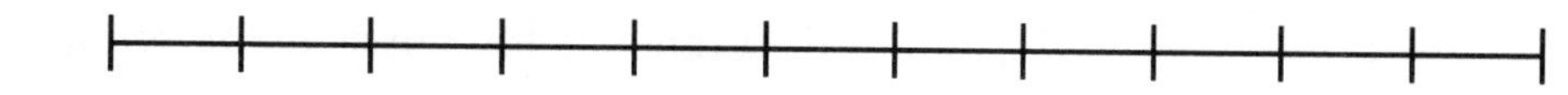

4. Locate the lower and upper extremes by marking a dot under their coordinates on the scale as in the example.

• **The third step is to find the median of the data.**

20 30 40 50 60

Median = 41

5. To find the median of the finger-snapping data, fold the strip in half by folding the ends together. The median is the number the fold passes through or, if the fold falls on the line between two numbers, it is the mean of the numbers.

6. Record the median and mark its location by making a dot under its coordinate on the scale.

   Median = ___________

• **The fourth step is to find the quartiles for the data.**

The *lower quartile* (LQ) is the median of the data values that are less than the median.

Data less than median / Data greater than median

20 35 | 37 39 41 45 45 | 47 59

LQ = 36 UQ = 46

7. Look at the part of the strip that contains data values that are less than the median. Find the median of these values by folding this part of the strip in half. This is the lower quartile of the data.

   Lower Quartile = ___________

The *upper quartile* (UQ) is the median of the times that are greater than the median of the data.

8. Look at the part of the strip that contains data values that are greater than the median. Find the median of these values by folding this part of the strip in half. This is the upper quartile of the data.

   Upper Quartile = ___________

20 30 40 50 60

LQ UQ

9. Mark the locations of the upper and lower quartiles by making a dot for each below its coordinate on the scale.

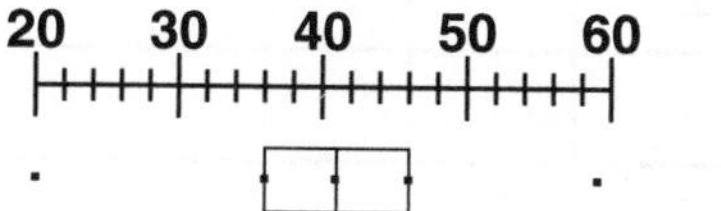

10. Form a box by drawing vertical segments through the dots for the upper and lower quartiles and connecting the endpoints of the segments as in the example. Then draw a vertical segment through the median as shown.

- **The final step is to identify and plot any outliers in the data.**

IQR = 46 – 36 = 10

11. Find the difference between the upper and lower quartiles. This difference is known as the *interquartile range* (IQR).

    Interquartile Range = UQ – LQ = ________

If a data point is more than 1.5 interquartile ranges above the upper quartile or more than 1.5 interquartile ranges below the lower quartile, it is called an *outlier*.

In the example, $LQ - 1.5 \times IQR = 36 - 1.5 \times 10 = 36 - 15 = 21$.

Thus 20 is an outlier, since $20 < LQ - 1.5 \times IQR$.

12. Identify any outliers in your data. Mark their locations by making a dot for each one below its coordinate on the scale.

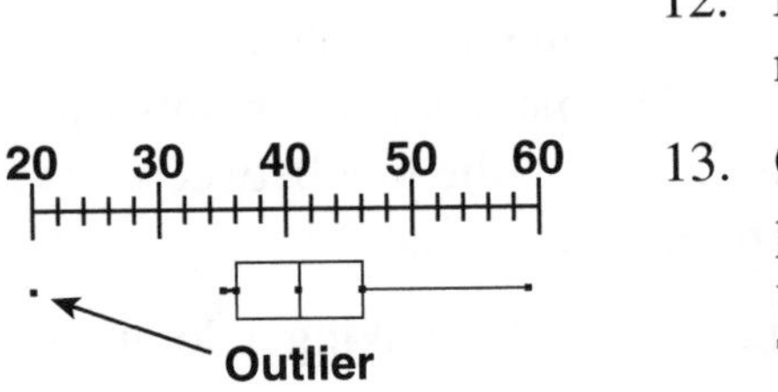

13. Complete the plot by drawing segments from the smallest data point that is not an outlier to the lower quartile and from the upper quartile to the largest data point that is not an outlier. These segments are the whiskers.

14. Study the completed box-and-whisker plot. About what percent of the data points lie between the

    a. lower extreme and the lower quartile?

    b. upper extreme and the upper quartile?

    c. lower quartile and the median?

    d. median and the upper quartile?

15. About what percent of the data points lie in the box?

16. Repeat the finger-snapping experiment, only this time snap your fingers as fast as you can for 30 sec. Divide the number of snaps by 2 to get your rate per 15 sec. Record your rate, your partner's rate, and the rates for 13 other people in the table below.

Finger Snaps

Person	Sex	Finger Snaps per 15 sec	Person	Sex	Finger Snaps per 15 sec	Person	Sex	Finger Snaps per 15 sec
You			4			9		
Partner			5			10		
1			6			11		
2			7			12		
3			8			13		

17. a. Construct a box-and-whisker plot for the new data.

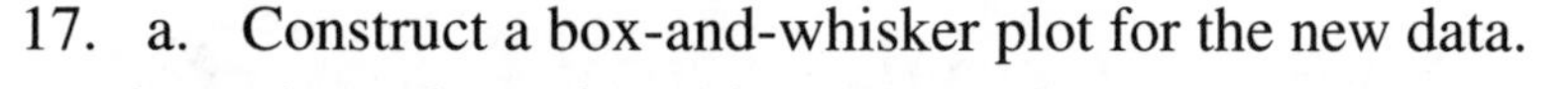

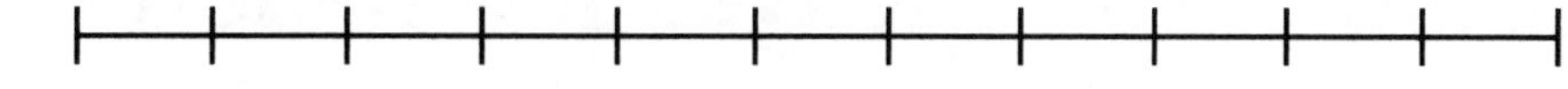

b. Use the scale above to construct a box-and-whisker plot for the original data in the space below the plot for the new data.

18. a. How do the medians of the two sets of data compare?

b. The extremes?

c. The interquartile ranges?

d. The upper quartiles?

e. The lower quartiles?

19. Based on your observations in Exercise 18, how do the finger-snapping rates in the first experiment compare to the rates in the second experiment? How would you explain any similarities or differences?

EXTENSIONS

1. a. Separate the data in the table for the second experiment into rates for males and rates for females. Construct box-and-whisker plots for the male and female rates using the same scale for both.

b. Based on your graphs, do you think there is any difference between the finger-snapping rates for males and for females? Explain your answers.

2. a. Construct a stem-and-leaf plot for the data from the first finger-snapping experiment.

b. What can you learn about the distribution of the data (gaps, clusters, extremes, outliers, etc.) and the averages **from both** the box-and-whisker plot and the stem-and-leaf plot of the data?

c. What information can you get about the distribution of the data and the averages from a box-and-whisker plot **but not from** a stem-and-leaf plot?

d. What information can you get about the distribution of the data and the averages from a stem-and-leaf plot **but not from** a box-and-whisker plot?

Activity 6: The Weather Report

PURPOSE Apply the concepts of data analysis, evaluate statements based on data, and analyze the advantages and disadvantages of using different displays of data.

MATERIALS A calculator and graph paper

GROUPING Work individually or in pairs.

Normal Daily Temperature for Some U.S. Cities

Based on the Period 1950–1980

	Jan	Feb	Mar	Apr	May	Jun	July	Aug	Sep	Oct	Nov	Dec
Los Angeles, CA	56	57	57	60	62	66	69	70	70	66	61	57
San Francisco, CA	49	52	53	55	58	61	62	63	64	61	55	49
Wichita, KS	30	35	44	56	66	76	81	80	71	59	44	34
Portland, ME	22	23	32	43	53	62	68	67	59	48	38	26
Portland, OR	39	43	46	50	57	63	68	67	63	54	46	41
Charlotte, NC	41	43	50	60	68	75	79	78	72	61	51	43
Seattle, WA	39	43	44	49	55	60	65	64	60	53	45	41
Spokane, WA	26	32	38	46	54	62	70	69	59	48	35	29

SOURCE: U.S. National Oceanic and Atmospheric Administration

1. Construct a stem-and-leaf plot for the weather data given for San Francisco and Wichita.

San Francisco

4	
5	
6	

Wichita

2	
3	
4	
5	
6	
7	
8	

2. Compute the mean, median, and mode for the data given for San Francisco and Wichita.

	San Francisco	Wichita
Mean		
Median		
Mode		

3. Construct line graphs for the temperatures for San Francisco and Wichita on the grid below.

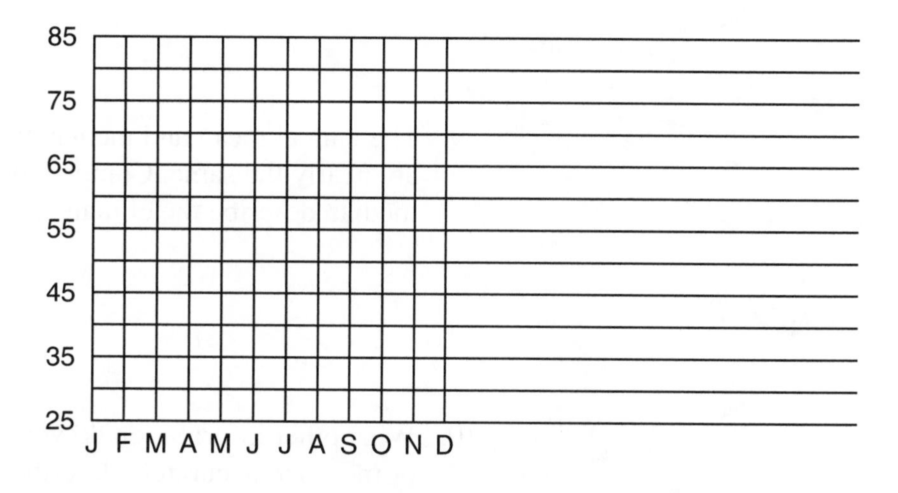

4. Using the scale on the grid above, construct vertical box-and-whisker plots for the temperatures for San Francisco and Wichita to the right of the line graphs.

5. Draw a horizontal line through the graphs to show the median for each set of data.

6. For how many months of the year is the temperature in Wichita within the range of the temperatures in San Francisco?

7. Which display of data can most easily be used to answer Exercise 6? Explain why or why not for each display.

8. The Wichita Chamber of Commerce could advertise that "the annual average temperature in Wichita is the same as that in *balmy* San Francisco." Evaluate this claim on the basis of the actual temperature data.

9. The annual mean and median temperatures for these two cities are nearly the same. Compare how accurately the mean and the median describe the climate in each city.

10. What other information do you need in addition to the medians or means to accurately describe the annual temperatures?

11. Find the latitude of Wichita: ____; and of San Francisco: ____.

    Which city is farther north?

12. If the difference in latitude is not significant, explain how the geographical location of each city affects its annual temperature.

13. Construct a back-to-back stem-and-leaf plot for Portland, ME, and Portland, OR.

Portland, ME		Portland, OR
	2	
	3	
	4	
	5	
	6	
	7	
	8	

14. On the grid below, construct line graphs and accompanying vertical box-and-whisker plots for the weather data for the two cities. Then draw the median lines. Explore questions similar to Exercises 6–12 for these two sets of data.

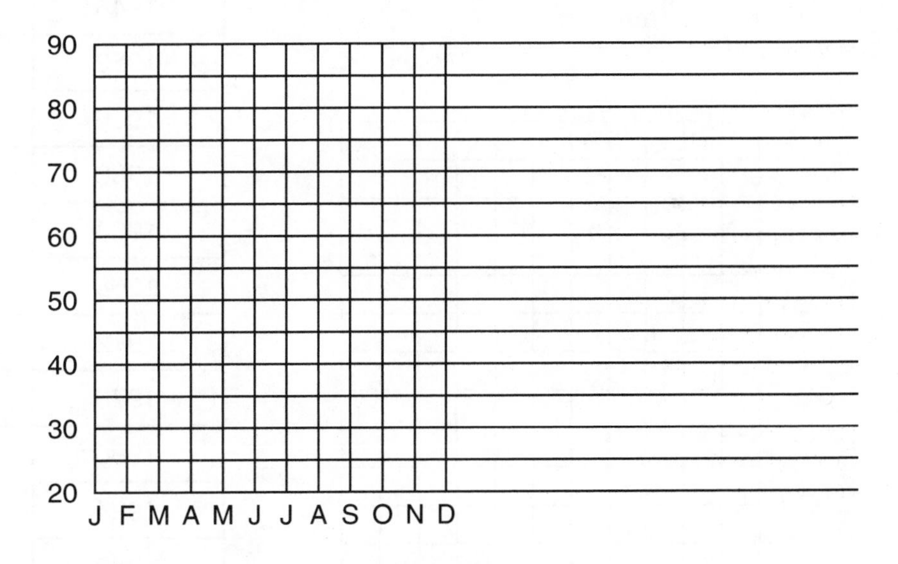

15. Choose other pairs of cities from the table and construct various displays for the weather data. Explore similarities and differences in the data. Determine the latitude and geographical location of each city. Explain how these affect the similarities or differences in the climates of the chosen cities.

Activity 7: Are Women Catching Up?

PURPOSE Display and analyze data using line graphs, scatter plots, and median fit lines. Make conjectures based on data and defend the conjectures.

GROUPING Work individually or in pairs.

GETTING STARTED In 1992, researchers at the University of California, Los Angeles, School of Medicine published an analysis of data indicating that within 65 years, top female and male runners might perform equally well in the 200-m, 400-m, 800-m, and 1500-m runs. Other researchers quickly responded that this was not possible. What do you think? Let's look at some of the data.

Olympic Track Records: 200-m Run

Year	Men's Time (sec)	Women's Time (sec)
1948	21.1	24.4
1952	20.7	23.7
1956	20.6	23.4
1960	20.5	24.0
1964	20.3	23.0
1968	19.83	22.5
1972	20.00	22.40
1976	20.23	22.37
1980	20.19	22.03
1984	19.80	21.81
1988	19.75	21.34
1992	19.73	21.72

SOURCE: *The World Almanac and Book of Facts 1992*

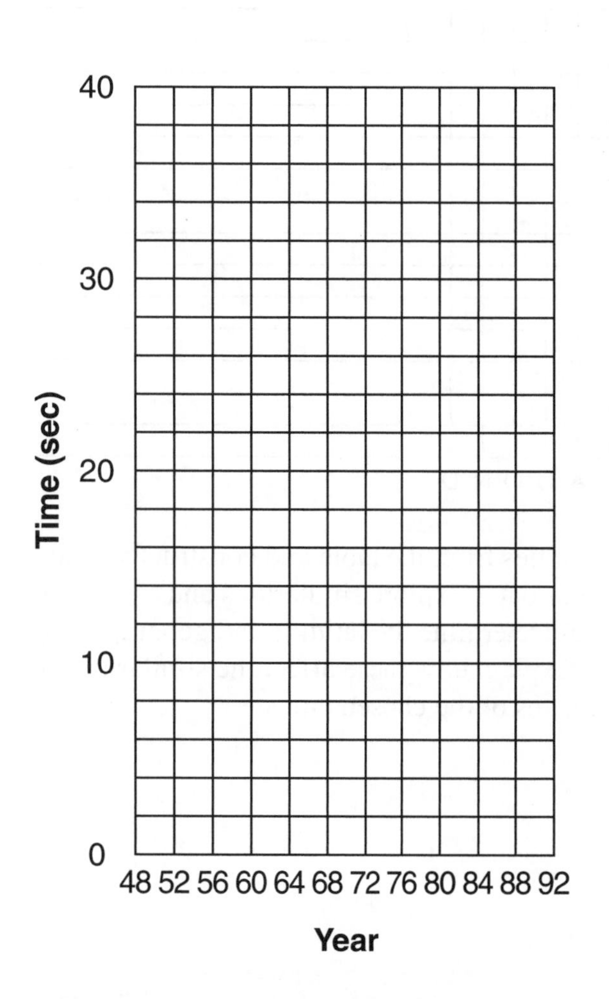

1. On the grid at the left, draw line graphs for the men's and the women's record times for the 200-m run.

2. Based on the graphs, draw a conclusion about the men's and women's record times for the 200-m run.

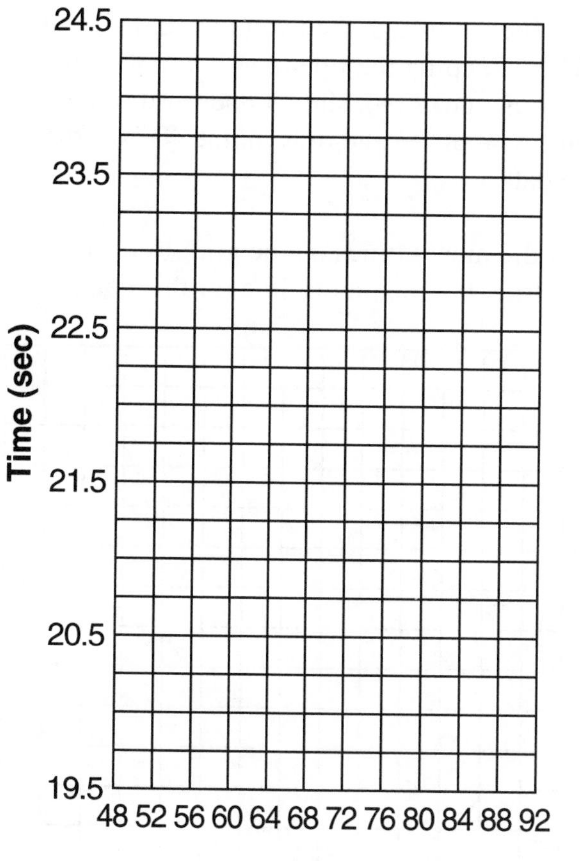

3. On the grid at the left, redraw the line graphs for the men's and the women's record times for the 200-m run.

4. Based on these graphs, draw a conclusion about the men's and women's record times for the 200-m run.

5. Is your conclusion in Exercise 4 the same as that in Exercise 2? Explain why or why not.

6. Both pairs of graphs present the same data on the same-sized grid. Why do they look so different?

When data appear to lie roughly along a straight line, it is often possible to fit a line to the data and use the line to make predictions. The following exercises develop one technique for doing this. *Note:* You must always be cautious about fitting a line to data because the data may actually fall near a curve or be clustered in two or more regions.

7. The grid on page 196 contains a scatter plot of the women's record times for the 200-m run. Draw two vertical lines that divide the data into three groups with approximately the same number of data points in each group. If the data cannot be divided evenly, the two outer groups should contain the same number of data points.

8. The group on the left contains four data points: (1948, 24.4), (1952, 23.7), (1956, 23.4), and (1960, 24.0). The median of the years is 1954, and the median of the times is 23.85 sec. Use a plus symbol (+) to mark the point (1954, 23.85) on the grid.

9. Find the median of the years and the median of the times for the data points in each of the other two groups, as in Exercise 8. Use a plus symbol (+) to mark the corresponding points on the grid.

10. Place a ruler so that it passes through the two plus marks (+) in the outside groups. Then, keeping the ruler parallel to the line through the two outer plus marks (+), slide it one-third of the way to the middle plus symbol (+) and draw a line. This is the *median fit line* for the data.

11. Make a scatter plot for the men's times on the grid. Repeat Exercises 7–10 to construct the median fit line for this data.

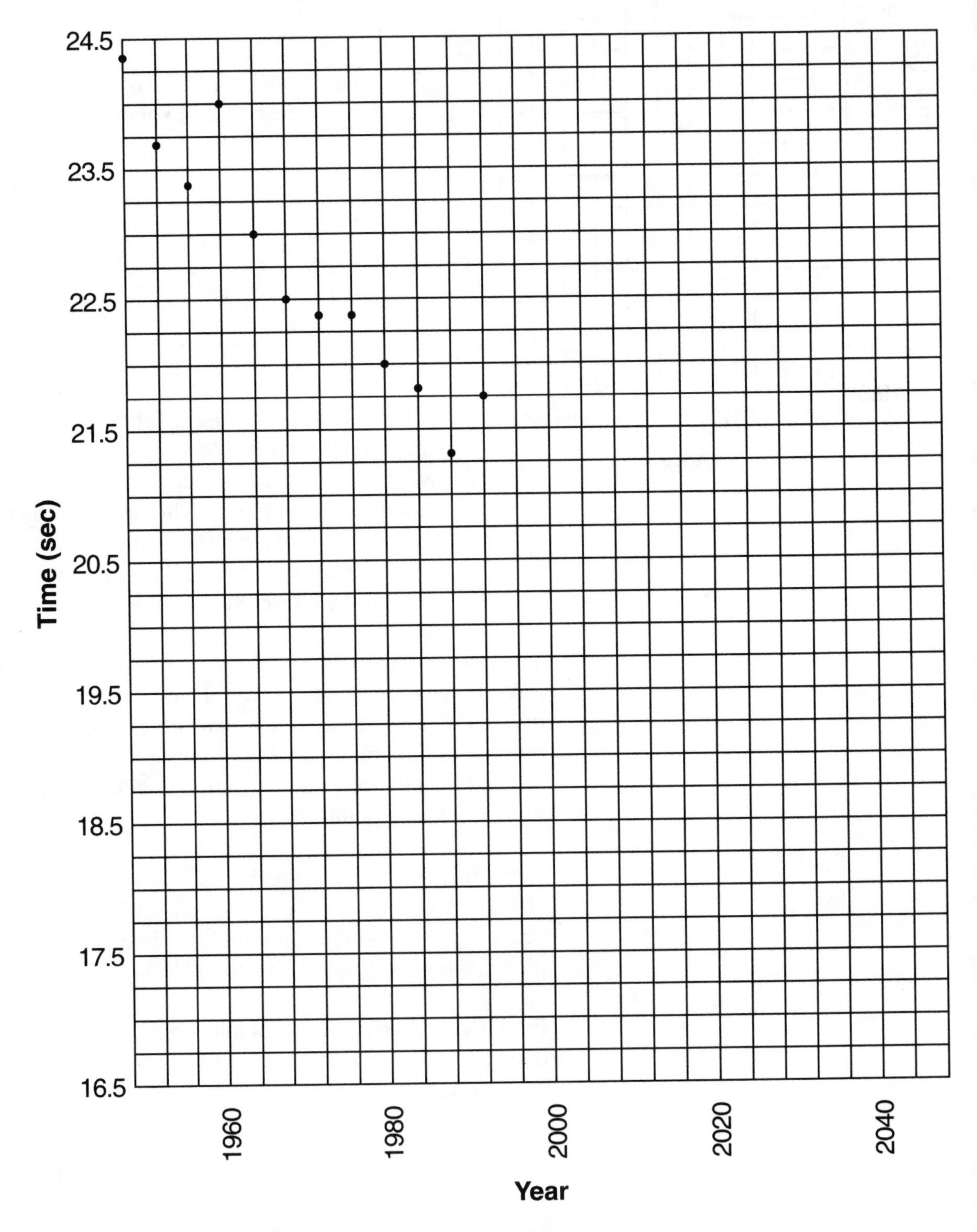

12. a. Use the median fit lines to predict the year that women will have the same time in the 200-m run as men.

    b. How does your prediction in Part (a) compare with that made by the University of California researchers?

    c. Approximately what will be the record for the 200-m run when men's and women's times are equal?

13. Why are women's times for the 200-m run decreasing at a faster rate than men's times? Explain.

14. Prepare an argument that supports the prediction that in the future, women's times for the 200-m run will equal men's times.

15. Prepare an argument to support the position that women's times for the 200-m run will never equal men's times.

EXTENSIONS Find the men's and women's winning times for the Boston Marathon over the last twenty years.

1. Use line graphs to construct two displays of the data, one that could be used to argue that in the future, women's times for the marathon will be equal to the men's, and one that could be used to argue that they will not.

2. Construct scatter plots of the data. Find the median fit lines for both sets of data. Use the lines to predict the year that the women's time for the Boston Marathon will equal the men's.

3. Compare the data using box-and-whisker plots. Explain how the two different visual presentations of the data can lead to different conclusions.

Activity 8: To Change or Not to Change

PURPOSE Analyze a set of data and prepare a proposal based on the results.

GROUPING Work individually.

GETTING STARTED There has been a great deal of debate in recent years about the need to lengthen the school day or to extend the school year for students in the United States. What's your opinion?

The following table contains data on the average mathematics scores on the International Assessment of Educational Progress for 13-year-old students in various countries. It also includes information on the length of the school year and the length of the school day in each country.

Country	Average Percent Correct	Average Days of Instruction per Year	Average Minutes of Instruction per School Day
Canada	62	188	304
Emilia-Romagna, Italy	64	204	289
France	64	174	370
Hungary	68	177	223
Ireland	61	173	323
Israel	63	215	278
Jordan	40	191	260
Korea	73	222	264
Scotland	61	191	324
Slovenia	57	190	248
Soviet Union	70	198	243
Spain	55	188	285
Switzerland	71	207	305
Taiwan	73	222	318
United States	55	178	338

[**Source:** Educational Testing Service, *Learning Mathematics,* Feb. 1992]

Use the statistical concepts you have studied to organize and analyze the data in the table. Based on the results, prepare a report recommending what changes, if any, should be made in the length of the school year and the school day for students in the United States. Whenever appropriate, use averages and statistical displays to support and clarify your position.

Chapter Summary

Activities 1, 3, and 5 introduced a variety of graphs, tables, and plots that can be used to organize and display data. The focus of the activities was on the potential uses and limitations of each statistical display, not just the construction of each display.

In general, *real graphs* do not provide a practical method for displaying statistical data. *Pictographs* are common in the popular media, probably because of their aesthetic appeal, but they are difficult to interpret and construct. *Bar graphs,* on the other hand, display exactly the same information and have the advantage of being easy to construct and to interpret.

Because *line plots* are quick and easy to construct, they provide a very useful tool for preliminary data analysis. Maximum and minimum values, clusters and gaps in the data, outliers, the median, and the mode are all easily identified on a line plot. The major limitation is that line plots are convenient to use only with relatively small sets of data.

Stem-and-leaf plots provide information about how data are distributed. Maximum and minimum values, the median, the mode, clusters and gaps in the data, and outliers are all easily identified in a stem-and-leaf plot. *Box-and-whisker plots* focus on the range and distribution of the data. Clustering of the data can sometimes be identified, but gaps in the data cannot. Box-and-whisker plots are very useful when working with large sets of data and for comparing different sets of data.

Activity 4 introduced the concept of an average and developed the meanings of the mean, median, and mode. The development focused on (a) how each measure provides an indication of where data are centered or concentrated, (b) how each measure is affected by extremes, and (c) how the physical modeling is related to the algorithm for computing the mean.

The activities illustrated that the *mean* has a leveling or smoothing effect. Another way of expressing this is, if all the data had the same value, the data points would all equal the mean. The mean is the most commonly used average—in fact, many people erroneously use the terms average and mean synonymously. However, the value of the mean may be affected by extremes in the data, and therefore it may not be the most representative value to use for an average.

The *median* is the middle data point or the mean of the two middle data points. The value of the median is not significantly affected by unusually large or small data points; thus the median is a more appropriate average than the mean when there are extremes in the data. However, the median may not accurately reflect concentrations in the data.

The *mode*, the data point that occurs most often, is a measure of where the data are concentrated. Its usefulness is limited, however, since either the frequency of occurrence of the mode may not be significantly different than that of other data points, the mode may be an outlier, or the data may have more than one mode.

Activities 1 and 2 explored the relationship between a sample and the population. Predicting the characteristics of a population from a sample is an important concept in statistics. The reliability of such predictions is affected by two factors: the sample size and the randomness of the sample. Assuming that the samples are selected randomly, predictions generally become more accurate as the size of the sample increases. Randomness was demonstrated in Activity 1 by the fact that there was a great deal of variation in the individual samples of *m&m's*®.

The focus of Activity 6 was on analyzing various plots to determine what information can be more easily derived from one display versus another and to evaluate the use of the mean or median as a good descriptor of average.

Advertisers, policy makers, and decision makers constantly use the term *average*, but the average reported may not accurately describe the entire set of data. It is very important to know the range of data to fully understand what the mean or median really indicates. For example, the mean and median temperatures for San Francisco and Wichita differ by only a degree. However, the range of the data varies considerably.

The stem-and-leaf plots illustrate the difference in the range of data, but the difference is much more visually apparent in line graphs or in side-by-side box-and-whisker plots. The median and the upper and lower quartiles of data can be identified in a stem-and-leaf plot; however, the relationships among these measures are much clearer in box-and-whisker plots. Each display has its own unique characteristics, and each offers a different insight into the data.

Since each display provides a different insight, the way data is displayed can affect the conclusions drawn from it. As illustrated in Activity 7, the conclusions can also be influenced by how the display is constructed. The choice of scales for a line graph, for example, can dramatically alter the visual impact of the data and result in two different interpretations. Activity 8 provided an opportunity for you to apply statistical concepts in a real-world context to display and interpret data, to reason from the data, to make a decision based on your interpretation, and to defend your decision.

Chapter 10
Introductory Geometry

"The study of geometry should provide experiences that deepen students' understanding of shapes and their properties, with an emphasis on their wide application in human activity. . . . Physical models and other real-world objects should be used to provide a strong base for the development of students' geometric intuition so that they can draw on those experiences in their work with abstract ideas."

—*Curriculum and Evaluation Standards for School Mathematics*

In his book, *A Mathematician's Delight,* the renowned mathematician W. W. Sawyer wrote:

"The best way to learn geometry is to follow the road which the human race originally followed:

Do things,
Arrange things,
Make things,
and only then
Reason about things."

The activities in this chapter provide a variety of opportunities to arrange, measure, and construct geometric shapes in two and three dimensions. Following the constructions, you will conjecture and reason about the figures to develop an understanding of their fundamental properties.

Pattern blocks, cubes, and geoboards are excellent tools for exploring geometric concepts. You will use these manipulatives, a computer, and dynamic geometry software to develop the conceptual understanding of triangles and selected quadrilaterals, their properties, and the relationships among triangles and among the quadrilaterals. This informal exploration of shapes and their properties builds the foundation that is needed for the study of formal, deductive geometry.

Activity 1: What's the Angle?

PURPOSE Develop the concept of measurement of angles using the central angle in a circle and its intercepted arc.

MATERIALS A circular geoboard or pages A-38 and A-39 and geo-bands

GROUPING Work individually.

GETTING STARTED The number of degrees in an angle is the measure of the opening between the two rays or the amount of rotation as one ray turns from coinciding with one side of the angle coinciding with the other side. A complete rotation of a ray about a point results in an angle with a measure of **360°**. One degree is $\frac{1}{360}$ of a complete rotation.

A **central angle** is an angle whose vertex is the center of a circle and whose sides contain radii.

Example:

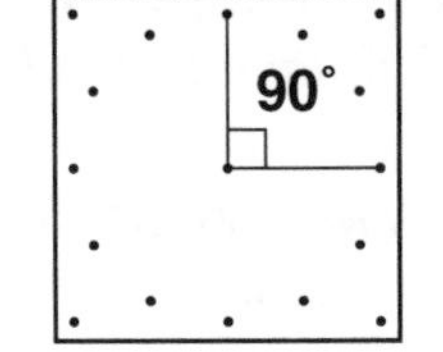

For each central angle, record its measure and its classification: *acute, right,* or *obtuse.*

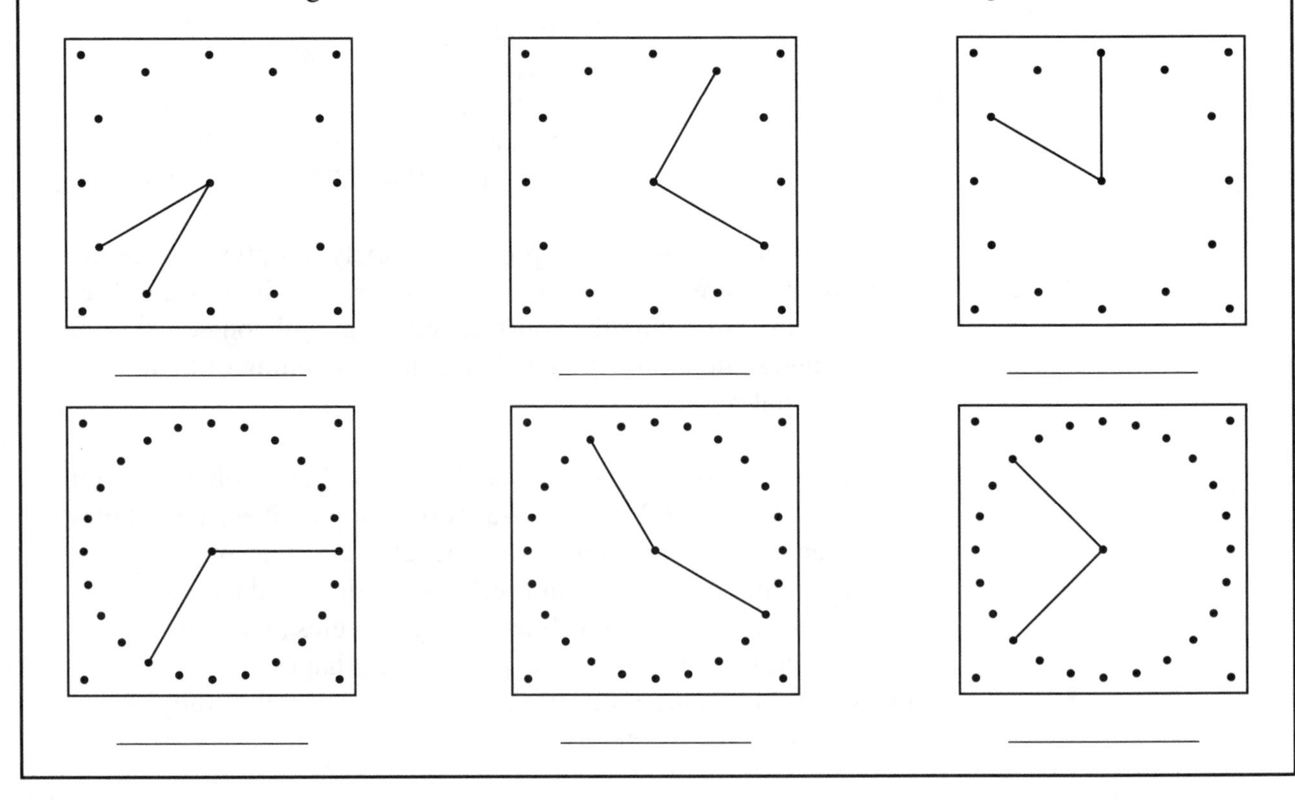

For each exercise, construct an angle that has the given measurement. In some exercises, one side of the angle is given.

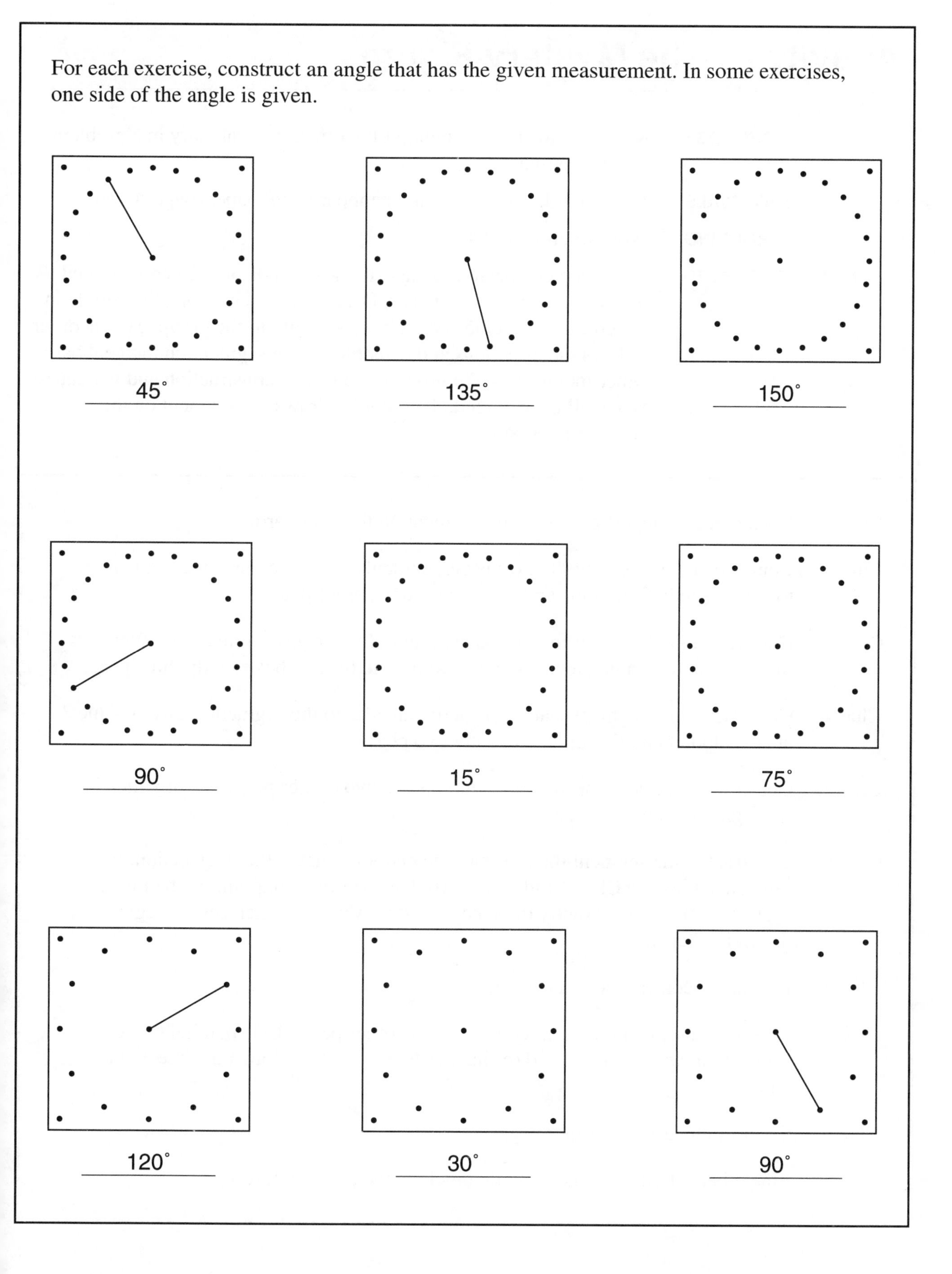

Activity 2: The Mystery Figure

PURPOSE Reinforce the understanding of geometric vocabulary in a problem-solving setting.

MATERIALS A geoboard, geo-bands, and geoboards dot paper (page A-40)

GROUPING Work in groups of 4–5.

GETTING STARTED Designate one person to place the geo-bands on the geoboard and record the work on dot paper. Divide the clues among the people in the group. Each clue should be read aloud to the group. The recorder follows the directions and constructs each segment on the geoboard. Other members of the group review each construction and validate it. When all have agreed, the recorder draws the segment on the geoboard dot paper.

Clue 1. Construct the longest possible line segment on the geoboard.

Clue 2. Construct a line segment that: (a) is perpendicular to the longest line segment, (b) touches exactly four pegs, and (c) touches one corner peg.

Clue 3. Construct a line segment that: (a) is parallel to the longest segment, (b) intersects the segment drawn in Clue 2 at an endpoint, and (c) touches exactly three pegs.

Clue 4. Construct a line segment that: (a) is perpendicular to the segment drawn in Clue 3 at an endpoint and (b) touches exactly two pegs.

Clue 5. Construct a line segment whose endpoints are two corner pegs, one of which is on the segment drawn in Clue 2.

Clue 6. Construct a line segment that: (a) has one endpoint that is the intersection of the segments drawn in Clue 2 and Clue 3, (b) has the other endpoint on the longest segment, (c) touches exactly three pegs, and (d) does not intersect the segment drawn in Clue 4.

Clue 7. Repeat the construction in Clue 5.

Clue 8. Construct a segment that: (a) connects two corner pegs, (b) is parallel to the segment drawn in Clue 6, and (c) intersects the segment drawn in Clue 3 at an endpoint.

Clue 9. Connect the last two corner pegs.

Find other activities in the book that include the mystery figure.

Activity 3: Classifying Triangles

PURPOSE Reinforce the classification of triangles according to their side and angle relationships.

MATERIALS A centimeter ruler, a protractor, scissors, and copies of the triangles on page A-41

GROUPING Work individually or in pairs.

GETTING STARTED Work with your partner to share information on the measurements and to discuss sorting the triangles into various groups. Remember that measurement is not exact and that this may result in differences between you and your partner in classifying the triangles. These differences should be resolved so that you agree on one classification for each triangle.

For each triangle on page A-41, find the measure of each angle and each side. Write the measures inside each figure and then cut out each triangle.

Triangles may be classified in two ways:

A. By the number of congruent sides:

1. A *scalene* triangle has no congruent sides.
2. An *isosceles* triangle has at least two congruent sides.
3. An *equilateral* triangle has three congruent sides.

B. By the types of angles:

1. An *acute* triangle has three acute angles.
2. A *right* triangle has one right angle.
3. An *obtuse* triangle has one obtuse angle.

1. Sort the triangles using the previous definitions. In the blanks below, write the letter of each triangle that belongs in each group. A triangle may belong to more than one group.

   a. Scalene ____________________ b. Equilateral ______________

   c. Acute ______________________ d. Isosceles ________________

   e. Right ______________________ f. Obtuse __________________

2. Is it possible to sort the triangles using

   a. two classifications by sides (e.g., isosceles and scalene)? Explain.

   b. two classifications by angles (e.g., right and obtuse)? Explain.

3. List all the possible ways to sort triangles using two classifications, one by sides and one by angles. Then sort the triangles according to your groups of the two classifications and indicate the letters of the triangles that belong in each group.

4. Which combinations of two classifications, one by sides and one by angles, are not possible? Why?

EXTENSIONS Research material on the van Hiele levels that describe the developmental levels associated with the understanding of geometry. At what level would you classify the questions in this activity?

Activity 4: Toothpick Triangles

PURPOSE Reinforce the classification of triangles by sides and by angles in a problem-solving setting.

MATERIALS Toothpicks that are all the same length and a ruler or straight edge

GROUPING Work individually or in pairs.

GETTING STARTED Use each of the numbers of toothpicks shown in the table below to form triangles. All toothpicks in a figure must be the same length and must be placed end to end. If necessary, lay them along a ruler to form a straight line segment.

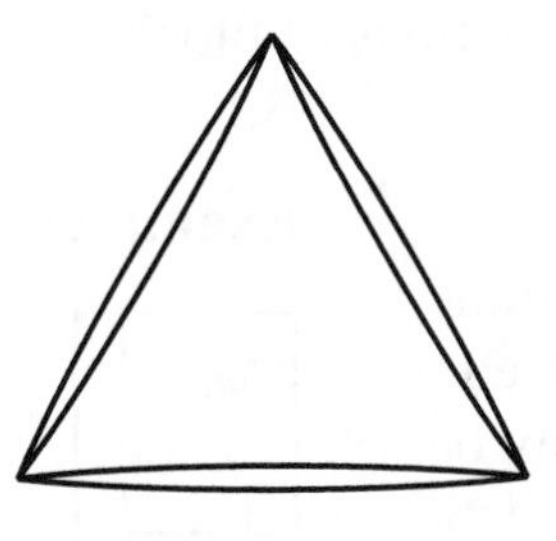

Determine the kind of triangle that can be constructed and complete the table. When listing the triangles that can be formed, list only those that are unique. A triangle with sides of length 5, 8, and 11 is the same as one with sides of length 8, 11, and 5.

Number of Toothpicks	Possible Triangles	Type of Triangle
3	1-1-1	Acute Equilateral
4		
5		
6		
7		
8		
9		
10		
11		

Given three sets containing *a* toothpicks, *b* toothpicks, and *c* toothpicks, what must be true about *a, b,* and *c* in order to construct a triangle in which *a, b,* and *c* represent the lengths of the sides?

Activity 5: Angles on Pattern Blocks

PURPOSE Determine the sum of the measures of the interior angles of a polygon and the measure of each angle of a regular polygon.

MATERIALS One set of pattern blocks

GROUPING Work individually.

Determine the measure of each interior angle of each pattern block. You may use **only** the fact that the square has four right angles. **HINT:** You may place combinations of blocks on top of a block to assist in determining the measures of the angles.

Indicate the measure of each angle inside the figures as shown in the example. For each pattern block, explain the method you used to determine the measure of each angle. You may draw a sketch of the blocks you used to illustrate your explanations. The measures you find on one block may be used to determine the measures of the angles of other blocks.

Example:

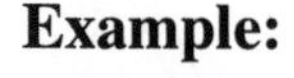

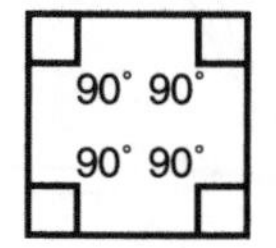

a.

b.

c.

d.

e.

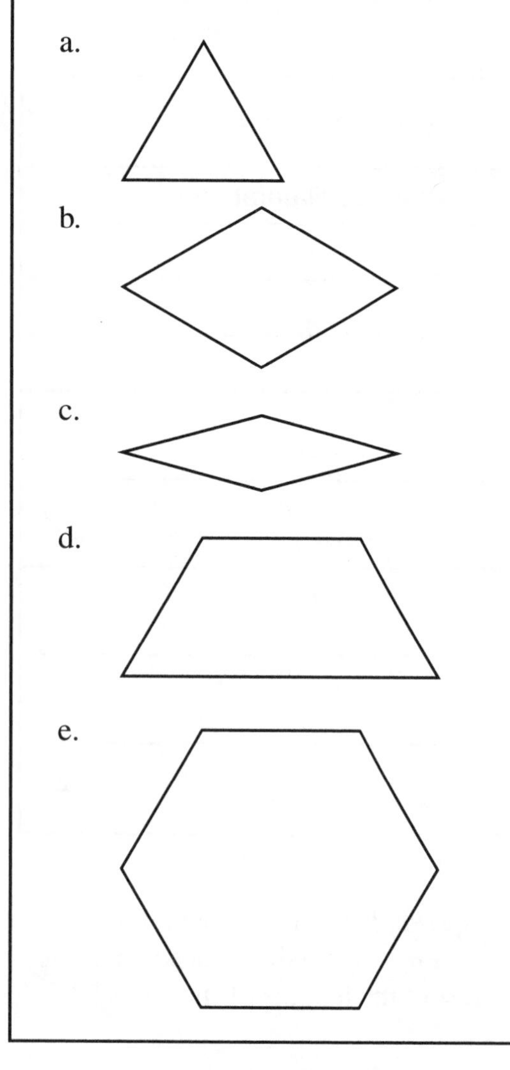

1. Use pattern blocks to construct a convex pentagon. Determine the measure of each interior angle of the pentagon. Sketch your pentagon and indicate the measure of each angle.

2. Use pattern blocks to construct a convex heptagon (seven sides). Determine the measure of each interior angle of the heptagon. Sketch your heptagon and indicate the measure of each angle.

3. Use the measures of each of the interior angles of the pattern blocks and the polygons you constructed to complete the following table.

Number of Sides	Sum of the Measures of the Angles
3	
4	
5	
6	
7	
8	
9	
n	

4. What is the relationship between the increase in the number of sides and the increase in the sum of the measures of the interior angles?

5. Given the number of sides of a polygon, n, how would you determine the **sum** of the measures of the interior angles of the polygon?

6. If a polygon is **regular**, how would you determine the measure of **each** interior angle?

Activity 6: Classifying Quadrilaterals

PURPOSE Reinforce the definitions of special quadrilaterals and their classifications and develop their distinguishing properties.

MATERIALS A centimeter ruler, a protractor, scissors, and copies of the quadrilaterals on page A-42

GROUPING Work in pairs.

GETTING STARTED Measure each side and angle of each quadrilateral. Work with your partner to share the information on the measurements and to discuss the sorting of the quadrilaterals into various groups. Draw both diagonals in each figure and measure their lengths and the angles formed at the point of intersection. Write all measures inside each figure and then carefully cut out the quadrilateral.

1. Sort the quadrilaterals into groups according to the following properties. Figures may be placed into more than one group. Indicate the letters of the quadrilaterals that have each property.

   a. All sides are congruent ____________
   b. At least one pair of sides are parallel ____________
   c. Has more than one right angle ____________
   d. Diagonals are congruent ____________
   e. Two pairs of opposite sides are parallel ____________
   f. Diagonals bisect each other ____________
   g. Two pairs of opposite sides are congruent ____________
   h. Diagonals are perpendicular ____________
   i. Two pairs of opposite angles are congruent ____________
   j. Has one or more lines of symmetry ____________
   k. Has rotational symmetry (turn must be less than 360°) ____________
   l. Adjacent sides are congruent ____________

2. Identify the letters of the figures that are

   a. Squares ____________ b. Rectangles ____________
   c. Parallelograms ____________ d. Trapezoids ____________
   e. Rhombuses ____________ f. Kites ____________

1. Which properties named in Exercise 1, on page 210, do each of the following quadrilaterals have? Write the correct letters of the properties in the blanks below.

   a. Rectangle__________________ b. Square __________________

   c. Parallelogram__________________ d. Rhombus __________________

   e. Trapezoid__________________ f. Kite __________________

2. On the basis of your answers, write definitions for *square, parallelogram, rectangle, trapezoid,* and *rhombus.* Use the least number of properties possible to define each polygon.

   a. A *square* is a quadrilateral with

   b. A *parallelogram* is a quadrilateral with

   c. A *rectangle* is a quadrilateral with

   d. A *trapezoid* is a quadrilateral with

   e. A *rhombus* is a quadrilateral with

3. Let *SQ, PR, RC, TR,* and *RH* designate the sets of quadrilaterals you defined in Parts (a)–(e), respectively. Which sets are subsets of another set?

EXTENSIONS Construct a tree diagram showing the relationships among the following quadrilaterals: trapezoid, parallelogram, rectangle, rhombus, kite, and square. Start with *quadrilateral* at the top; as you proceed down, each line segment connecting two figures will indicate that the figures below are subsets of the figures above.

Activity 7: Name That Polygon

PURPOSE Reinforce geometric vocabulary and the properties of polygons.

MATERIALS A length of rope or heavy string (2–3 m) tied at the ends to form a large loop

GROUPING Work in groups of four.

GETTING STARTED Each person in the group should hold the rope at two different points. One person reads the description of the polygon. The group then decides how each person must slide his or her hands along the rope to form the given polygon. Once the group has correctly formed each polygon and decided on the correct name, each person should name the polygon and make a drawing of it at the right of each statement. Find as many different kinds of polygons as possible for each description.

1. A four-sided figure with exactly two right angles
2. An equilateral quadrilateral
3. An equilateral quadrilateral with one right angle
4. A polygon with at least two pairs of parallel sides
5. A polygon with one pair of parallel sides and two right angles
6. A quadrilateral with congruent diagonals
7. A quadrilateral with two pairs of congruent angles
8. A quadrilateral with two pairs of congruent sides

EXTENSION Each person in the group should write a statement that describes some polygon. Students in the group should then pick up the rope. The person who wrote the description should read it, and the group should try to form the given polygon and name it correctly.

Activity 8: Mysterious Midpoints

PURPOSE Apply coordinate geometry techniques to reinforce understanding of the properties of quadrilaterals.

MATERIALS Graph paper, ruler, and a computer with a dynamic geometry software package (e.g., Geometer's Sketchpad, Cabri, etc.)

GROUPING Work individually or in groups of 2–3.

1. Plot and label each of the following sets of points on a separate pair of coordinate axes. Draw four quadrilaterals by drawing segments connecting the points of each set in order.

   a. $P(2, 5)$, $I(4, 2)$, $N(13, 1)$, $K(7, 8)$
   b. $B(^-1, ^-5)$, $R(4, 2)$, $O(0, 0)$, $W(^-4, 5)$
   c. $R(5, ^-1)$, $O(^-3, ^-1)$, $S(^-6, ^-5)$, $E(5, ^-5)$
   d. $P(11, 3)$, $O(11, 8)$, $L(2, 4)$, $Y(0, 0)$

2. Mark the midpoints of the sides of each quadrilateral and label them consecutively *M, A, T,* and *H*. Connect the points in order to form another quadrilateral. What appears to be true about each of the polygons *MATH*?

3. Explain how you can use coordinate methods to check your conjecture.

---

1. Plot and label the points in each of the following sets on separate coordinate axes. Connect the points in each set in order to form the following special quadrilaterals: a parallelogram, a rhombus, a rectangle, and a square.

   a. $D(4, 8)$, $U(1, 3)$, $C(10, 6)$, $K(13, 11)$
   b. $B(2, ^-2)$, $I(9, 1)$, $K(2, 4)$, $E(^-5, 1)$
   c. $D(2, 3)$, $A(^-3, 3)$, $V(^-3, ^-4)$, $E(2, ^-4)$
   d. $P(^-2, ^-5)$, $I(^-7, 0)$, $C(^-2, 5)$, $K(3, 0)$

2. Locate the midpoints of the sides and label them consecutively *M, O, N,* and *T*. Connect the points in order to form quadrilaterals.

   a. Are any of the quadrilaterals *MONT* that were formed special quadrilaterals?

   b. Are any of the quadrilaterals *MONT* the same type of quadrilateral as the original figure?

3. In each of the quadrilaterals *MONT* in Exercise 2, locate the midpoints of the sides, label them consecutively *W, X, Y,* and *Z,* and connect them in order to form new quadrilaterals. How are the new quadrilaterals related to the original figures drawn in Exercise 1?

1. Complete each of the following statements.
   a. If you connect the midpoints of the sides of a quadrilateral in order, the resulting figure is a ________________________.
   b. If you connect the midpoints of the sides of a parallelogram in order, the resulting figure is a ________________________.
   c. If you connect the midpoints of the sides of a rectangle in order, the resulting figure is a________________________.
   d. If you connect the midpoints of the sides of a rhombus in order, the resulting figure is a ________________________.
   e. If you connect the midpoints of the sides of a square in order, the resulting figure is a ________________________.
2. Explain which properties of the original quadrilateral determine which special quadrilateral is formed by connecting the midpoints of the sides.

EXTENSIONS

1. Use a dynamic geometric drawing program to construct a quadrilateral. Locate the midpoints of the four sides and connect them in order as you did to form the polygon *MATH*. Use the dynamic feature of the program to alter the quadrilateral in various ways to verify that your first conjecture was correct.
2. On separate screens, construct a trapezoid, parallelogram, rhombus, rectangle, kite, and square. In each figure, locate the midpoints of the sides and connect them in order. Then locate the midpoints of the sides of the new figure and connect them in order to form a second quadrilateral. Use the dynamic feature of the program to alter each of the original figures in various ways.

   Given that you start with a special quadrilateral, explain the relationship between the original quadrilateral and the quadrilateral that is formed by connecting the midpoints, and between the original quadrilateral and the second quadrilateral formed. Check your conjectures using the measuring capabilities of the software.

Activity 9: A View from the Top

PURPOSE Develop spatial perception by using various views to construct models of buildings.

MATERIALS Cubes, at least 16 per person or pair

GROUPING Work individually or in pairs.

GETTING STARTED Architectural plans may include various views of a building: top, front, back, left, and right. By viewing a building from the top and sides, you can determine its shape. Each number on a building mat tells you the number of stories (cubes) in that section of the building.

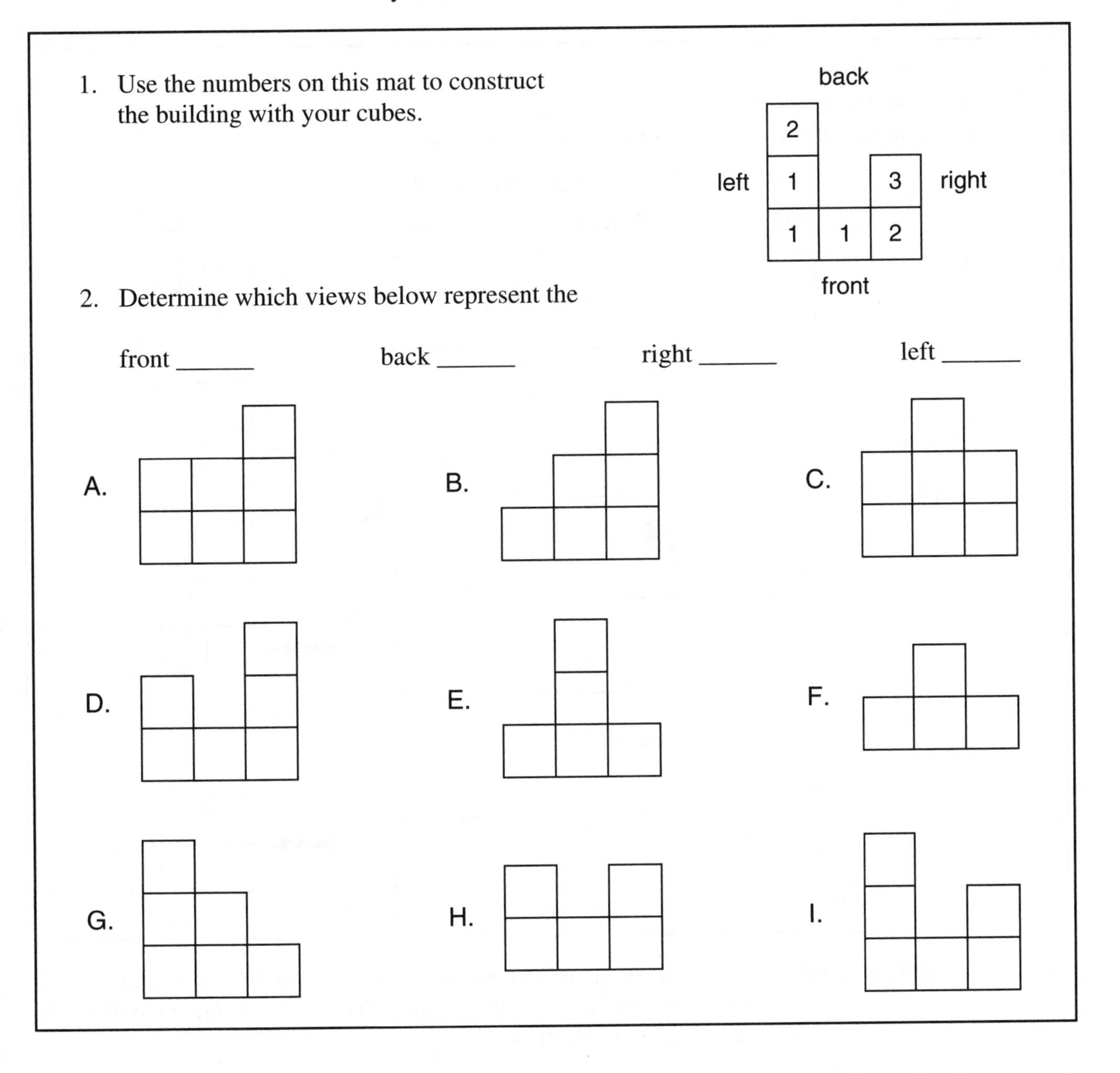

1. Use the numbers on this mat to construct the building with your cubes.

2. Determine which views below represent the

front ______ back ______ right ______ left ______

1. Use your cubes to construct the building represented by the following mats:

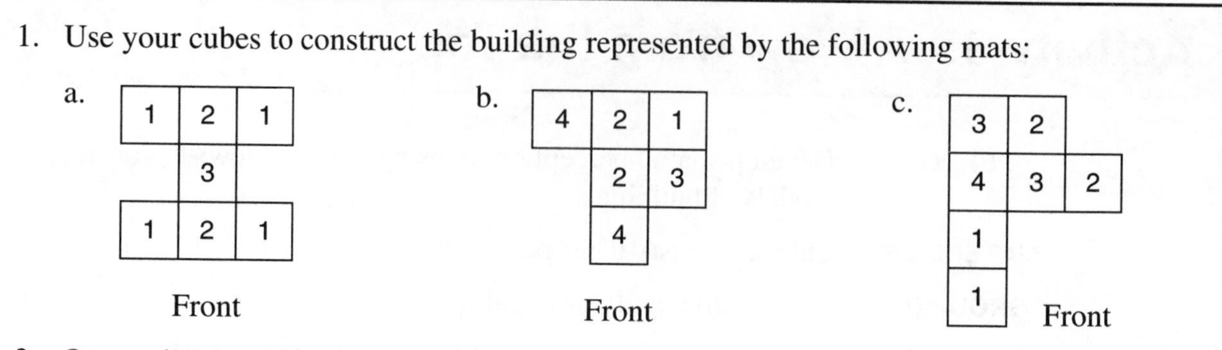

2. On centimeter grid paper, draw the architectural plans for each building. Label the top, front, back, left, and right view for each.
3. What is the relationship between the front and back views? Between the left and right views?

Use the plans below to construct each building.
Record the height of each section of the building on the mat.

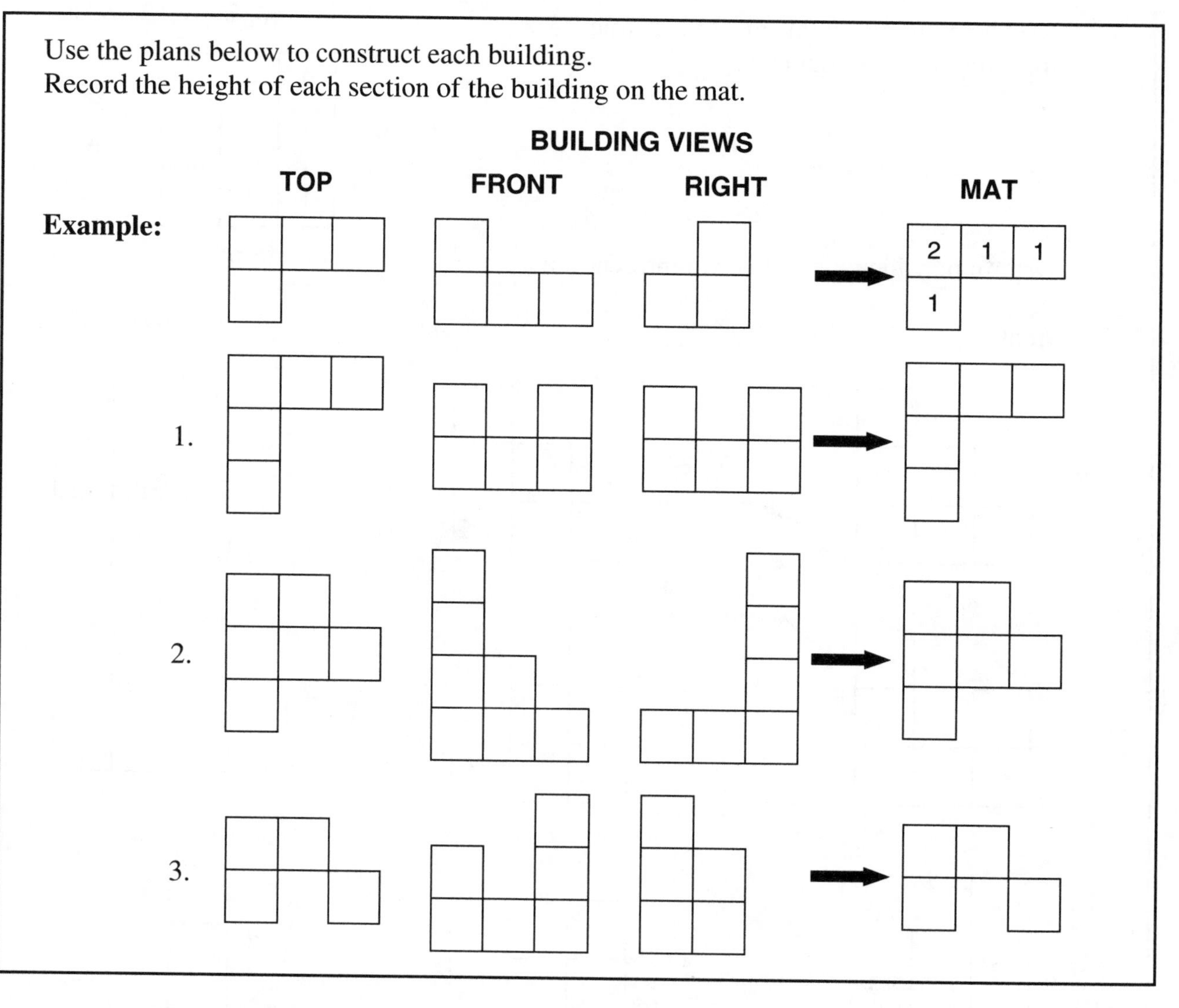

EXTENSIONS Draw a set of plans for a building showing the top, front, and one side view. The building can use no more than 20 cubes. Give the plans to a classmate to construct.

Chapter Summary

Geometric shapes are a part of our everyday life. Throughout the chapter, the use of manipulatives allowed hands-on exploration of geometric figures and promoted the understanding of the underlying concepts that are used in the study of geometry. The activities introduced you to several important geometric figures. In Activity 1, geoboards were used to explore angle measurement. Classifying the angles as acute, right, or obtuse laid the foundation for later exercises involving triangles. Linear and angle measurements were important in the development of the properties of both triangles and quadrilaterals.

Geometric vocabulary was reinforced in Activity 2. A cooperative group setting allowed for shared problem solving as you worked together to define and possibly redefine vocabulary related to the placement of the geo-bands to complete the Mystery Figure.

Triangles and their properties were the subject of Activities 3 and 4. You discovered that triangles can be classified by the number of congruent sides, the types of angles, or a combination of both sides and angles. The Toothpick Triangle activity reinforced the classification of triangles using the combination of sides and angles. The Triangle Inequality (that is, the sum of the measures of any two sides of a triangle must be greater than the measure of the third side) was also introduced.

Activities 6–8 developed the properties of selected quadrilaterals and the relationships among the quadrilaterals. Following a thorough exploration of the properties of the figures, you developed a definition of each quadrilateral using the least number of properties. This is exactly what mathematicians do when they construct precise definitions of terms.

Activity 9 helped to develop spatial visualization in various ways. First, you had to determine two-dimensional views of a building as seen from different sides. Then you were given the structure of a model and required to draw it as seen from four views. Finally, you had to construct a model based on three different views. This activity has wide application in making architectural drawings and interpreting plans.

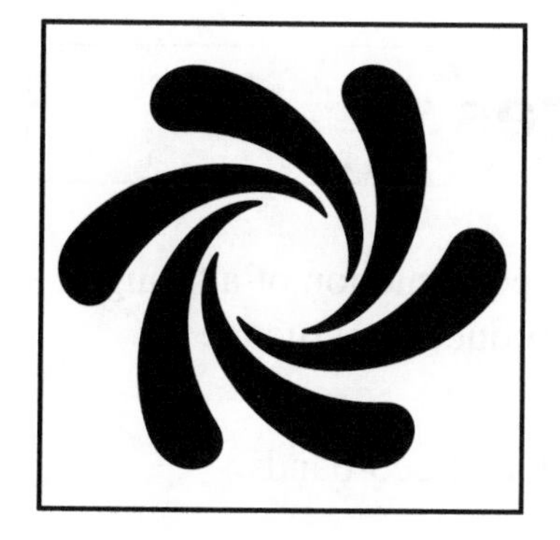

Chapter 11
Constructions and Similarity

Students discover relationships and develop spatial sense by constructing, drawing, measuring, visualizing, comparing, transforming, and classifying geometric figures. Discussing ideas, conjecturing, and testing hypotheses precede the development of more formal summary statements. . . . geometry should focus on investigating and using geometric ideas and relationships rather than on memorizing definitions and formulas.

—*Curriculum and Evaluation Standards for School Mathematics*

The activities in this chapter will engage you in all of the problem-solving situations described in the *Standards*. After you construct a given shape and compare the result with a classmate's, you will be asked to make and test conjectures based on your observations. In one activity, you will use a computer software program to complete a construction and then alter the figure dynamically to illustrate that a conclusion is true for any triangle.

In the final activity, you will use the concepts that were developed in the previous activities on similar triangles to measure inaccessible heights. This activity illustrates the use of similar triangles in a real-world application. It provides a good answer for the age-old question, "When are we ever going to use this?"

Activity 1: Triangle Properties—Sides

PURPOSE Develop the Triangle Inequality, reinforce construction of a triangle using a compass and straightedge, and introduce congruence of triangles.

MATERIALS A centimeter ruler, a compass, a geoboard, and geo-bands

GROUPING Work in pairs.

GETTING STARTED Make all constructions on a separate piece of paper. As you construct the triangles in each section, compare your results with those of your partner.

1. Use a ruler to construct two different triangles in which one side is 6 cm long and another side is 9 cm long.
   a. Compare your triangles with your partner's triangles. What do you notice?
   b. How many different triangles could you construct given the lengths of two sides?
2. Use a ruler and a compass to construct triangles with sides of the following lengths.
   a. 9 cm, 7 cm, 5 cm b. 7 cm, 7 cm, 10 cm c. 8 cm, 8 cm, 8 cm
   d. 6 cm, 5 cm, 12 cm e. 7 cm, 12 cm, 11 cm f. 10 cm, 6 cm, 4 cm
3. Which sets of measures **did not** result in a triangle? Why is it impossible to construct triangles with sides of these lengths?
4. Which sets of measures **did** result in a triangle? In these cases, what is true about the sum of the lengths of the two shorter sides?
5. List measures for three sets of three lengths of sides that **can** be used to construct a triangle. Do not use a set in which all the lengths are equal.
   a. ____________ b. ____________ c. ____________
6. List measures for three sets of three lengths of sides that **cannot** be used to construct a triangle.
   a. ____________ b. ____________ c. ____________
7. What can you conclude about the lengths of the sides of a triangle?

1. For each set of three measures in Exercise 2 that resulted in a triangle:

   a. Classify the triangle using a combination of classifications, one by sides and one by angles.

   b. Compare each of your triangles to the corresponding triangle of your partner. What do you notice?

2. With your partner, decide on two additional sets of measures for three segments that will form a triangle. Each person should construct a triangle using these numbers. Compare the triangles as before. What do you notice?

How many unique triangles can be constructed on a 3×3 portion of a geoboard? Build your triangles on a geoboard and record the results on the 3×3 grids below. Classify each triangle that you construct. It may be necessary to put more than one triangle on a grid.

Activity 2: Triangle Properties—Angles

PURPOSE Reinforce the theorem on the angle sum of a triangle, introduce the concept of similarity of triangles, and reinforce the construction of a triangle using a protractor and straightedge.

MATERIALS A centimeter ruler and a protractor

GROUPING Work in pairs.

GETTING STARTED Make all constructions on a separate piece of paper. As you construct the triangles in each section, compare your triangle with your partner's triangle.

Use a ruler and a protractor to construct a triangle that has two angles with the following measures. Record your results in the table.

a. 30°, 50° b. 40°, 50° c. 90°, 95°

d. 60°, 60° e. 110°, 70° f. 80°, 80°

Problem	Sum of the Given Angles	Is a Triangle Possible?	If Yes, What Is the Measure of the Third Angle?
a.			
b.			
c.			
d.			
e.			
f.			

In each Part (a)–(f) that **did not** result in a triangle, what is true about the sum of the measures of the two given angles?

In each Part (a)–(f) that **did** result in a triangle, what is true about the sum of the measures of the two given angles?

1. List the measures for three pairs of angles that **can** be used to construct a triangle.

   a. ____________ b. ____________ c. ____________

2. List the measures for three pairs of angles that **cannot** be used to construct a triangle.

   a. ____________ b. ____________ c. ____________

3. What can you conclude about the sum of the measures of the three angles of a triangle?

4. For each exercise that resulted in a triangle

   a. identify the type of triangle that was constructed.

   b. compare your triangle with your partner's triangle. What do you notice?

Measure the lengths of the sides of each of your triangles. Then find the ratio of the length of each side of your triangle to the length of the corresponding side of your partner's triangle and record the ratios in the table below.

Problem	Ratios of Lengths of Sides					
	Shortest Side to Shortest Side		Middle Side to Middle Side		Longest Side to Longest Side	
	Fraction	Decimal	Fraction	Decimal	Fraction	Decimal

For each pair of triangles, what can you conclude about the ratios of the corresponding sides?

Activity 3: To Be or Not to Be Congruent?

PURPOSE Use constructions to develop the triangle congruence axioms.

MATERIALS A compass, a protractor, and a ruler

GROUPING Work in pairs.

GETTING STARTED In Activity 1, you learned that two triangles are congruent if the three sides of one triangle are congruent respectively to the three sides of the other. Are there other combinations of corresponding parts of two triangles that will ensure that the triangles are congruent?

1. Use a compass and ruler to construct a triangle with the parts given in each exercise.

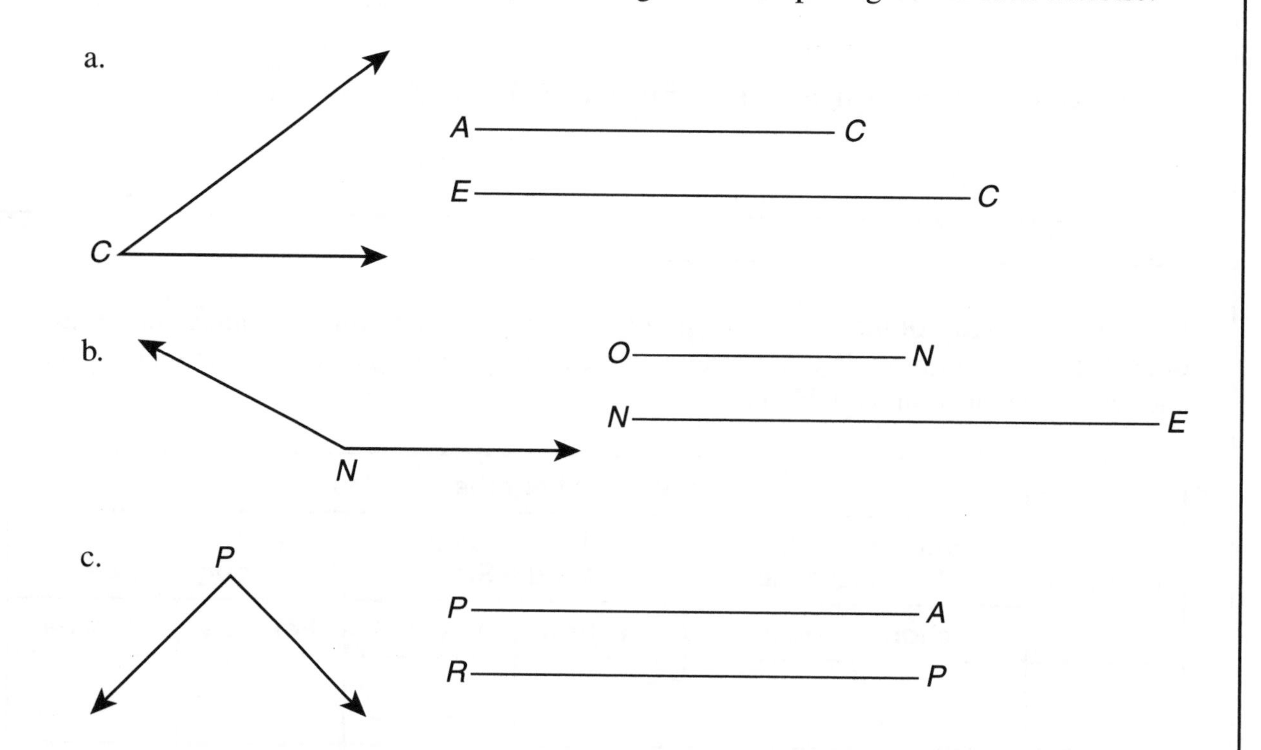

2. In each part of Exercise 1, how were the given sides and the angle of the triangle related?

3. Compare each triangle you constructed to the corresponding triangle of your partner. What do you notice?

4. What can you conclude from your answers to the above questions?

1. In the following problems, use a ruler and compass to construct a triangle with the given parts. Label the third angle of each triangle, Z.

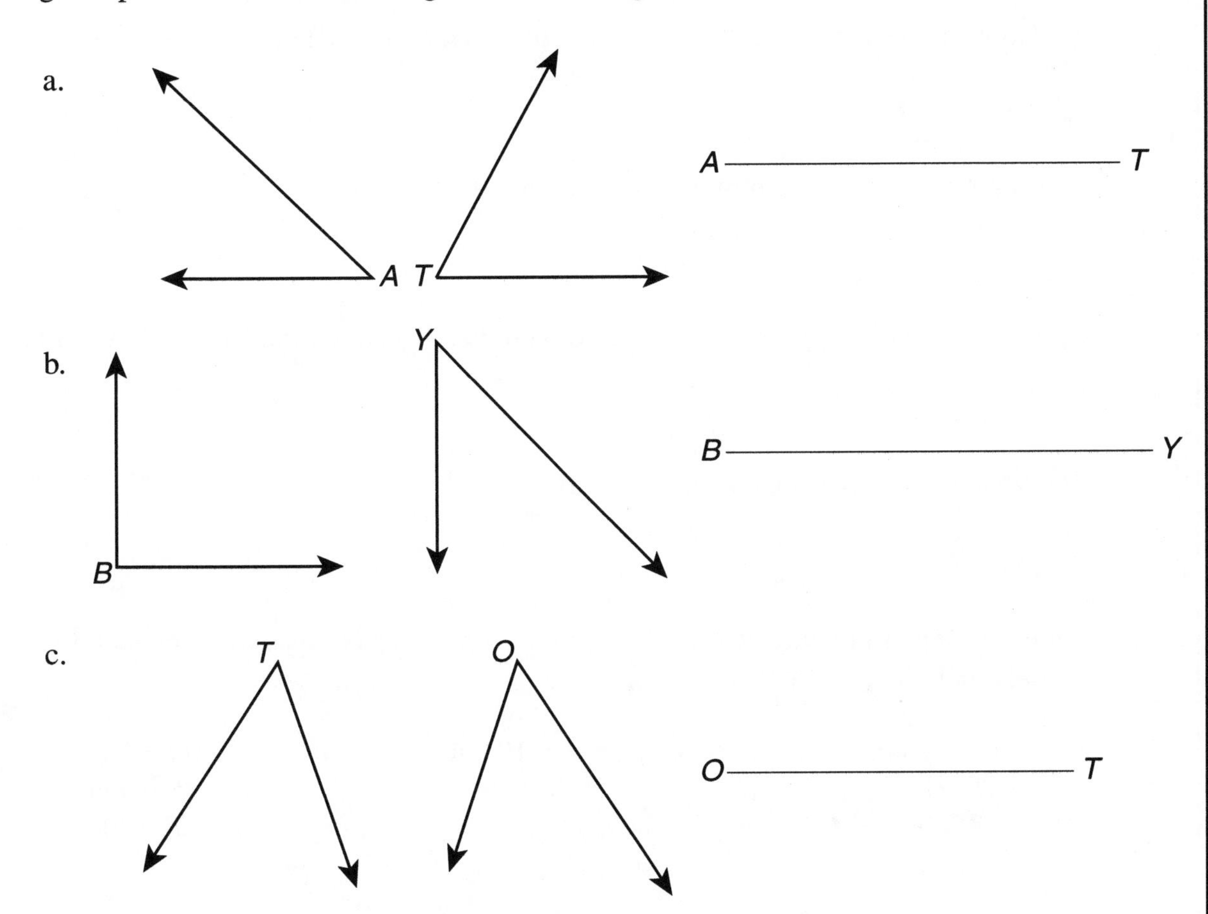

2. In each part of Exercise 1, how is the side related to the given angles?

3. Compare each of your triangles to the corresponding triangle of your partner. What do you notice?

4. What can you conclude from your answers to the above questions?

1. Use a ruler and a protractor to construct a triangle *BAD* in which $AB = 10$ cm, $m \angle A = 40°$, and $m \angle D = 80°$.

   a. How are the angles and the given side of the triangle related?

   b. Describe step by step how you constructed the triangle.

   c. Compare the triangle you constructed with that of your partner. What do you notice?

   d. What can you conclude from Parts (a) and (c)?

2. For each of the following, use a ruler, a protractor, and a compass to construct a triangle with the given parts.

   a. $TO = 5$ cm
$OP = 8$ cm
$m \angle T = 30°$

   b. $A = 10$ cm
$AT = 6$ cm
$m \angle H = 37°$

   c. $TI = 9$ cm
$IE = 4$ cm
$m \angle T = 40°$

3. For each part in Exercise 2, how are the given sides and angle related?

4. Which exercises did not result in a triangle?

5. Did any of the exercises result in more than one triangle? If so, which one(s)?

6. If any exercise resulted in only one triangle, what type of triangle was it?

7. What can you conclude from your answers in Exercises 3–6?

Activity 4: Matching Triangles

PURPOSE Reinforce the concepts of congruence and similarity of triangles and reinforce the classification of triangles by angles.

MATERIALS A pair of scissors and copies of pages A-43–A-45

GROUPING Work in pairs or in groups of four. This can also be done as a class activity directed by the instructor.

On a sheet of paper, make a copy of the Triangle Sorting Board shown below. Cut out the triangles on the matching triangles sheets and distribute them evenly among the members of your group.

1. Alternating turns, place each triangle in the appropriate column on the Triangle Sorting Board. Justify the placement of each triangle to the group.

Triangle Sorting Board

Acute	Right	Obtuse

2. After all the triangles have been correctly placed on the board, choose one triangle from one column. Your partner must then find another triangle in that column that is the same shape as yours. Place one triangle on the other to determine if:

   a. the corresponding sides have the same length,

   b. the corresponding angles have the same measure, or

   c. both the corresponding sides and the corresponding angles have the same measure.

Repeat the previous steps, switching roles each time, until all triangles have been matched.

1. If the three sides of the triangles match, do the three angles also match?

2. If the three angles of the triangles match, do the three sides also match? Will this always be true? Explain.

Two triangles are *congruent* if all the corresponding sides and the corresponding angles are congruent.

Two triangles are *similar* if all the corresponding angles are congruent and the corresponding sides are proportional.

1. List the pairs of triangles that are *congruent*.

2. List the pairs of triangles that are *similar*. Find the ratio of the corresponding sides in each pair of similar triangles.

Activity 5: Paper Folding Construction

PURPOSE Use paper folding to construct the perpendicular bisector of a line segment, the bisector of an angle, and the centroid, incenter, and circumcenter of a triangle.

MATERIALS Squares of waxed paper (10–12 cm on a side) or paper squares used to separate hamburger patties, compass, ruler, protractor, and computer with dynamic geometry software

GROUPING Work in pairs.

GETTING STARTED One student should read the directions and the other should do the folding. Both should review the results of the folding, answer any questions, and make a conjecture based on the findings. The conjecture should be compared with the conjecture of another pair of students and discussed so as to arrive at a mutually accepted conclusion. All of the drawing and folding should be done on the waxed paper.

1. Draw $\angle ABC$.

2. Fold $\overrightarrow{BA}$ onto $\overrightarrow{BC}$ and crease the paper.

3. Open the paper and mark a point D on the line formed by the crease.

4. Measure $\angle ABD$ and $\angle CBD$.
   What is true about the measures of $\angle ABD$ and $\angle CBD$?
   What is the relationship between $\overrightarrow{BD}$ and $\angle ABC$?

Recall that the distance from a point to a line is the length of the perpendicular segment drawn from the point to the line.

5. Place the edge of another piece of paper on side $\overrightarrow{BC}$ of the angle as shown so that the adjacent side of the paper passes through point D. Label the point E on $\overrightarrow{BC}$ and mark the length DE on the edge of the second sheet as shown. Then repeat the steps to find a point F on $\overrightarrow{BA}$. Compare DF and DE. What did you find? Place the point of your compass at D, open to E, and draw a circle. What is the relationship between the circle and the sides of $\angle ABC$?

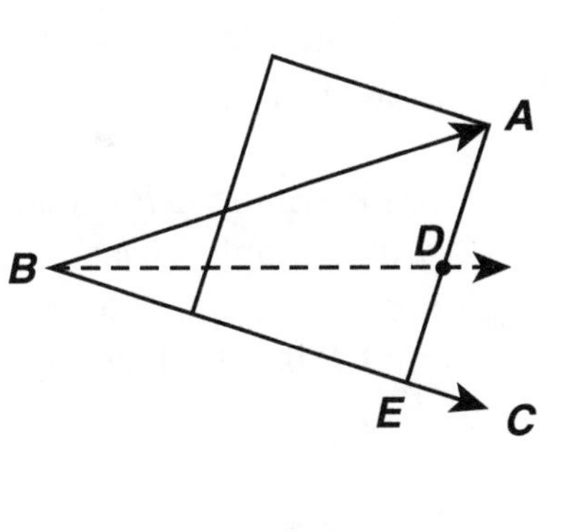

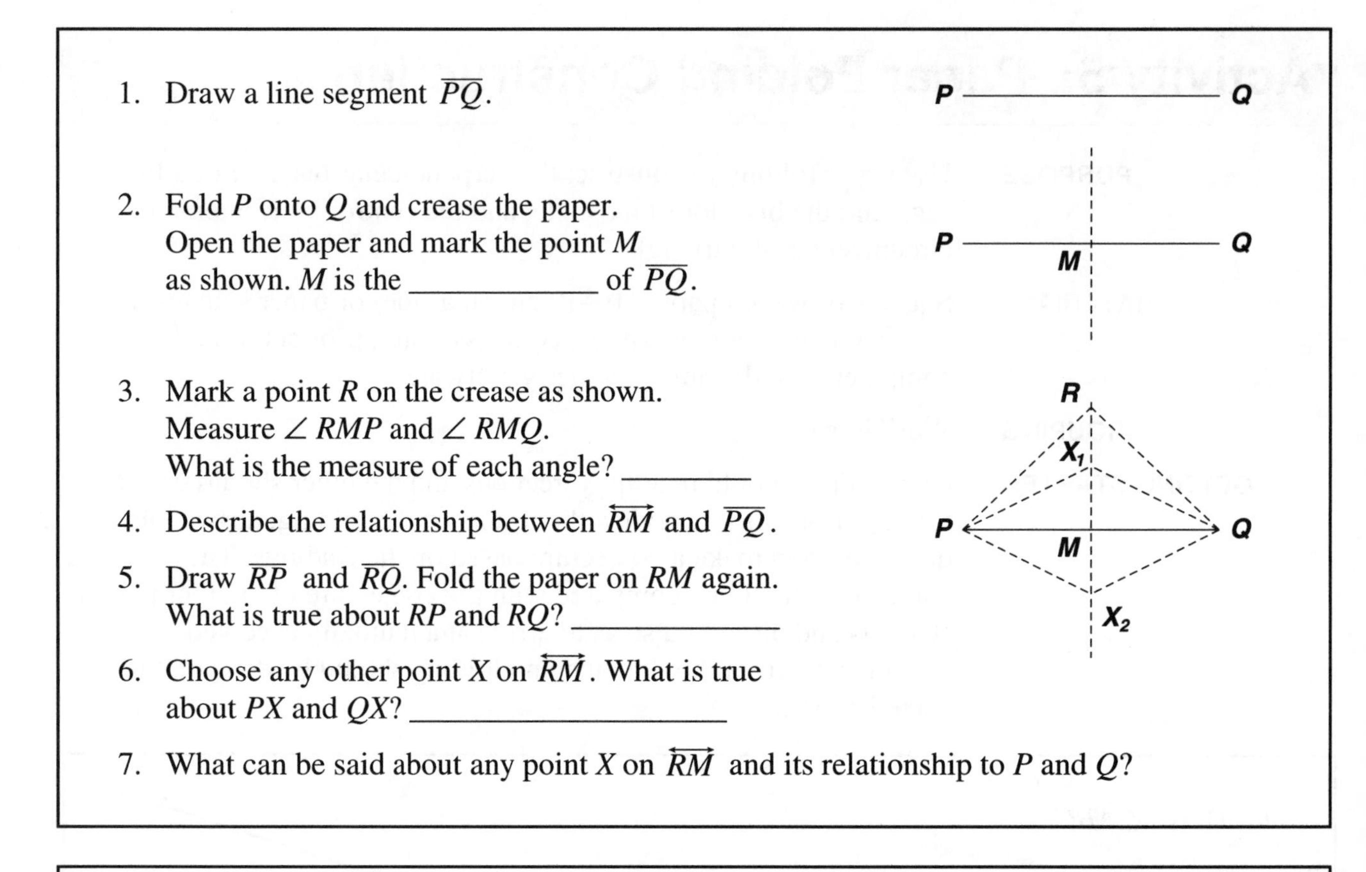

1. Draw a line segment $\overline{PQ}$.
2. Fold P onto Q and crease the paper. Open the paper and mark the point M as shown. M is the ____________ of $\overline{PQ}$.
3. Mark a point R on the crease as shown. Measure $\angle RMP$ and $\angle RMQ$. What is the measure of each angle? ________
4. Describe the relationship between $\overleftrightarrow{RM}$ and $\overline{PQ}$.
5. Draw $\overline{RP}$ and $\overline{RQ}$. Fold the paper on RM again. What is true about RP and RQ? ____________
6. Choose any other point X on $\overleftrightarrow{RM}$. What is true about PX and QX? ____________________
7. What can be said about any point X on $\overleftrightarrow{RM}$ and its relationship to P and Q?

For the remaining constructions, you may use dynamic geometry software or paper folding. If you use a drawing program, complete the construction and then alter the original triangle with the dynamic feature of the program to verify that your conjecture works for *any* triangle.

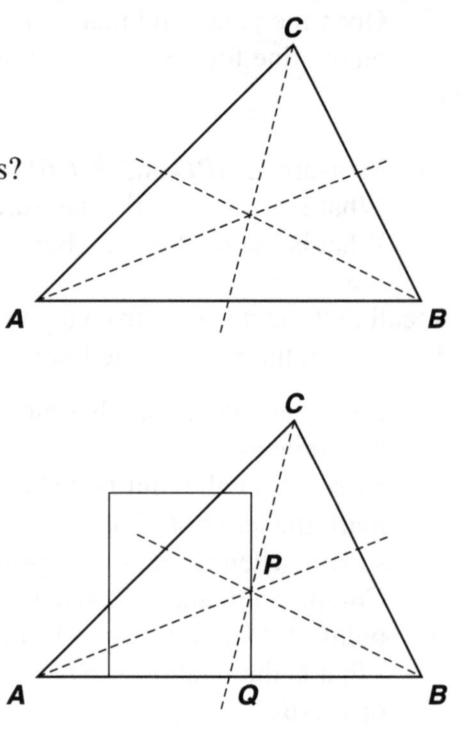

1. Draw ΔABC on your paper. Fold the paper to bisect each angle.
2. What appears to be true about the three angle bisectors? Check your results with those of another group.
3. Label the point where the bisectors intersect P. Place the edge of a second piece of paper along one side of the triangle as shown so that the adjacent side passes through P. Mark the point Q on $\overline{AB}$.
4. To determine if P is equidistant from the three sides of the triangle, place the point of your compass at P, open the compass to point Q, and draw a circle. The point P is called the *incenter* of the triangle.

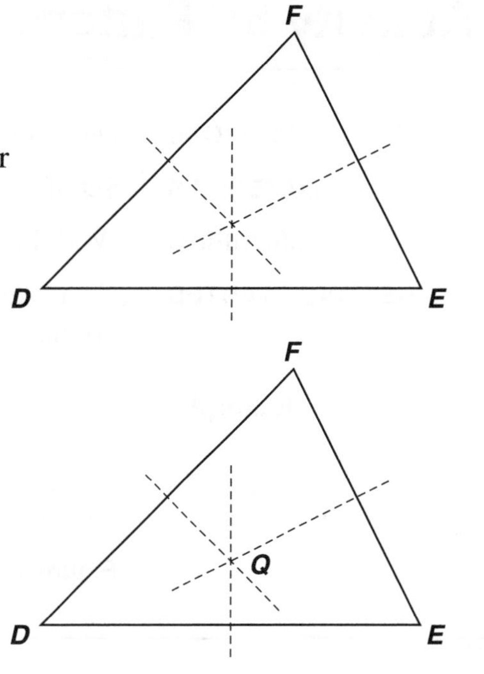

1. Draw an acute ΔDEF.
2. Fold the paper to construct the perpendicular bisector of each side of the triangle. What appears to be true about the perpendicular bisectors?
3. Mark the point of intersection Q. Measure DQ, EQ, and FQ. What did you find?
4. Place the point of your compass at Q, open it to D, and draw a circle. Q is called the *circumcenter* of the triangle.
5. Repeat the steps above for an obtuse triangle and a right triangle. How are the results the same? How are they different?

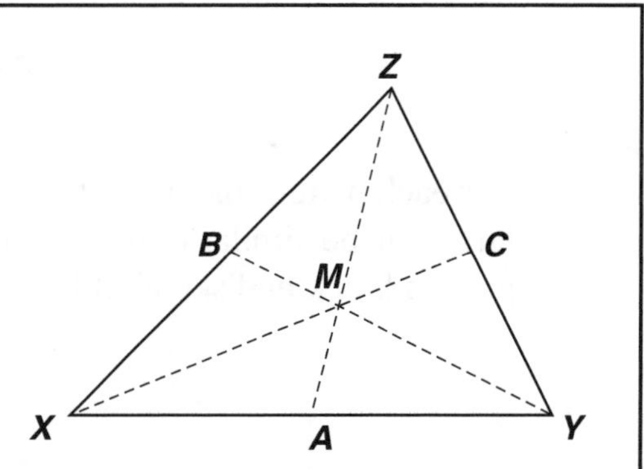

1. Draw ΔXYZ. Fold the paper to locate the midpoints of the sides of the triangle and label them A, B, and C as shown.
2. Draw the *medians* $\overline{XC}$, $\overline{YB}$, and $\overline{ZA}$. What appears to be true about the medians?
3. Mark the point of intersection, called the *centroid, M*. Measure the two segments on each median and find the ratio of the length of the shorter segment to the length of the longer segment. The *centroid* divides each median into two segments so that the ratio of their lengths is __________.

EXTENSIONS In a new triangle, construct the medians, angle bisectors, altitudes, and perpendicular bisectors of the sides. The points at which these sets of line segments intersect are the *centroid*, *incenter*, *orthocenter*, and *circumcenter* respectively. Research the *Euler Line* and find it in your triangle. Which of these points lie on the *Euler line*? Using the dynamic feature of your software, verify that the *Euler Line* contains the same points in any triangle.

Activity 6: Pattern Block Similarity

PURPOSE Develop the concept of similarity using pattern blocks.

MATERIALS Set of pattern blocks

GROUPING Work in pairs or individually.

GETTING STARTED Using squares, you can construct larger squares **similar** to the original one.

Example:

Figure 1 **Figure 2** **Figure 3**

1. Use triangles to construct the next larger triangle similar to the green triangle. Sketch the similar triangle at the right.

2. For each pattern block, is it possible to use only blocks of the same shape to construct a larger shape similar to the original block? If so, construct the similar shape for each pattern block and sketch it below.

3. For each shape in Exercise 2, use the least number of blocks possible to construct the next larger similar shape. Sketch the figures below.

4. How many blocks were needed for each figure?

Repeat the process once more using more blocks. Record the number of blocks that are used to construct each similar figure for each pattern block in the table below.

Figure	1	2	3	4	5	6	7	8	n
Number of Blocks									

1. Describe the set of numbers in the **Number of Blocks** row of the table.

2. Describe the *numerical pattern of the differences* in the number of blocks in each successive similar figure.

Construct two different parallelograms using four red trapezoids each.

1. Are the measures of the angles of the red parallelograms equal to the measures of the corresponding angles of the blue rhombus?

2. Are the corresponding sides proportional?

3. Are the red parallelograms similar to the blue rhombus? Explain.

Construct a trapezoid using three red trapezoids.

1. Are the measures of the angles of the small trapezoid equal to the measures of the corresponding angles of the large trapezoid?

2. Are the corresponding sides proportional?

3. Are the trapezoids similar? Explain.

Construct a rhombus using four tan rhombuses. Compare this figure to the blue rhombus.

1. Are the corresponding sides of the two rhombuses proportional? If so, what is the ratio of the corresponding sides?

2. Are the measures of the corresponding angles equal?

3. Are the two rhombuses similar? Explain.

Activity 7: Outdoor Geometry

PURPOSE Apply the properties of similar triangles to indirect measurement.

MATERIALS A mirror, a five to ten meter measuring tape, a straw, a small washer, and a 40-cm length of string or thread

GROUPING Work in groups of four.

GETTING STARTED Identify several tall objects to be measured. Divide the tasks among the members of the group and switch roles for each object to be measured. One student should sketch a drawing of the method of solution (see example) and record the data. Another student can provide the shadow or do the siting, as in the mirror method. The other two students can do the measuring.

SHADOW METHOD

Measure *NA*, the height of a person; *AD*, the length of the person's shadow; and *JI*, the length of the shadow of the object for which the height is being determined.

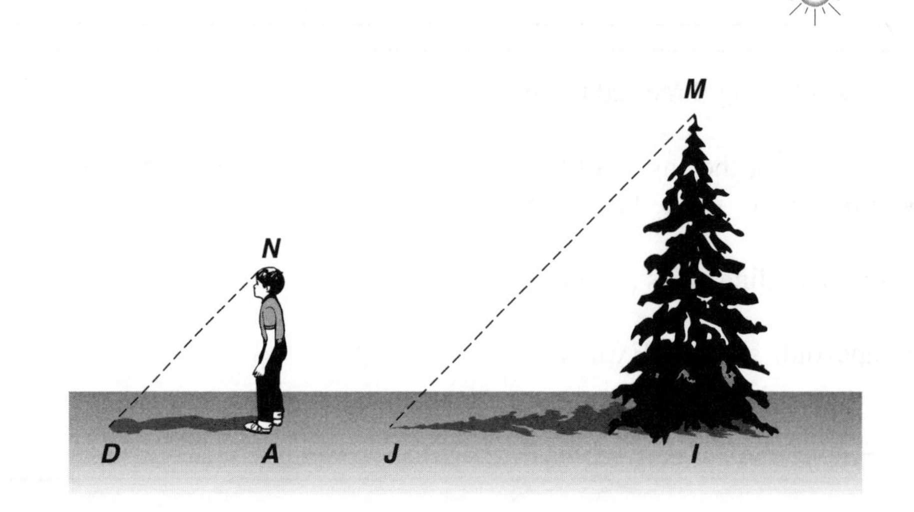

1. Explain why $\Delta DAN \sim \Delta JIM$.

2. Write a proportion that can be used to determine the height of the object, *MI*. Solve the proportion to find *MI*.

MIRROR METHOD

With a felt marker, draw a segment connecting the midpoints of one pair of opposite sides of a mirror. Place the mirror on the ground so that the segment on the mirror is parallel to a line determined by the tips of the toes of the shoes of a person facing the object to be measured. The person should look into the mirror and align the reflection of the top of the object to be measured (point *J* on the house) with the line on the mirror represented by $\overline{RS}$. Point *M* is the intersection of $\overline{AI}$ and $\overline{RS}$.

Measure *AM, MI,* and *EI*. Note that *EI* is the eye-to-ground distance, not the height of the person. The point *I* should be vertically below the person's eye (point *E*), approximately at the toes of the shoes.

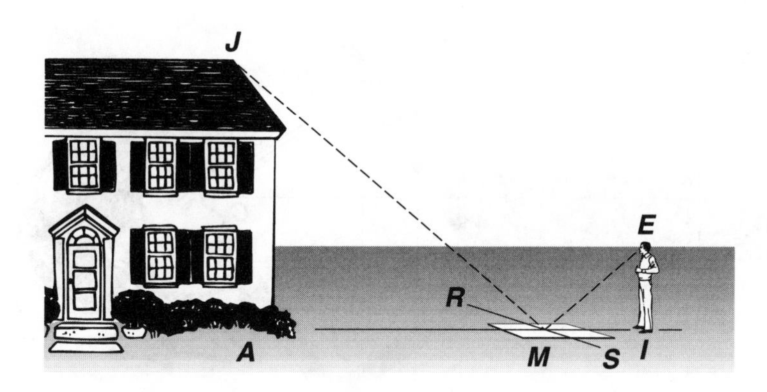

1. Explain why $\Delta JAM \sim \Delta EIM$.

2. Write a proportion that can be used to determine *JA*. Solve the proportion and find the height of the object being measured.

HYPSOMETER METHOD

Use a clipboard with a pad of paper attached to it. Pin a drinking straw along the top of the pad at *C* and *B*. Attach a small weight to one end of a 40-cm length of thread and tie the other end of the thread to the pin at *B*. One person should hold the clipboard and sight through the straw until the top of the object to be measured is sighted. A second person should mark the point *F* on the edge of the pad to determine the ΔBAF. Measure *DC*, *BA*, *AF*, and the distance from eye level to the ground.

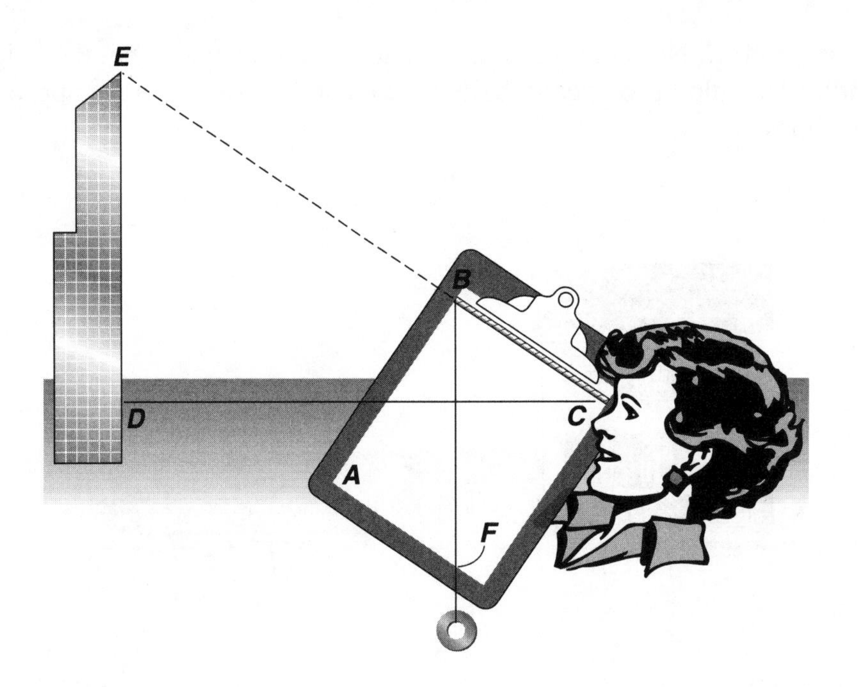

1. Explain why $\Delta CDE \sim \Delta BAF$.

2. Write a proportion that could be used to determine *DE*. Solve the proportion and find *DE*.

3. Is *DE* equal to the measure of the height of the object being measured? What must be done to *DE* to determine the height of the object?

Chapter 11 Summary

In Activity 1, the Triangle Inequality theorem was reinforced as you constructed triangles given the measures of their sides. When a triangle could be constructed, comparison with a classmate's triangle introduced the idea that triangles are congruent if the corresponding sides are congruent.

The theorem on the sum of the measures of the angles of a triangle was reinforced in Activity 2. When you constructed a triangle given the measures of the angles, you discovered through comparison of your triangle to that of a classmate, that the triangles were not congruent; rather they were similar.

Activity 3 developed the Side-Angle-Side, Angle-Side-Angle, and Angle-Angle-Side axioms for congruence of triangles. These axioms were further extended in the Matching Triangles activity, as was the concept of similarity.

Activity 5 developed concurrence theorems involving the angle bisectors, medians, altitudes, and perpendicular bisectors of the sides of a triangle. The incenter and circumcenter were introduced, as was the fact that these two points and the orthocenter lie on the *Euler Line*. By using dynamic geometry software, you were able to make the constructions on one triangle and then alter the completed figure to show that the constructions worked for any triangle.

The Outdoor Geometry activity allowed you to apply the concepts developed in earlier activities in a real-world problem setting. Geometry surrounds us in our world and applications of geometry abound.

The introduction to geometry provided in this chapter will enhance your understanding of the topics presented in the next two chapters, where you will explore additional applications of geometry in the real world.

Chapter 12
Concepts of Measurement

"Length, area, and volume of one-, two-, and three-dimensional figures are especially important. For example, once students have discovered that it is possible to find the area of a rectangle by covering a figure with squares and then counting, they are ready to explore the relationship between the areas of rectangles and the areas of other geometric figures. . . . This exploration gives students an opportunity to reason deductively and see how mathematical ideas relate to one another. . . . These connections require students to understand that the area of a figure does not change if it is partitioned and rearranged. It is also important that students understand the association between multiplication and determining the area of a rectangle. The formula is not a 'black box'."

—*Curriculum and Evaluation Standards for School Mathematics*

The activities in this chapter follow the recommendation of Sawyer to arrange things and make things before reasoning about them. Through exploration with a variety of manipulatives, you will develop an understanding of the concept of area and the formulas for the areas of certain polygons.

All of the formulas will be developed sequentially, beginning with the relationship to the product arrays used to illustrate multiplication. Each new formula will be related to a previously developed one by comparing the actual areas of the two polygons. A similar sequence of activities develops the rules for determining the volumes of certain polyhedra.

In the final activity, you will explore the relationship between surface area and volume and apply this knowledge to the process of manufacturing boxes. As you collect and analyze the data, you will simulate an industrial application of geometry, where decisions are based on minimizing costs and maximizing profits.

Activity 1: Regular Polygons in a Row

PURPOSE Reinforce the concept of perimeter and use the pattern strategy for developing a rule to determine the perimeter of any number of regular polygons placed end to end.

MATERIALS Pattern blocks can be used for part of the activity.

GROUPING Work individually.

GETTING STARTED Fill in the blanks in each section. Draw a model for the figures that are not given and make a table to help determine the pattern.

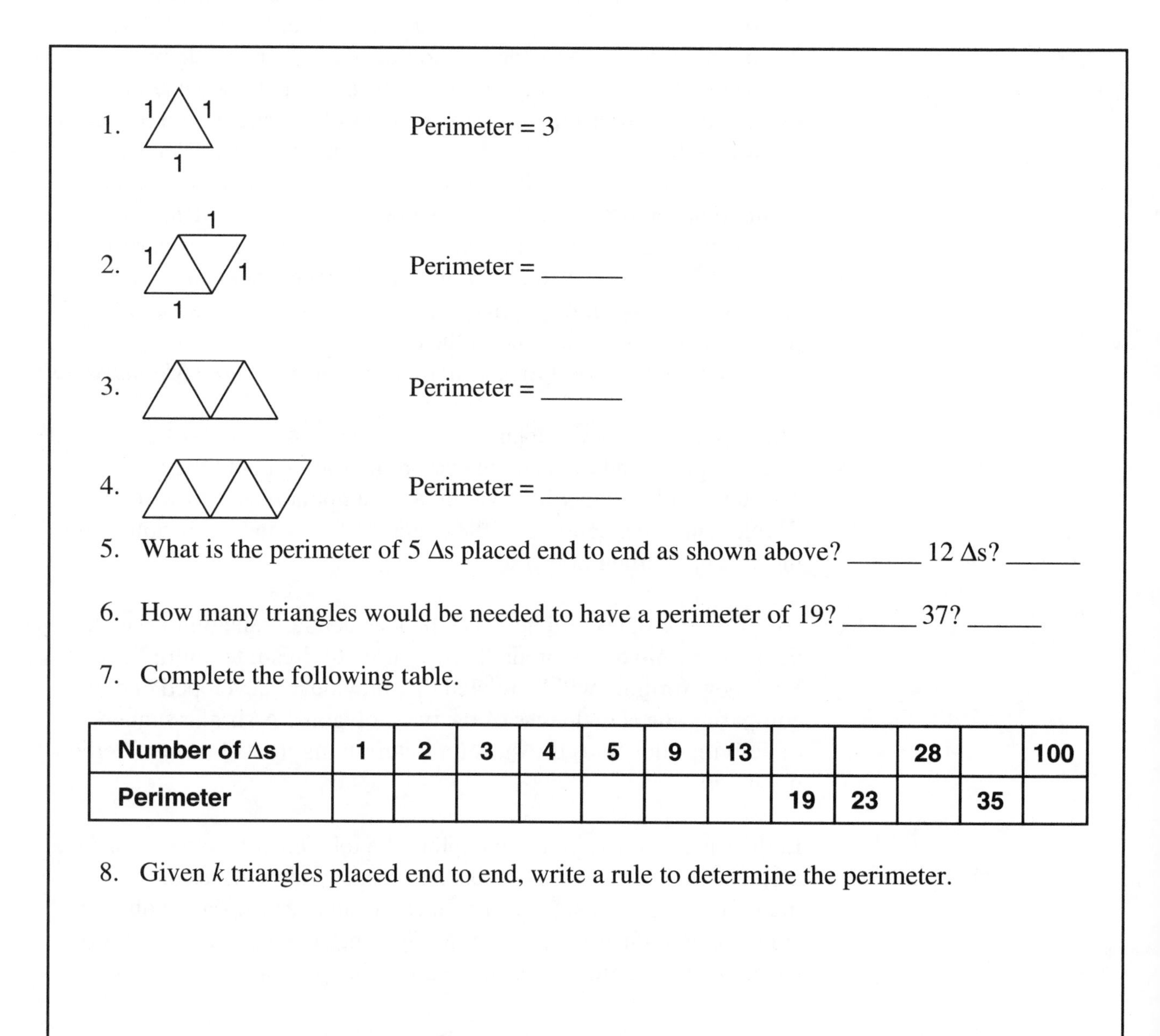

1. Perimeter = 3
2. Perimeter = ______
3. Perimeter = ______
4. Perimeter = ______
5. What is the perimeter of 5 Δs placed end to end as shown above? ______ 12 Δs? ______
6. How many triangles would be needed to have a perimeter of 19? ______ 37? ______
7. Complete the following table.

Number of Δs	1	2	3	4	5	9	13			28		100
Perimeter								19	23		35	

8. Given k triangles placed end to end, write a rule to determine the perimeter.

1. 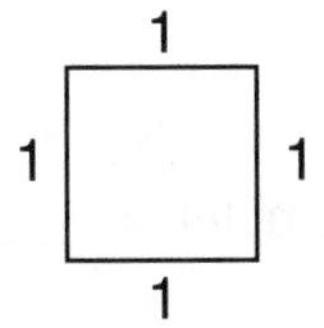

Perimeter = 4

2. 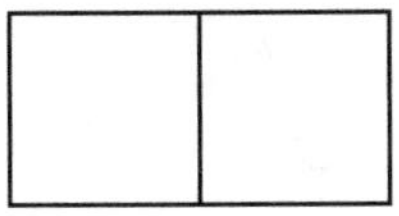

Perimeter = ______

3. 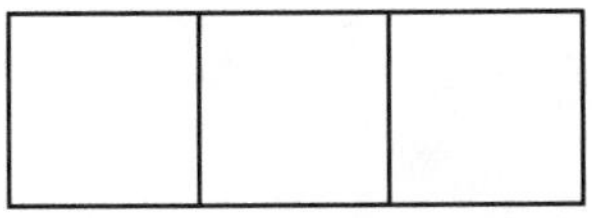

Perimeter = ______

4. What is the perimeter of 5 □s placed end to end? ______ 9 □s? ______

5. Complete the following table.

Number of □s	**1**	**2**	**3**							**100**
Perimeter										

6. What is the perimeter of 13 □s? ______ 23 □s? ______

7. How many squares would be needed to have a perimeter of 38? ______ 66? ______

8. Given *k* squares placed end to end, write a rule to determine the perimeter.

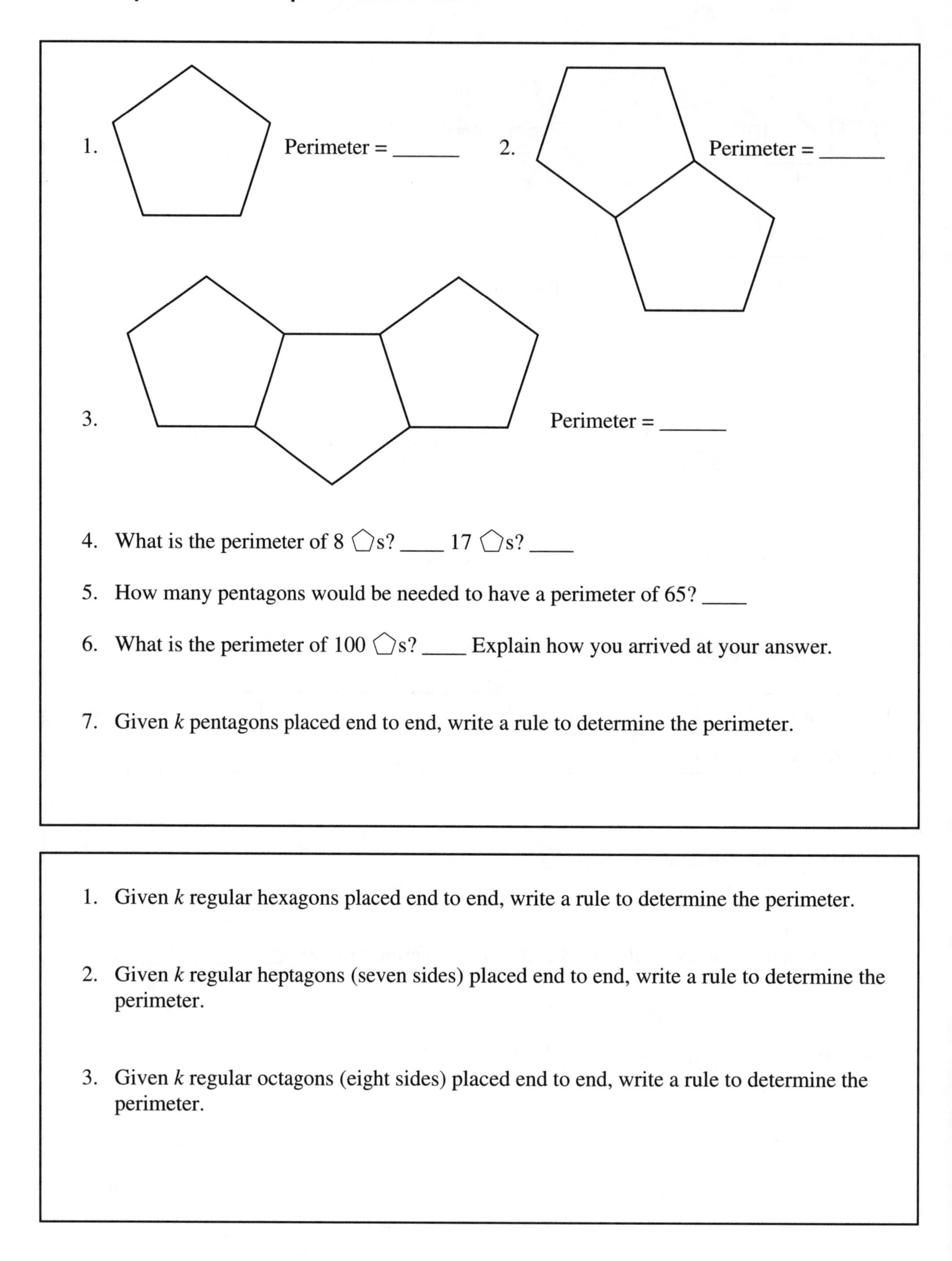

1. Perimeter = ______

2. Perimeter = ______

3. Perimeter = ______

4. What is the perimeter of 8 ⬠s? ____ 17 ⬠s? ____

5. How many pentagons would be needed to have a perimeter of 65? ____

6. What is the perimeter of 100 ⬠s? ____ Explain how you arrived at your answer.

7. Given k pentagons placed end to end, write a rule to determine the perimeter.

1. Given k regular hexagons placed end to end, write a rule to determine the perimeter.

2. Given k regular heptagons (seven sides) placed end to end, write a rule to determine the perimeter.

3. Given k regular octagons (eight sides) placed end to end, write a rule to determine the perimeter.

Use the pattern in the rules you developed to help write a general rule for finding the perimeter of any number k of regular polygons with n sides that are placed end to end as in the previous exercises.

Regular Polygons	Number of Sides in Each Polygon	Rule
Triangles	3	$k + 2$
Squares		
Pentagons		
Hexagons		
Heptagons		
Octagons		
⋮		
n-gons	n	

Activity 2: What's My Area?

PURPOSE Develop the concept of area through estimation and measurement of area using nonstandard units.

MATERIALS One set of tangram pieces (page 247)

GROUPING Work individually or in pairs.

GETTING STARTED Use the small triangles from the tangram set.

If the area of the triangle is one square unit, estimate the area of each figure on pages 244–246. Record your estimates. Then use the triangle to measure the area of each figure and record your answers.

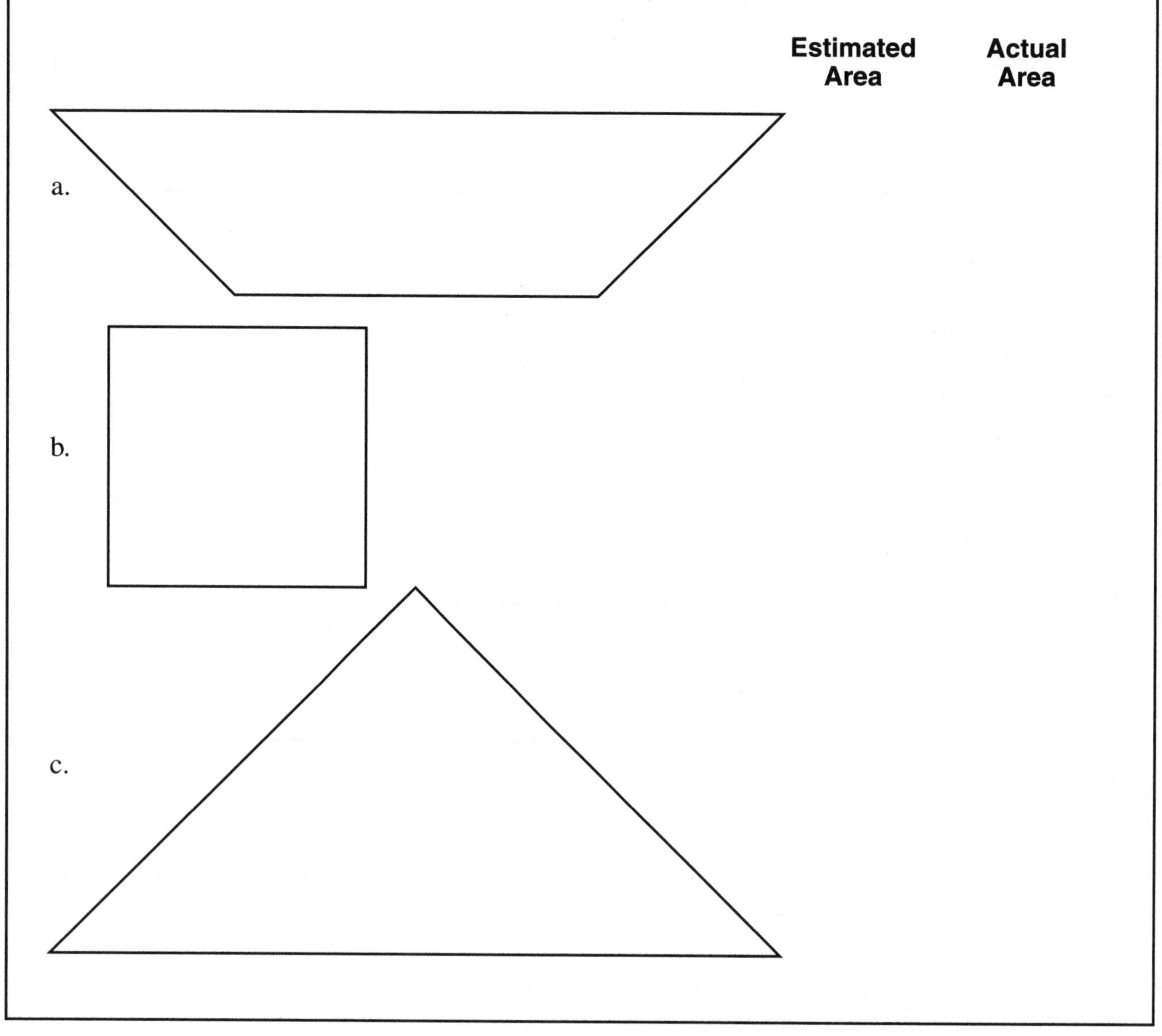

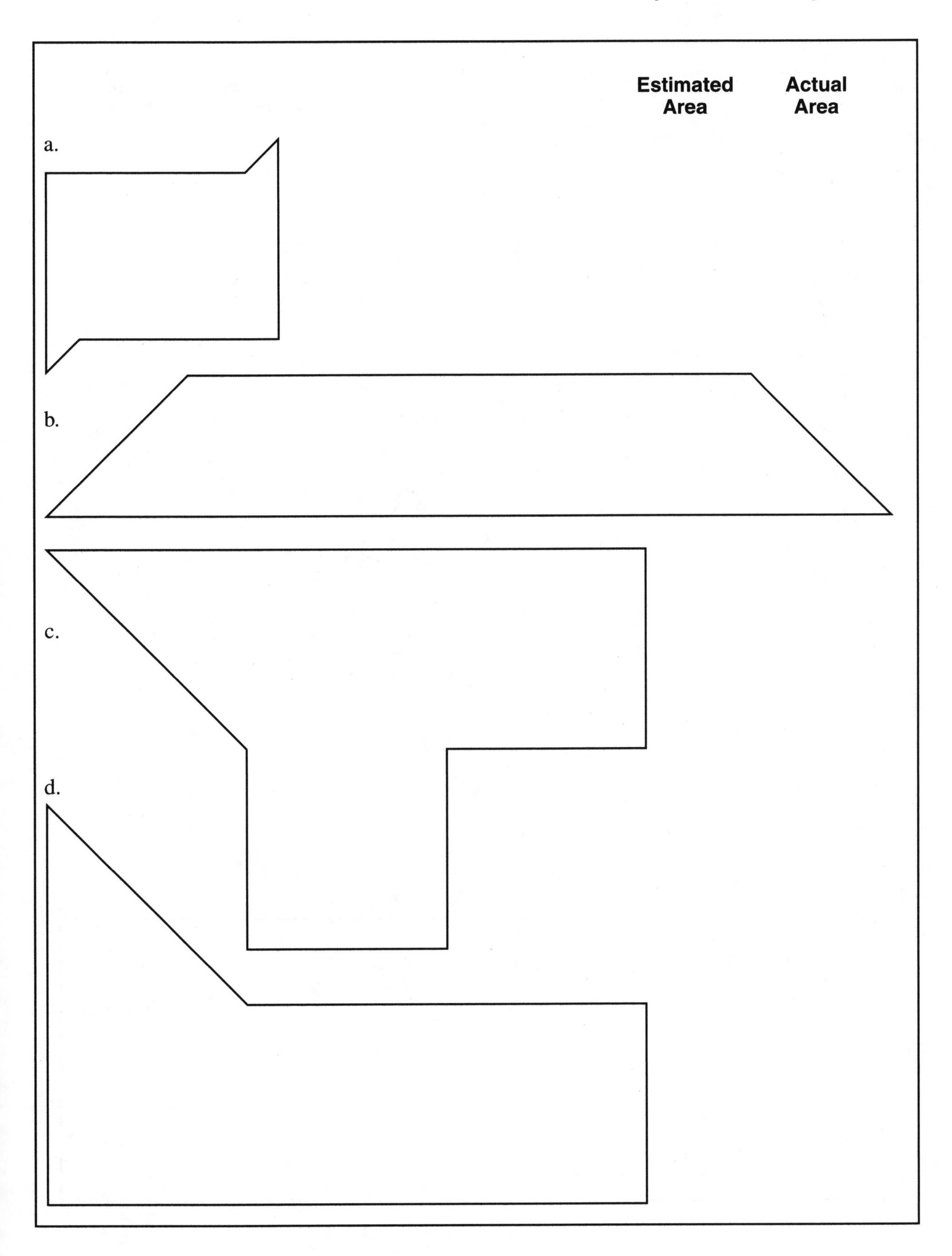
Estimated Area
Actual Area
a.
b.
c.
d.

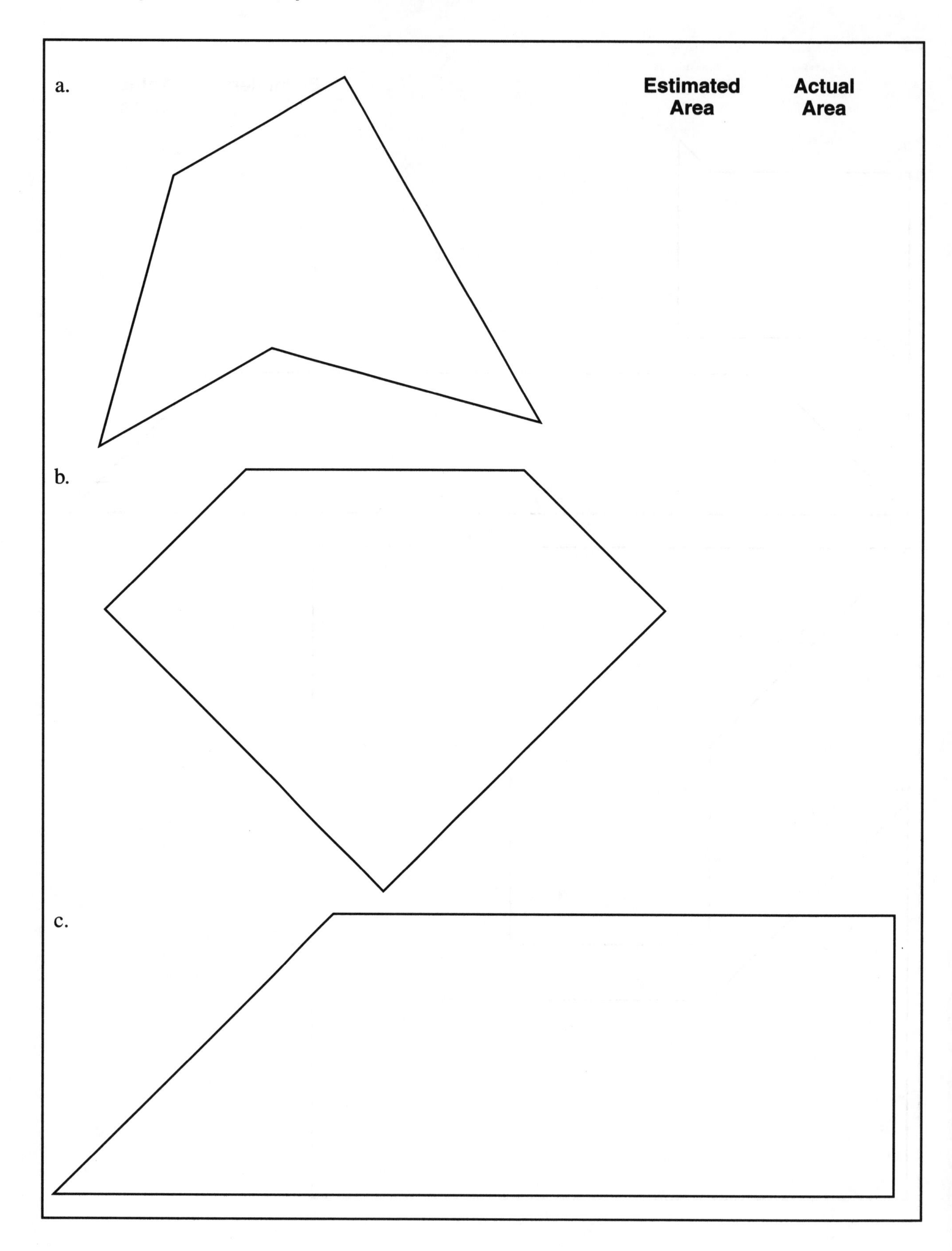
a.
Estimated Area
Actual Area
b.
c.

Activity 3: Areas of Polygons

PURPOSE Develop the concept of the area of a polygon through the composition and decomposition of non-overlapping parts of a figure and reinforce the concept of convex and concave polygons.

MATERIALS A geoboard, geo-bands, and dot paper for recording

GROUPING Work individually.

GETTING STARTED The smallest square that can be constructed on your geoboard has an area of one square unit.

1. On your geoboard, construct **10 or more** polygons that have an area of $2\frac{1}{2}$ square units and record your results on dot paper. Carefully check to be sure that each new figure is not congruent to a previous figure.
2. Sort the figures into sets of concave and convex polygons.

1. On your geoboard, construct **10 or more** polygons with an area of 3 square units and 10 or more polygons with an area of $3\frac{1}{2}$ square units. Record your results on dot paper. Carefully check to see that each new polygon is not congruent to a previous figure.
2. Sort the figures as above.
3. Look at the figures that you constructed with an area of 3 square units. Are the perimeters of all of the polygons equal? **NOTE:** The distance between two adjacent horizontal or vertical pegs is one unit. The distance between any other pair of pegs is more than one unit.
   a. What is the smallest perimeter?
   b. What is the greatest perimeter (approximately)?

c. If you had a 1000×1000 peg geoboard, could you construct a polygon with an area of 3 square units and a perimeter of 100? 1000? Explain.

Activity 4: Now You See It, Now You Don't

PURPOSE Develop the concept of conservation of area in a problem-solving setting.

MATERIALS Graph paper, a ruler, and scissors

GROUPING Work individually.

1. On a piece of graph paper, draw an 8×8 square and divide it into four parts as shown in the figure at the right. What is the area of the square?

2. Cut out the four parts and form a rectangle as shown. What is the area of the rectangle? Can you explain this?

3. Rearrange the four parts as shown. What is the area of this figure? Is this possible?

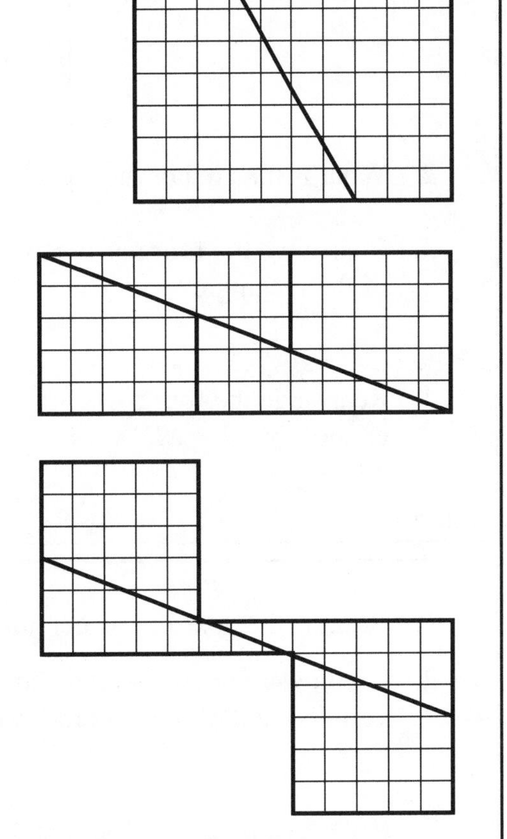

The Fibonacci numbers are found in a famous sequence of numbers: 1, 1, 2, 3, 5, 8, ... Each number in the sequence is the sum of the previous two numbers.

What are the next four Fibonacci numbers?

Note that the length of a side of the square was a Fibonacci number and that three sides were divided into two parts. The measure of each part was also a Fibonacci number.

1. On graph paper, draw a 13×13 square. Divide the square into four parts as shown. In this case, divide each side into two parts with measures 5 and 8. Note that, as in the previous exercise, the length of the side of the square (13) and the lengths of the two parts (5 and 8) are Fibonacci numbers.

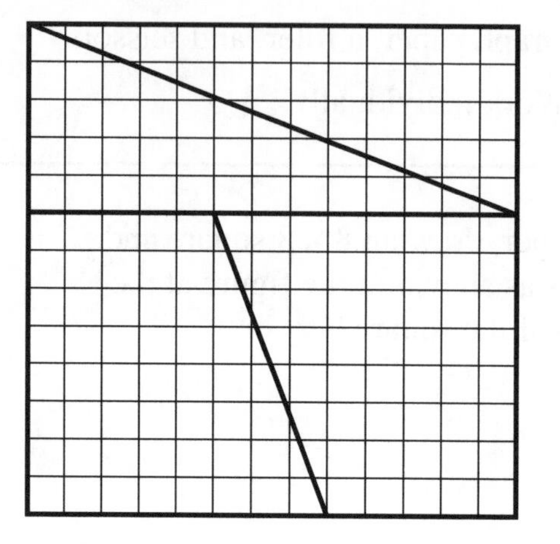

2. What is the area of the square?

3. Rearrange the four parts into a rectangle as in Exercise 2 on page 249. What is the area of the rectangle?

4. Rearrange the four parts into the figure shown in Exercise 3 on page 249. What is the area of this shape?

---

1. What is the next Fibonacci number after 13?

2. If you were to construct a square with that number as the length of the side, how would you divide the sides in order to make the four parts as in the previous exercises?

3. If you rearrange the four parts to form a rectangle and the other shape as before, what would be the area of the rectangle?

   of the other shape?

EXTENSIONS For any number in the Fibonacci sequence, what is the relationship between the area of the square with that number as the length of a side and the areas of the rectangle and the other shape that can be constructed if the original square is divided and rearranged as was done in this activity?

Activity 5: Pick's Theorem

PURPOSE Develop the concept of area, apply the patterns problem-solving strategy to determine a rule for finding the area of a polygon constructed on dot paper or a geoboard, reinforce the concept of dividing a figure into smaller non-overlapping parts to determine its area, and reinforce the concept of conservation of area.

MATERIALS A geoboard, geo-bands, and dot paper

GROUPING Work individually or in pairs.

GETTING STARTED The area of the smallest square that can be constructed on the geoboard or dot paper is **one** square unit.

For each polygon, count the number of points on the perimeter, find the area, and enter the results in the table below.

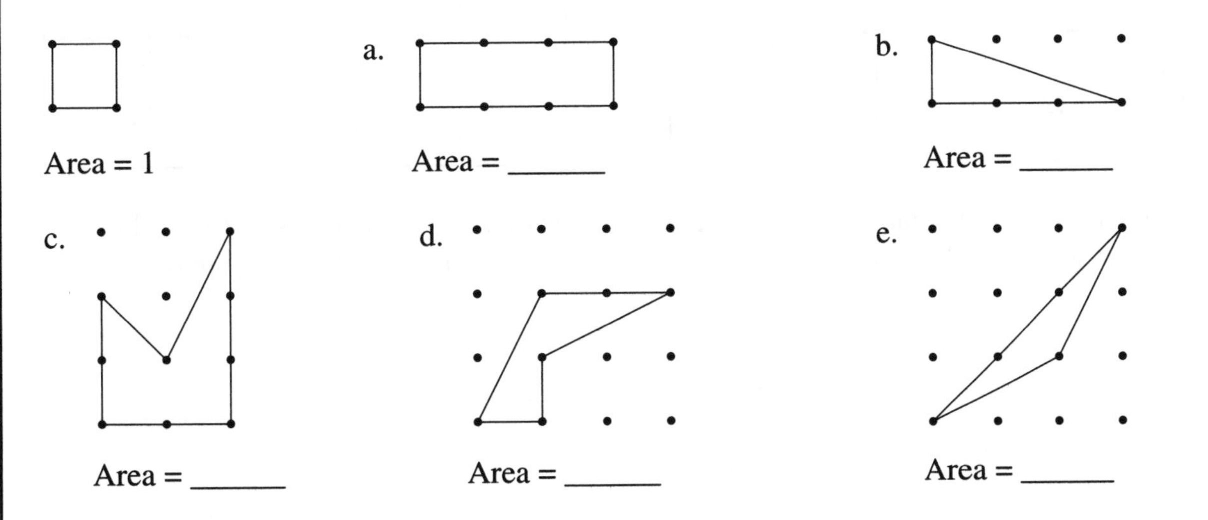

On your geoboard or on dot paper, construct several other polygons in which no points are in the interior. Determine the area of each polygon and record the results in the table.

Number of Points on the Perimeter (N_p)	**3**	**4**	**5**	**6**	**7**			
Area of the Polygon (A)		**1**						

Look for a pattern in the table. Write a rule that relates the area of the polygon (A) to the number of points on the perimeter of the polygon (N_p).

PICK'S THEOREM REVISITED

On your geoboard or dot paper, construct three different polygons that have 1 point in their interiors. Count the number of points (N_p) on the perimeter of each polygon and the number of points (N_i) inside the figure.

Determine the area (A) and record your data in the table below.

Construct several polygons that have 2 points in their interiors. Count the points as above and determine the area of each figure.

Continue this procedure, constructing polygons with 3 and 4 points in their interiors.

Determine the area of each polygon and record the data in the table.

Number of Points on the Perimeter (N_p)								
Number of Points in the Interior (N_i)								
Area of the Polygon (A)								

Examine the data in the table. Write a rule that relates the area of the polygon (A) to the number of points on the perimeter (N_p) and the number of points inside the figure (N_i).

Activity 6: From Rectangles to Parallelograms

PURPOSE Develop a formula for finding the area of a parallelogram by comparing the area of a parallelogram to the area of a related rectangle.

MATERIALS Dot paper, a ruler, and scissors

GROUPING Work individually.

Construct a parallelogram on your dot paper. Construct the altitude from one vertex of the upper base as shown. Cut out the parallelogram and then cut off the triangle. Move the triangle to the other end of the figure and match the vertices as shown.

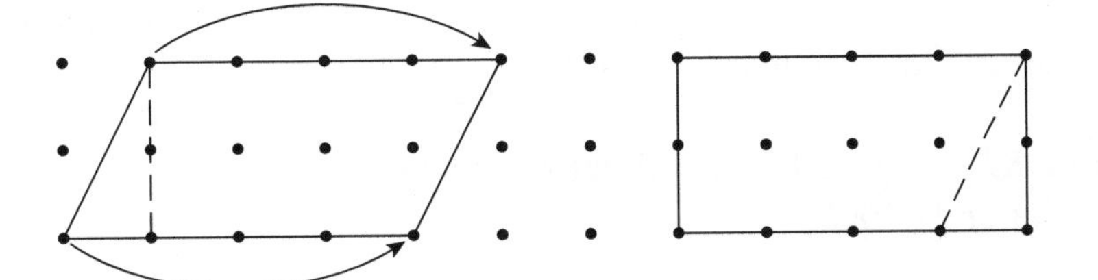

1. What kind of polygon is the new figure?
2. What is the relationship between the base and the altitude of the original parallelogram and those of the new polygon?
3. What is the area of the new polygon?
4. What is the relationship between the area of the original parallelogram and the area of the new polygon?
5. Describe two methods for finding the area of a rectangle.

Construct five additional parallelograms on your dot paper. Draw an altitude, cut out the figures, and construct a rectangle as shown above.

1. For each new parallelogram, determine the area of the related rectangle.
2. What is the relationship between the area of the parallelogram and the area of the related rectangle?
3. Write a rule to determine the area of a parallelogram.

Activity 7: From Parallelograms to Triangles

PURPOSE Develop a formula for finding the area of a triangle by comparing the area of a triangle to the area of a related parallelogram.

MATERIALS Dot paper, a ruler, and scissors

GROUPING Work individually.

Construct ΔKIM on your dot paper. Construct a segment $\overline{KE}$ that is parallel to $\overline{MI}$ and has length equal to the measure of $\overline{MI}$ as shown. Draw the segment $\overline{EM}$.

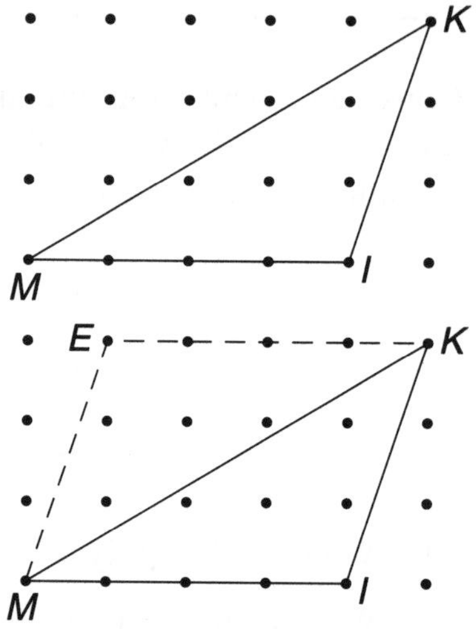

1. Polygon *MIKE* is what type of quadrilateral?
2. What is the area of quadrilateral *MIKE*?
3. The area of ΔKIM is what fractional part of the area of the quadrilateral *MIKE*?
4. What is the area of ΔKIM?

K
M
I
E
K
M
I

Construct five additional triangles on your dot paper. Then construct the related parallelogram as shown above.

1. For each new triangle, what is the relationship between the area of the triangle and the area of the related parallelogram?
2. For each new triangle, what is the relationship between the base and altitude of the triangle and those of the related parallelogram?
3. What is the formula for finding the area of a parallelogram?
4. Knowing the relationship between the area of a triangle and the area of a related parallelogram, write a rule to determine the area of a triangle.

Activity 8: From Parallelograms to Trapezoids

PURPOSE Develop a formula for finding the area of a trapezoid by comparing the area of a trapezoid to the area of a related parallelogram.

MATERIALS Dot paper, a ruler, and scissors

GROUPING Work individually.

Construct a trapezoid on your dot paper as shown. Duplicate the trapezoid, rotate the copy, and match the vertices with the original trapezoid as shown.

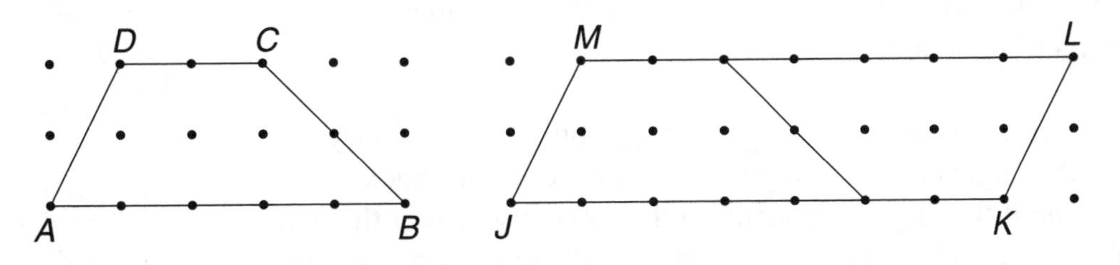

1. Polygon *JKLM* is what type of quadrilateral?
2. What is the area of quadrilateral *JKLM*?
3. What is the relationship between the area of trapezoid *ABCD* and the area of quadrilateral *JKLM*?
4. What is the area of trapezoid *ABCD*?

Construct five additional trapezoids on your dot paper. Then construct the related parallelogram.

1. For each trapezoid, what is the relationship between the area of the trapezoid and the area of the related parallelogram?
2. What is the relationship between the measure of the base of the parallelogram and the measures of the two bases of the trapezoid?
3. Knowing the relationship between the area of a trapezoid and the area of a related parallelogram, write a rule to determine the area of a trapezoid.

Activity 9: Right or Not?

PURPOSE Develop or reinforce the Pythagorean theorem and its converse and explore the relationship between the sides of a triangle and its classification as acute, right, or obtuse.

MATERIALS Centimeter graph paper (page A-47) and scissors

GROUPING Work individually or in pairs.

From a sheet of graph paper, cut out squares with areas 9, 16, 25, 36, 49, 64, 81, 100, 121, 144, and 169. Use three squares to construct a triangle as shown.

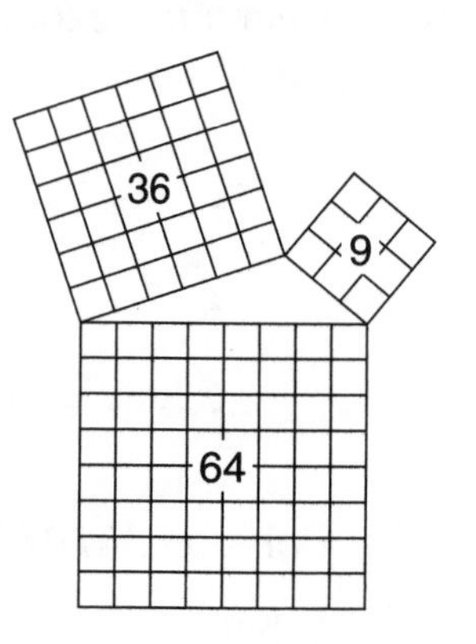

Determine if the triangle is acute, right, or obtuse. If necessary, compare the angles of the triangles to the corner of an index card to determine if the angles are acute, right, or obtuse. Enter the data in Table 1. Use additional sets of three squares to construct other triangles and enter the data in Table 1.

TABLE 1

Area of the Largest Square	Area of the Smallest Square	Area of the Third Square	Sum of the Areas of the Two Smaller Squares	Is the Triangle Acute, Right, or Obtuse?
64	9	36	45	obtuse
100	36	64		
36	16	25		
169				
121				
144				

Use the data from Table 1 to complete Table 2.

TABLE 2

Length of the Longest Side of the Triangle	Square of the Length of the Longest Side	Length of the Shortest Side of the Triangle	Length of the Third Side of the Triangle	Sum of the Squares of the Two Shorter Sides	Is the Triangle Acute, Right, or Obtuse?
8	64	3	6	45	obtuse

Use the data from Table 2 to complete the following statements.

a. If the square of the length of the longest side of a triangle is **less than** the sum of the squares of the lengths of the two shorter sides, the triangle is

a(n) ________________ triangle.

b. If the square of the length of the longest side of a triangle is **equal to** the sum of the squares of the lengths of the two shorter sides, the triangle is

a(n) ________________ triangle.

c. If the square of the length of the longest side of a triangle is **greater than** the sum of the squares of the lengths of the two shorter sides, the triangle is

a(n) ________________ triangle.

Activity 10: Volume of a Rectangular Solid

PURPOSE Develop the formula for finding the volume of a rectangular prism.

MATERIALS Unit cubes from the base-ten blocks or any set of cubes of the same size

GROUPING Work individually.

GETTING STARTED The numbers in the squares in each figure below indicate the number of cubes in a stack. Use your cubes to construct solids made up of these stacks of cubes.

4	3	1
2	1	5
1	2	2

Figure a

8	2
10	4
3	6

Figure b

4	3	6	5
2	5	2	3
4	3	6	4

Figure c

1. How many cubes are there in each layer of each solid and what is the total number of cubes in each solid?

TOTAL

a. ___ ___ ___ ___ ___ ________

b. ___ ___ ___ ___ ___ ___ ___ ___ ___ ___ ________

c. ___ ___ ___ ___ ___ ___ ________

2. How many more cubes must be added to each solid to construct a rectangular prism? The base and height of the prism must be the same as the base and the maximum height of the solid.

a. ____________ b. ____________ c. ____________

3. What is the volume of each rectangular prism in Exercise 2 (the total number of blocks in the solid)?

a. ____________ b. ____________ c. ____________

Use the information in the table to construct rectangular prisms with the given dimensions and complete the table.

Length	Width	Height	Number of Cubes
3	5	2	
4	6	5	
4	4	4	
7	8	3	
6	5.5	10	
14	3	6.5	
15	9	5	
8.7	13	5	

1. Write a formula to find the volume of a rectangular prism using the length, width, and height.

2. Write a rule for finding the total **number of cubes** for each prism using the area of the base and the height of the prism.

3. A rectangular prism with dimensions 2, 5, and 8 can have three different bases, each with different dimensions. Find the area of each base.

4. Use each base to find the volume of the rectangular prism.

5. Explain how the formula you wrote in Exercise 2 can be applied to determine the volume of a **cylinder**.

Activity 11: Roll a Cylinder

PURPOSE Develop the concept of the volume of a cylinder.

MATERIALS Two sheets of 8.5 in. × 11 in. paper, two sheets of 8.5 in. × 14 in. paper (construction paper or something stiffer than regular notebook paper works best), tape, and popcorn or rice

GROUPING Work in pairs.

GETTING STARTED Roll each sheet of 8.5 in. × 11 in. paper so that the edges meet. Tape the edges carefully to form two open-ended cylinders, one with a height of 11 in. and the other with a height of 8.5 in.

1. Do you think the cylinders have the same volume? If not, which one has the greater volume? Explain your reasoning.

2. Verify your estimate by placing the taller cylinder on a flat surface and filling it with popcorn or rice. Now place the shorter cylinder over the taller one. Lift the taller cylinder and allow the contents to fill the wider cylinder. What did you discover?

Repeat the activity using 8.5 in. × 14 in. paper. Now the height of the taller cylinder is almost twice that of the other cylinder.

1. Do you think the cylinders have the same volume? If not, which one has the greater volume? Explain your reasoning.

2. Verify your estimate as you did previously. What did you discover?

3. If you construct two different cylinders from two congruent sheets of paper, what dimension of the cylinder has the greatest effect on the volume? Explain.

EXTENSIONS Use the formula for the volume of a cylinder and compute the volumes of the two cylinders that can be made form 8.5 in. × 14 in. sheets of paper. What is the ratio of the two volumes?

Activity 12: Pyramids and Cones

PURPOSE Develop the relationship between the volume of a pyramid and its related prism and the relationship between a cone and its related cylinder.

MATERIALS Scissors, tape, rice or popcorn, and nets for the prism, pyramid, cylinder, and cone (pages 262–265).

GROUPING Work individually or in pairs.

GETTING STARTED Make copies of the nets for the prism, pyramid, cylinder, and cone on construction paper or tag board. Cut out each net and construct each model.

1. What is true about the areas of the bases of the prism and the pyramid?

2. What is true about the heights of the prism and the pyramid?

3. Estimate the number of pyramids full of rice it will take to fill the prism.

Fill the pyramid with rice. Pour the rice into the prism. Repeat the process until the prism is full. (Be sure to record each time you pour rice into the prism.)

4. One prism full of rice = ______ pyramids full of rice.

5. Write a ratio that compares the volume of the pyramid to the volume of the prism.

6. On the basis of this exploration and the ratio you wrote in Exercise 5, write a rule to determine the volume of a pyramid.

Repeat the experiment using the cylinder and the cone. Place the cylinder in a box or some container, since it is open at both ends.

1. Write a ratio that compares the volume of the cone to the volume of the cylinder.

2. On the basis of this exploration and the ratio you wrote in Exercise 1, write a rule to determine the volume of a cone.

NET FOR PRISM

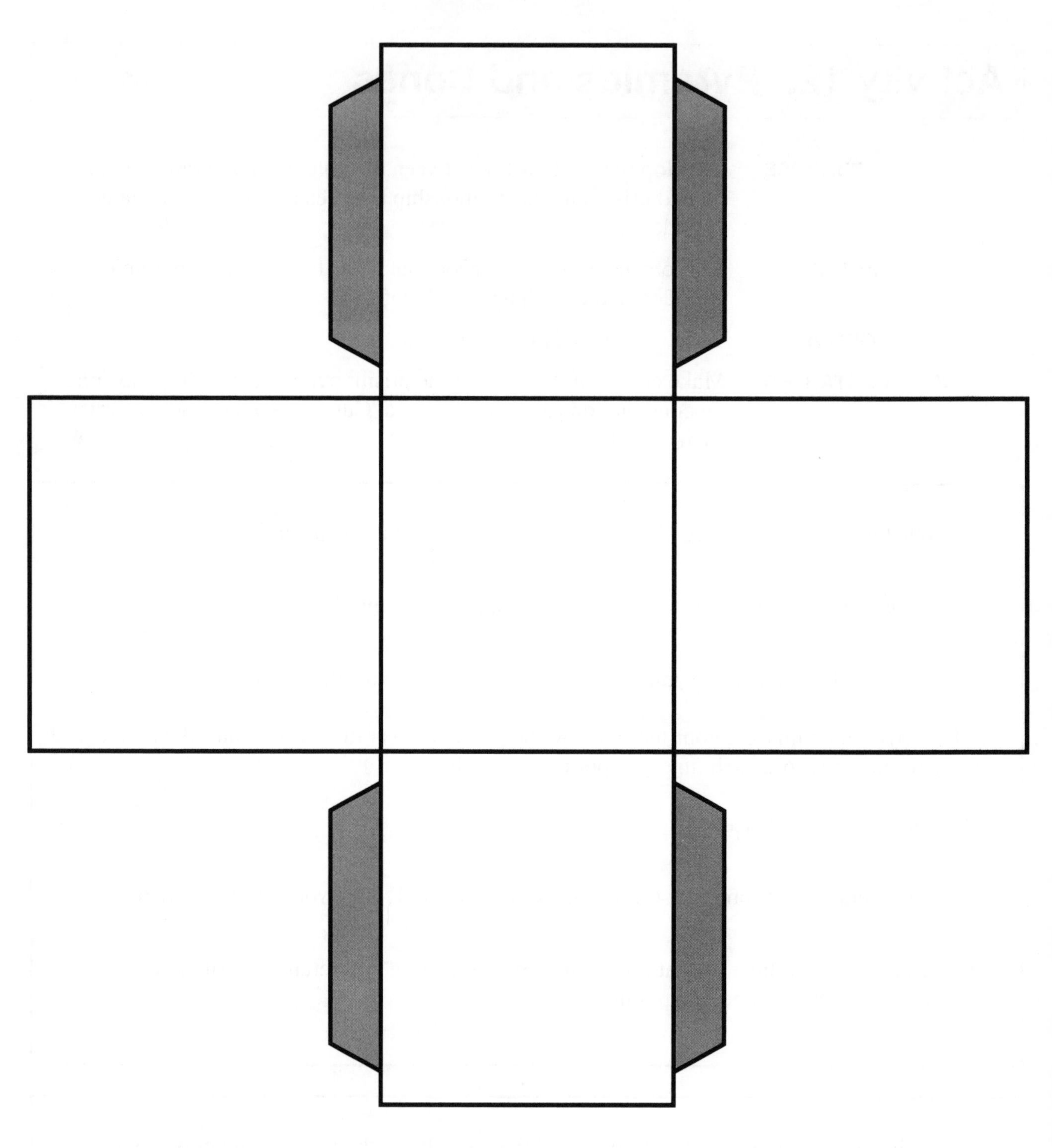

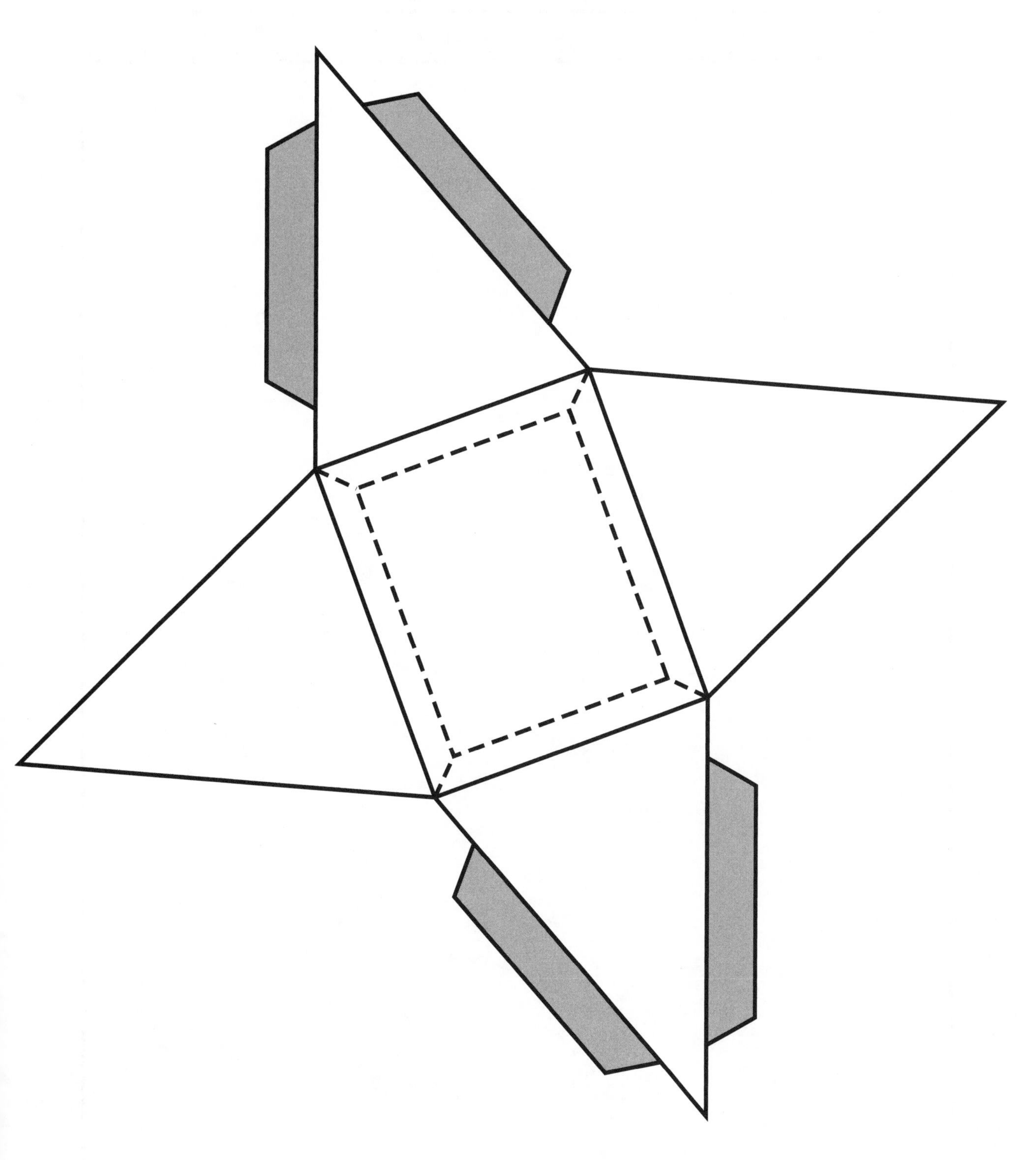

Cut on dotted line to open the pyramid.
Fold the tabs inside the pyramid.

NET FOR CYLINDER

Tab for overlap to tape

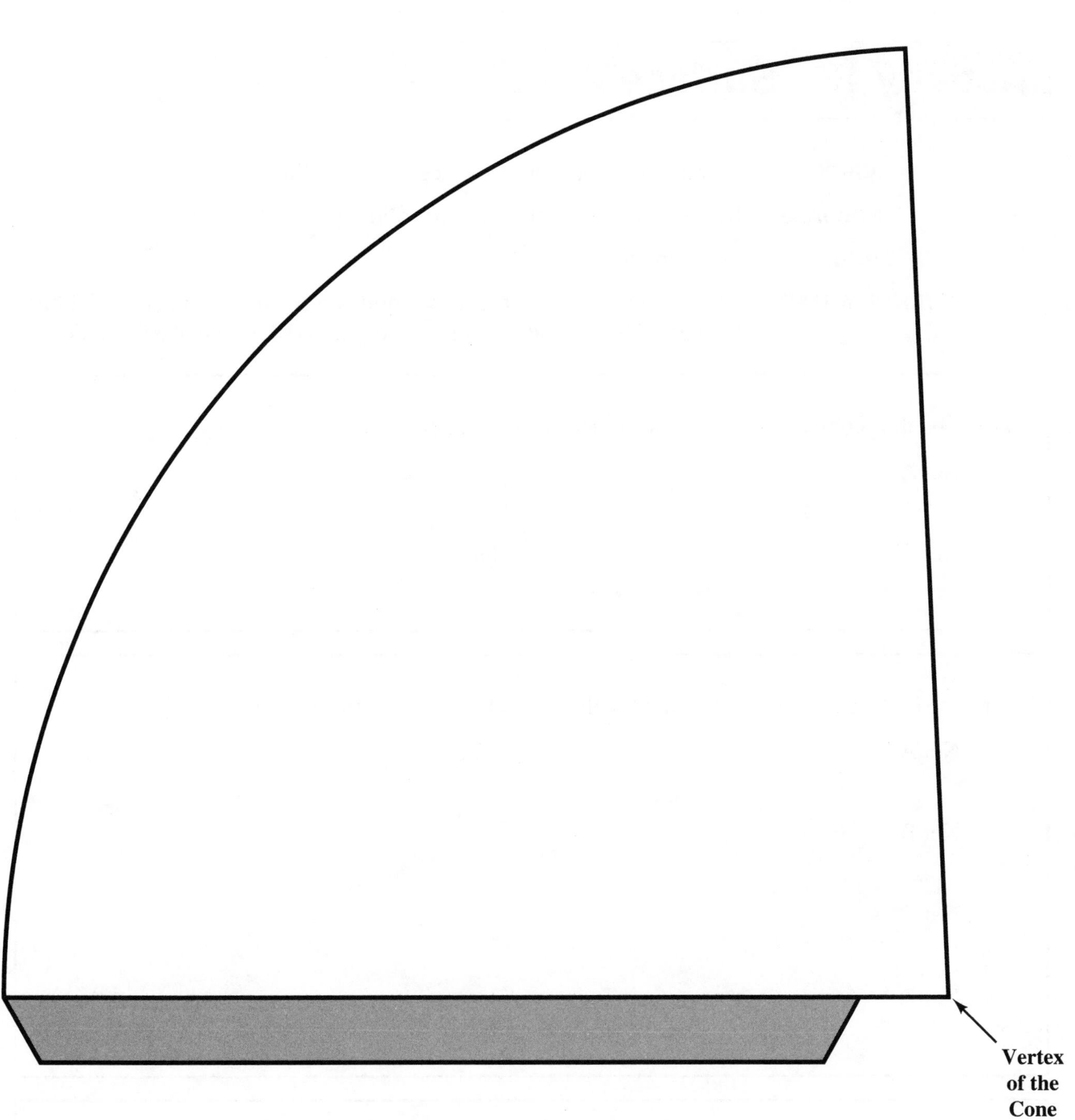
Vertex
of the
Cone

Activity 13: Surface Area

PURPOSE Develop the concept of surface area of solids.

MATERIALS Scissors, tape, and the nets for solids (pages 267–268)

GROUPING Work in pairs.

GETTING STARTED Make a copy of the nets for the solids on construction paper. Cut out the nets. Then fold them and tape them together to make the solids.

Describe the geometric solid that is formed with each net.

1. Net A
2. Net B
3. Net C
4. Net D

Find the surface area of solid formed with each net. Explain your method.

1. Net A
2. Net B
3. Net C
4. Net D

Use the formulas you developed in the previous activities to find the volumes of the solids. In some cases, a formula may not be apparent. Can you find the volumes for those solids in another way? Explain your method.

EXTENSIONS Make three copies of Net B. Cut them out. Then fold them and construct the pyramids. Put the three pyramids together to form a prism. What is the volume of the prism? What is the volume of each pyramid?

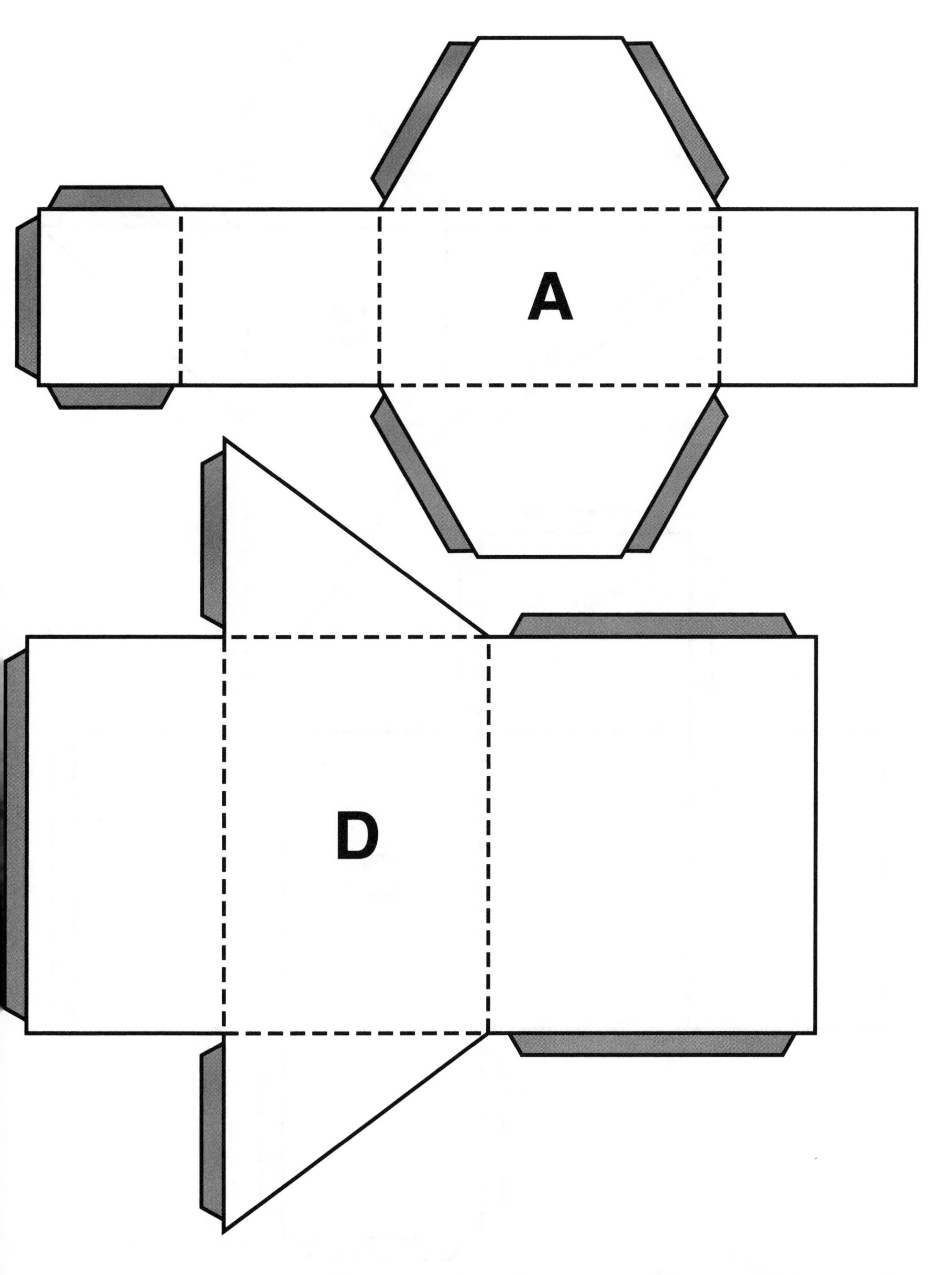
A
D

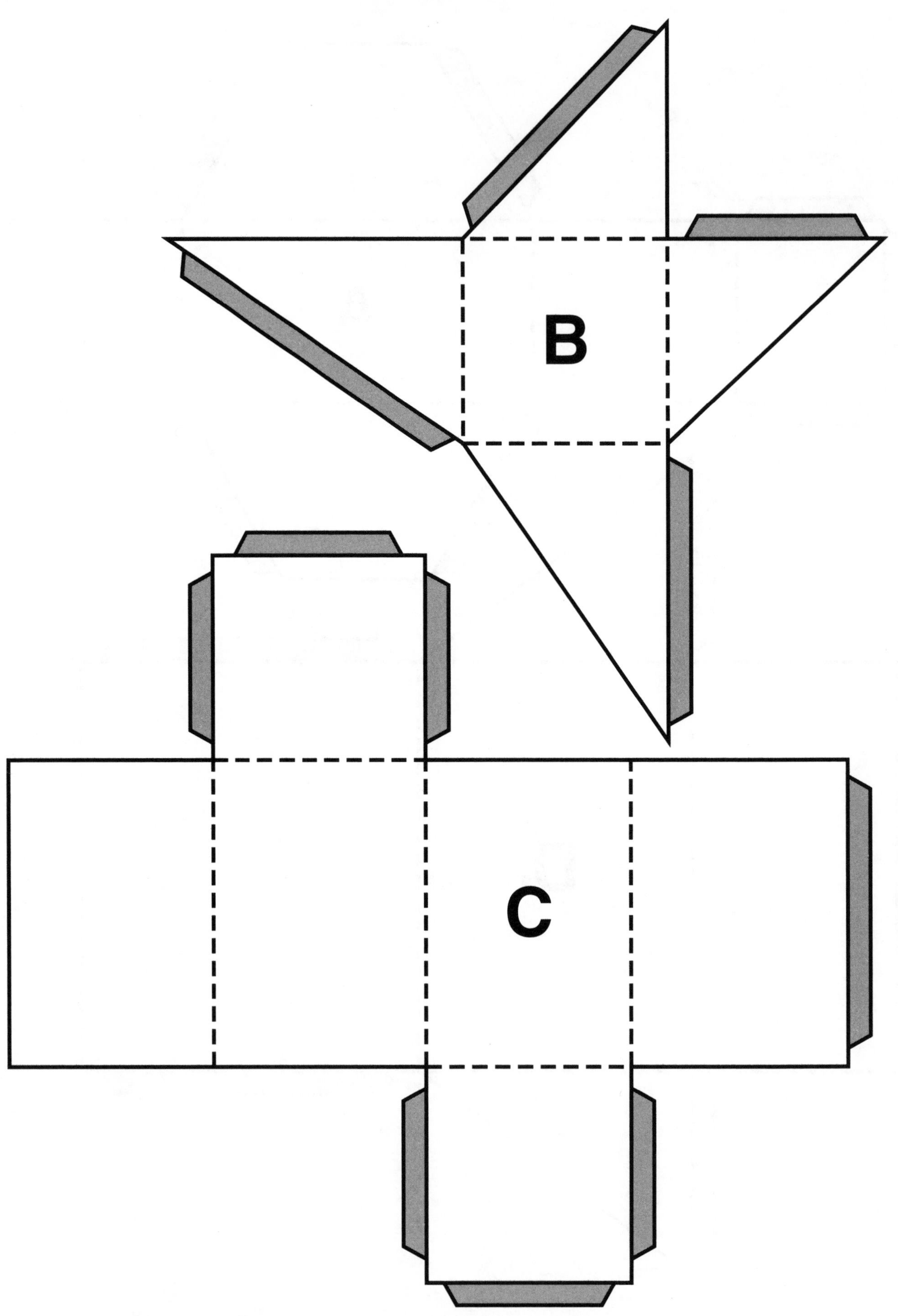
B
C

Activity 14: Compare Volume to Surface Area

PURPOSE Develop an understanding of the relationship between volume and surface area.

MATERIALS Scissors, tape, and several sheets of centimeter graph paper

GROUPING Work individually or in pairs.

GETTING STARTED Use a rectangle 17 cm × 24 cm. Cut one square from each corner of the paper. Fold up the sides and tape the edges to make an open-top box. Be careful not to overlap the paper. Use the squares on the paper to determine the area of the base, the altitude of the box, the surface area of the box, and the volume. Form new boxes by cutting 2 × 2, 3 × 3, etc., squares from each corner of a 17 cm × 24 cm rectangle of graph paper. Find the new measurements and enter your data for each box in the table below.

Dimensions of Squares Cut Off	Area of Base	Altitude	Surface Area	Volume
1 × 1				
2 × 2				
3 × 3				
4 × 4				
5 × 5				
6 × 6				
7 × 7				
8 × 8				

1. What were the dimensions of the box with the greatest volume?
2. Did that box also have the greatest surface area?
3. As surface area increases, does the volume also increase? Explain.

EXTENSIONS If you were a manufacturer of cereal boxes, how would you use the results of this activity to minimize costs and maximize profits?

Chapter Summary

Perimeter, area, and volume are important concepts in geometry and have wide application in the real world. In fact, of all the topics studied in geometry, they are applied more frequently in everyday life than any other topic.

The need for a developmentally appropriate instructional approach to these topics is outlined in the *Curriculum and Evaluation Standards for School Mathematics* which states:

> "Children need to understand the attribute to be measured as well as what it means to measure. . . . Premature use of instruments or formulas leaves children without the understanding necessary for solving measurement problems."

In Activity 1, perimeter was explored in a problem-solving setting in which you discovered a pattern to determine the perimeter of any given number of regular polygons laid end to end. No formulas were given or needed. Using the meaning of perimeter—the distance around a figure—and finding the rule to determine it helped you develop a better understanding of perimeter.

Activities 2 and 3 introduced area as covering with a given unit. Using a triangle or a square as the unit of area and determining the number of triangles or squares that would cover a shape helped you develop spatial visualization and estimation skills.

Activities 3 and 14 explored two concepts that are generally misunderstood: the relationship between perimeter and area and the relationship between surface area and volume. Asked about these relationships, most people will claim that as the area or volume of an object increases so does its perimeter or surface area respectively. You discovered that both of these assumptions are false by constructing shapes and measuring the attributes.

Activities 6 through 8 developed formulas for determining the areas of certain quadrilaterals. Each new quadrilateral was related to one previously studied. From these activities, you found that all the area formulas have their foundation in the rectangular arrays used to illustrate multiplication.

Activity 9 introduced one of the most important theorems of geometry: the Pythagorean theorem. In this activity, you constructed triangles by putting squares together. Thus the idea of "the square of the length of the side of the triangle" was firmly established. Classification of triangles by angles and the Triangle Inequality were also reinforced.

Activities 10 through 12 developed the concept of volume and the formulas for the volumes of certain solids. The process of counting cubes to determine volume was directly related to the process of counting squares used to determine area.

From the comparison of a pyramid to a prism and a cone to a cylinder, you developed a connection among these solids and the formulas to find their volumes. This comparison strategy is the same as that used to find area formulas, in which related polygons were compared.

Throughout this chapter, you were actively involved in doing mathematics. You discovered patterns, made and tested conjectures, and reasoned to develop formulas for perimeter, area, and volume. You also constructed new knowledge or reinforced prior knowledge about these important attributes. Your active involvement in developing the formulas helped to enhance your understanding of these concepts.

Chapter 13
Motion Geometry and Tessellations

"One of the most important connections in all of mathematics is that between geometry and algebra. Historically, mathematics took a great stride forward in the seventeenth century when the geometric ideas of the ancients were expressed in the language of coordinate geometry, thus providing new tools for the solution of a wide range of problems.

More recently, the study of geometry through the use of transformations—the geometric counterpart of functions—has changed the subject from static to dynamic, providing in the process great additional power. . . . Viewed as an algebraic system, transformations also provide . . . valuable experiences with properties of function composition and group structure."

—*Curriculum and Evaluation Standards for School Mathematics*

In this chapter, you will study the properties of a class of functions called isometries. An isometry is a mapping or transformation that preserves the distance between points. The transformations in this chapter include slides (translations), flips (reflections), turns (rotations), and glide reflections. You will also explore symmetry—line symmetry, point symmetry, and rotational symmetry. Isometries and symmetry are powerful new tools for studying congruence, similarity, and other geometric concepts. The chapter concludes with an investigation of transformations of geometric shapes on a coordinate system.

Activity 1: Cut It Out

PURPOSE Introduce line, rotational, and point symmetry.

MATERIALS Paper, scissors, ruler, protractor, and a straight pin or tack

GROUPING Work individually or in pairs.

GETTING STARTED Draw the angle with the given measure, fold the paper along the lines as shown, and cut out the shapes. Do the operations carefully, as accuracy in performing them is critical.

LINE SYMMETRY

1. On a separate sheet of paper, draw two lines, $\overleftrightarrow{AB}$ and $\overleftrightarrow{BC}$, that intersect at point B and such that $m\angle ABC = 60°$.
2. Draw an irregular polygon that has point B in its interior. 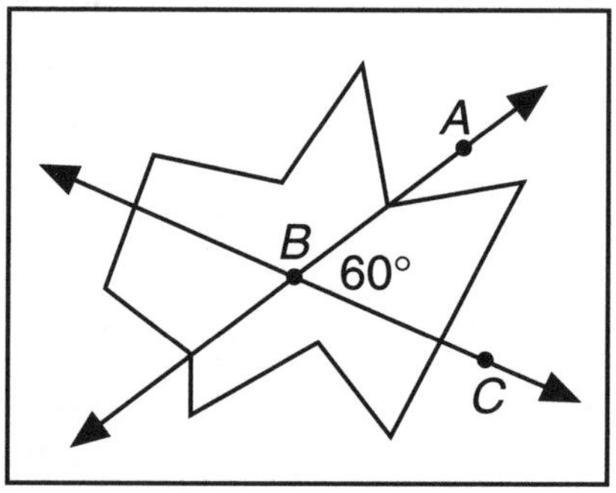
3. Carefully fold the paper along line $\overleftrightarrow{AB}$ so that the figure is on the outside of the paper and cut along the edges of the polygon. 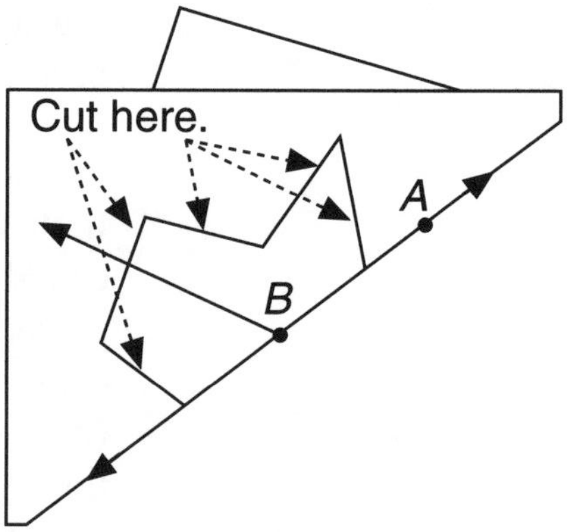
4. Turn the paper over and cut along any remaining edges of the polygon.
5. Open the paper. What do you observe?

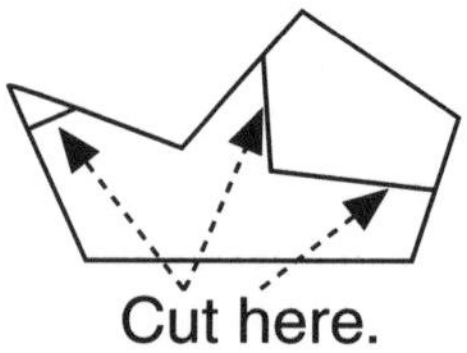

A geometric figure that can be folded along a line in such a way that the two halves are congruent and coincide is said to have *line symmetry.* The line along the fold is the *line of symmetry*.

6. What is the line of symmetry of the figure?

ROTATIONAL SYMMETRY

1. Fold the figure along the other line, $\overleftrightarrow{BC}$. **NOTE:** This is the second fold.
2. Cut away any portion of the figure where there is only one layer of paper. Turn the paper over and cut off the undoubled portion.
3. Open the paper. How many lines of symmetry does the figure have?
4. Continue folding the paper along lines $\overleftrightarrow{AB}$ and $\overleftrightarrow{BC}$ and cutting off the undoubled paper until there are no undoubled parts. If you have measured, folded, and cut accurately, this should require just one more fold (a total of three folds altogether).
5. Unfold the final figure. How many lines of symmetry does it have?
6. Flatten the figure on a piece of paper and trace its outline. Stick a pin through point B and rotate the figure about the pin. What do you observe?

A figure has rotational *symmetry* if rotating it through an angle less than 360° about some point makes the figure coincide with itself. The *angle of rotational symmetry* is the measure of the angle through which the figure was rotated.

7. What are the angles of rotational symmetry for the figure?

POINT SYMMETRY

1. On a separate sheet of paper, draw two lines, $\overleftrightarrow{AB}$ and $\overleftrightarrow{BC}$, that intersect at point B and such that $m\angle ABC = 90°$. Draw an irregular polygon that has point B in its interior.
2. Carefully fold the paper along line $\overleftrightarrow{AB}$ and cut along the edge of the polygon. Turn the paper over and cut along any of the remaining edges of the polygon. Continue the process of folding on lines $\overleftrightarrow{AB}$ and $\overleftrightarrow{BC}$ and cutting off the undoubled paper until there are no undoubled parts.
3. Unfold the final figure. How many lines of symmetry does it have?
4. What is the angle(s) of rotational symmetry for the figure?

Any figure that has 180° rotational symmetry is said to have *point symmetry* about the point of rotation.

5. Draw five segments that pass through point B and have their endpoints on the edge of the figure. How is point B related to the segments?

1. Repeat the experiment for two lines $\overleftrightarrow{AB}$ and $\overleftrightarrow{BC}$ that intersect to form a 45° angle. In the table below, record the maximum number of folds that were made before there was no excess paper to cut off.

Measure of $\angle ABC$	90°	60°	45°		30°		20°	18°	10°
Maximum Number of Folds	2	3	4	5					
Number of Lines of Symmetry	2	3	4			8			

2. a. Look for a pattern to find the missing entries in the table.

   b. Describe the pattern.

   c. Check your prediction for a 30° angle.

REGULAR POLYGONS

An n-gon is a polygon with n sides.

1. How many lines of symmetry does a regular n-gon have?
2. What are the rotational symmetries of a regular n-gon?
3. Which regular n-gons have point symmetry?

EXTENSIONS

1. a. Use the pattern in the table above. According to the pattern, if you start with two lines intersecting at an 80° angle, how many lines of symmetry should the resulting figure have?

   b. Is this possible?

   c. What do you think would happen if you started with an 80° angle?

   d. Try the experiment starting with an 80° angle. (**HINT:** Use a large piece of paper.) What happened? Why?

2. If a figure has exactly two lines of symmetry, what must be true about the lines of symmetry? Explain.

Activity 2: Draw It

PURPOSE Apply the concept of symmetry to drawing figures.

MATERIALS Graph paper and either a mirror or a Mira (see page 285)

GROUPING Work individually.

1. Use the grid to complete the drawing of the figure at the right. The final figure should be symmetric with respect to the vertical line.
2. Place a mirror or a Mira on the line so that you can see the reflection of the left half of the picture in it. What do you observe?

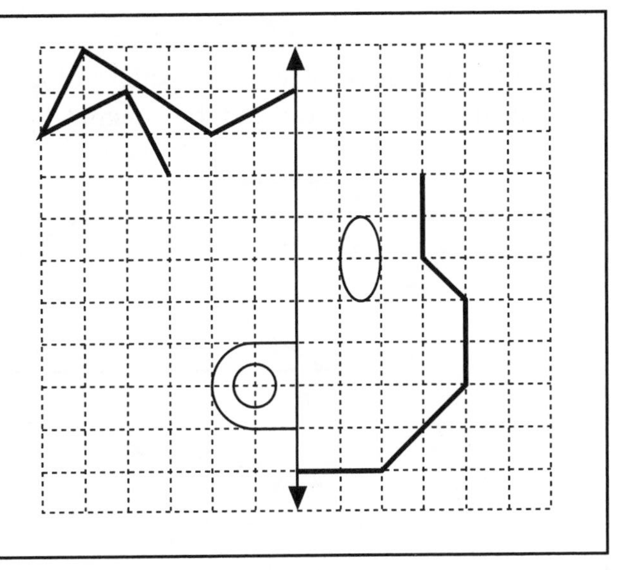

1. Use the grid to complete the drawing of the figure at the right. The final figure should be symmetric with respect to the point O.
2. Rotate the page 180° and look at the figure. What do you observe?

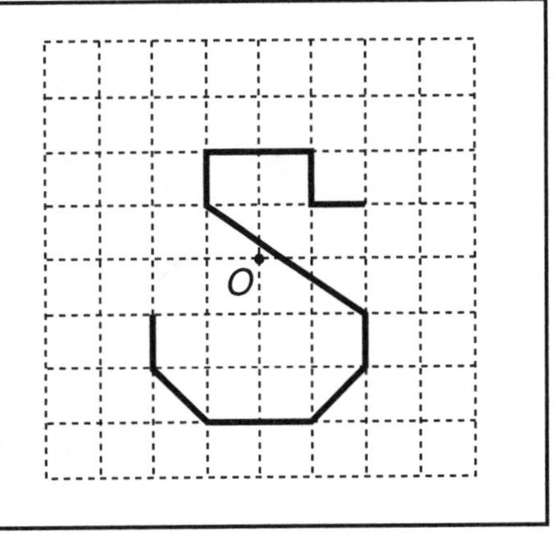

EXTENSIONS

1. Draw a figure that has line symmetry but no rotational symmetries.
2. Draw a figure that has point symmetry but no line symmetries.
3. Draw a figure that has rotational symmetry but neither point nor line symmetries.
4. Make up a design that has one or more lines of symmetry. Draw part of it on graph paper and have a classmate complete the design using symmetries as in the activity above.

Activity 3: Slide It

PURPOSE Use pattern blocks to introduce translations.

MATERIALS Pattern blocks

GROUPING Work individually or in pairs.

Cover each shape below with pattern blocks. Slide the blocks the distance and direction indicated by the arrow and trace them.

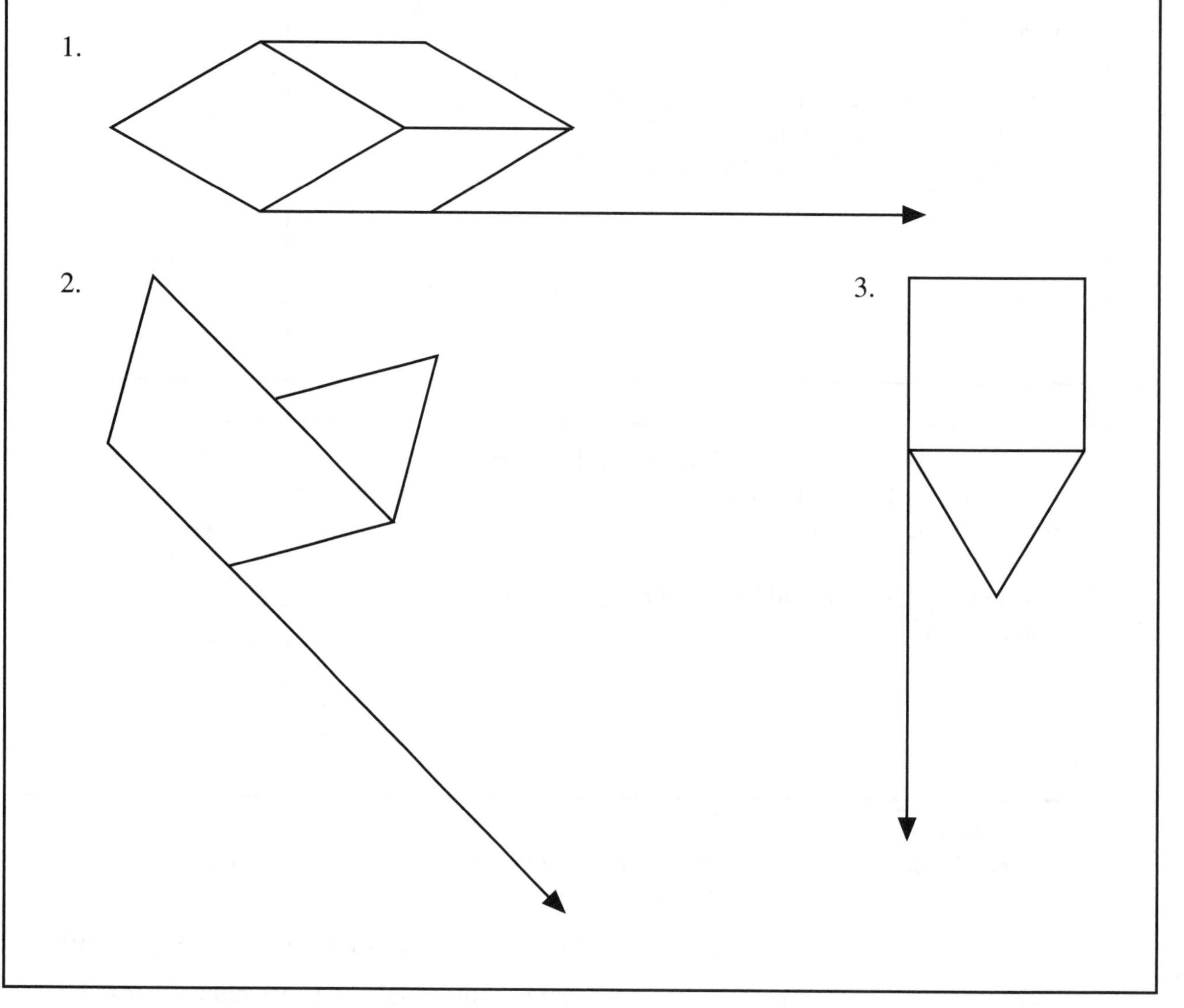

EXTENSIONS Make up a design using pattern blocks. Have a classmate slide it the distance and direction indicated by a given arrow and record the result.

Activity 4: Flip It

PURPOSE Use pattern blocks to introduce reflections.

MATERIALS Pattern blocks

GROUPING Work individually or in pairs.

Cover each shape below with pattern blocks. Flip the blocks across the line and trace them.

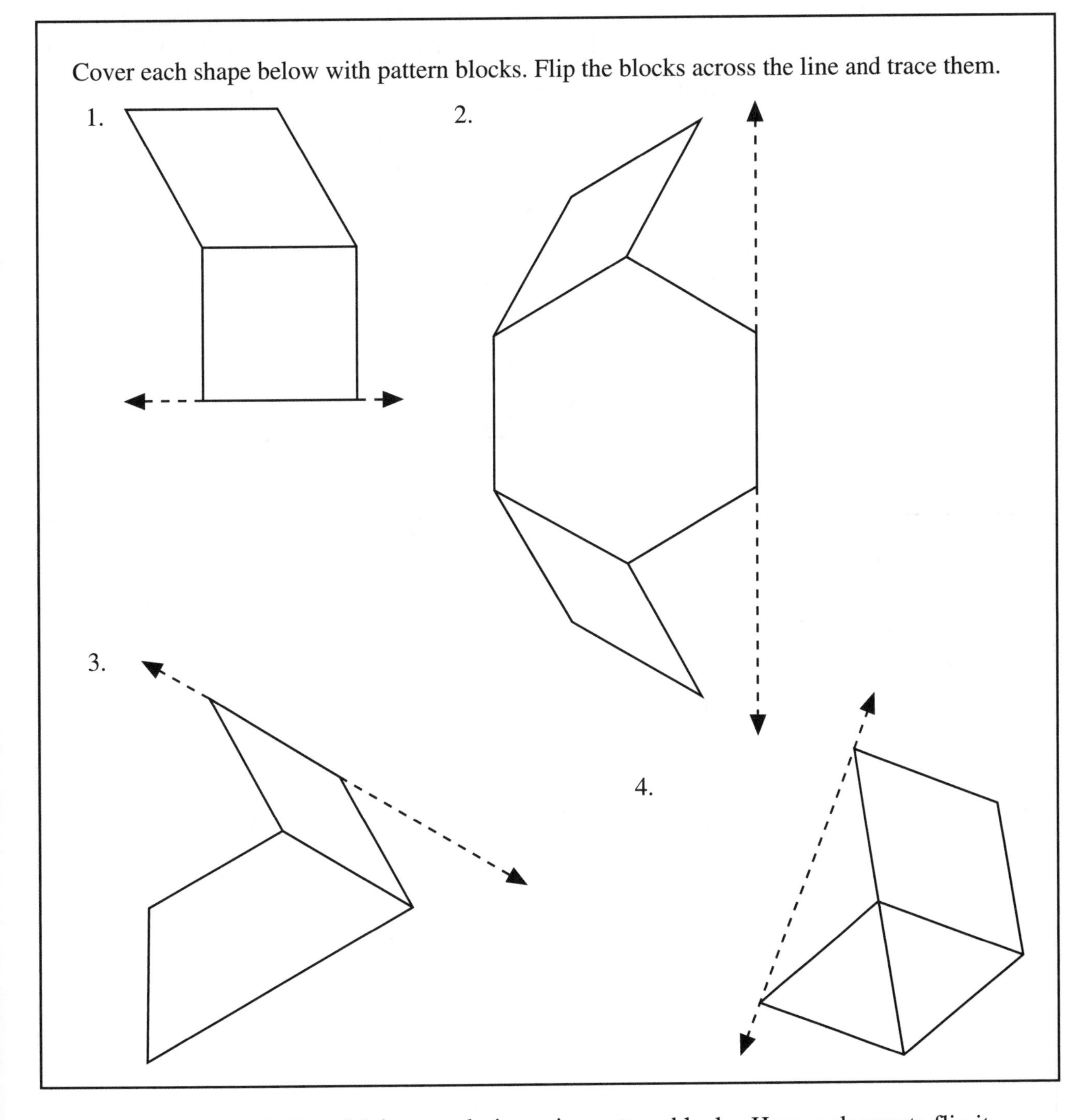

EXTENSIONS Make up a design using pattern blocks. Have a classmate flip it across a given line. Record the result.

Activity 5: Turn It

PURPOSE Use pattern blocks to introduce rotations.

MATERIALS Pattern blocks, a ruler, and a protractor

GROUPING Work individually or in pairs.

GETTING STARTED Cover each shape with pattern blocks. Turn the blocks through the given angle in the indicated direction about point *O*. Record the result.

Example: Turn the shaded block 60° clockwise about point *O*.

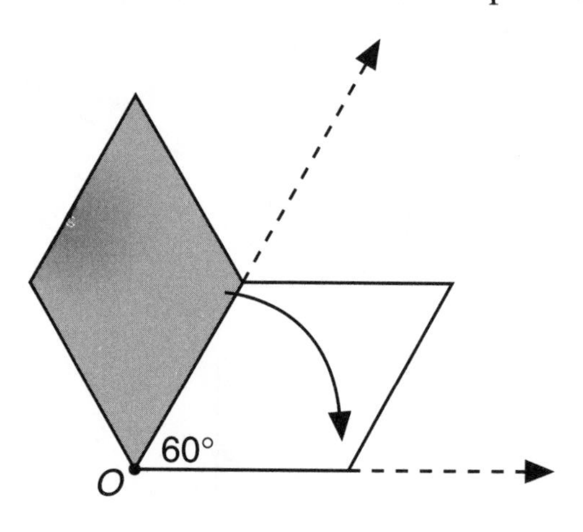

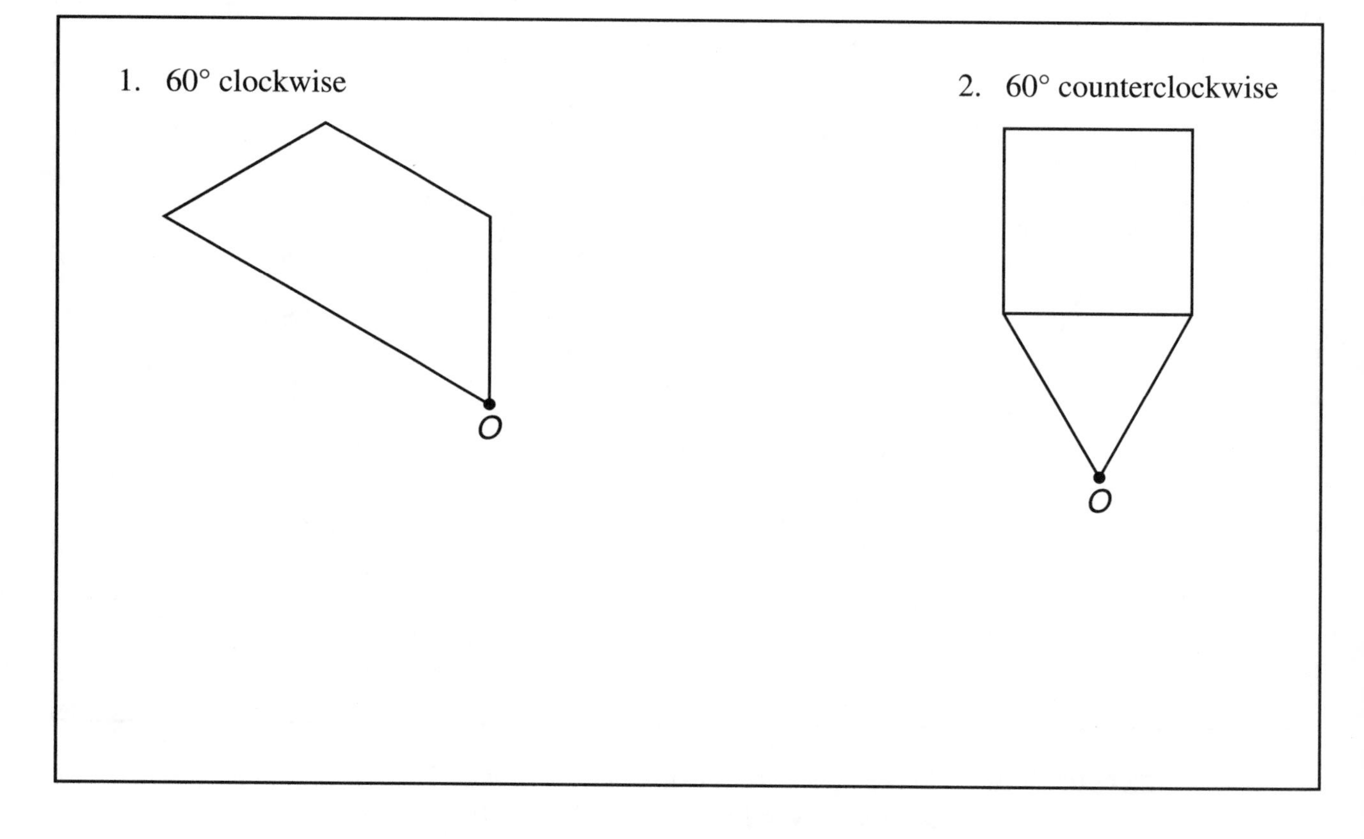

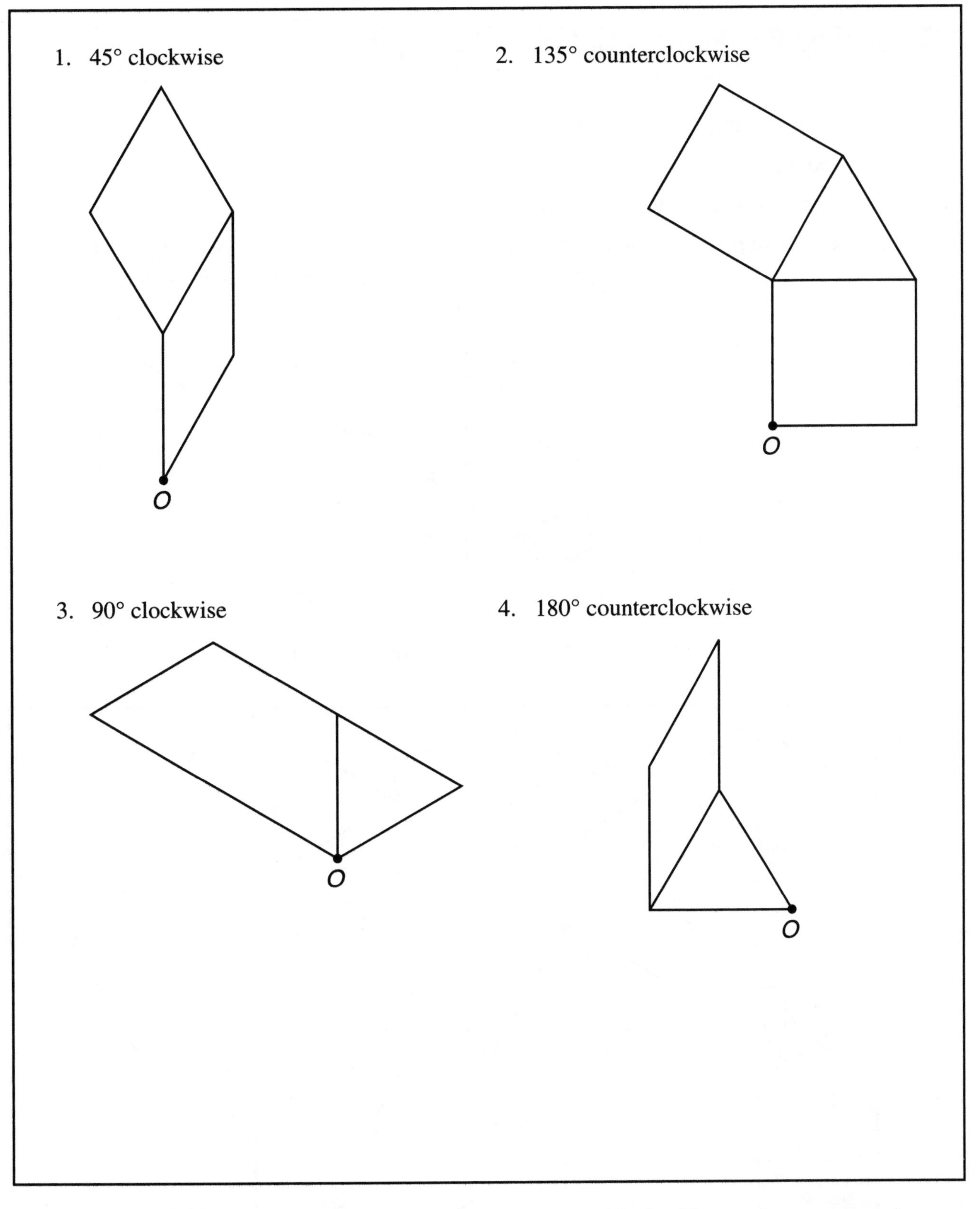

EXTENSIONS Make up a design using pattern blocks. Have a classmate turn it through a given angle either clockwise or counterclockwise about a point *O*. Record the result.

Activity 6: How Did You Do It?

PURPOSE Identify slides and flips.

MATERIALS Pattern blocks

GROUPING Work individually or in pairs.

GETTING STARTED Cover the shaded shapes with pattern blocks. Move the blocks to the next shapes by sliding or flipping them. Write the transformation that was used to make each move in the blank for each shape.

Example:

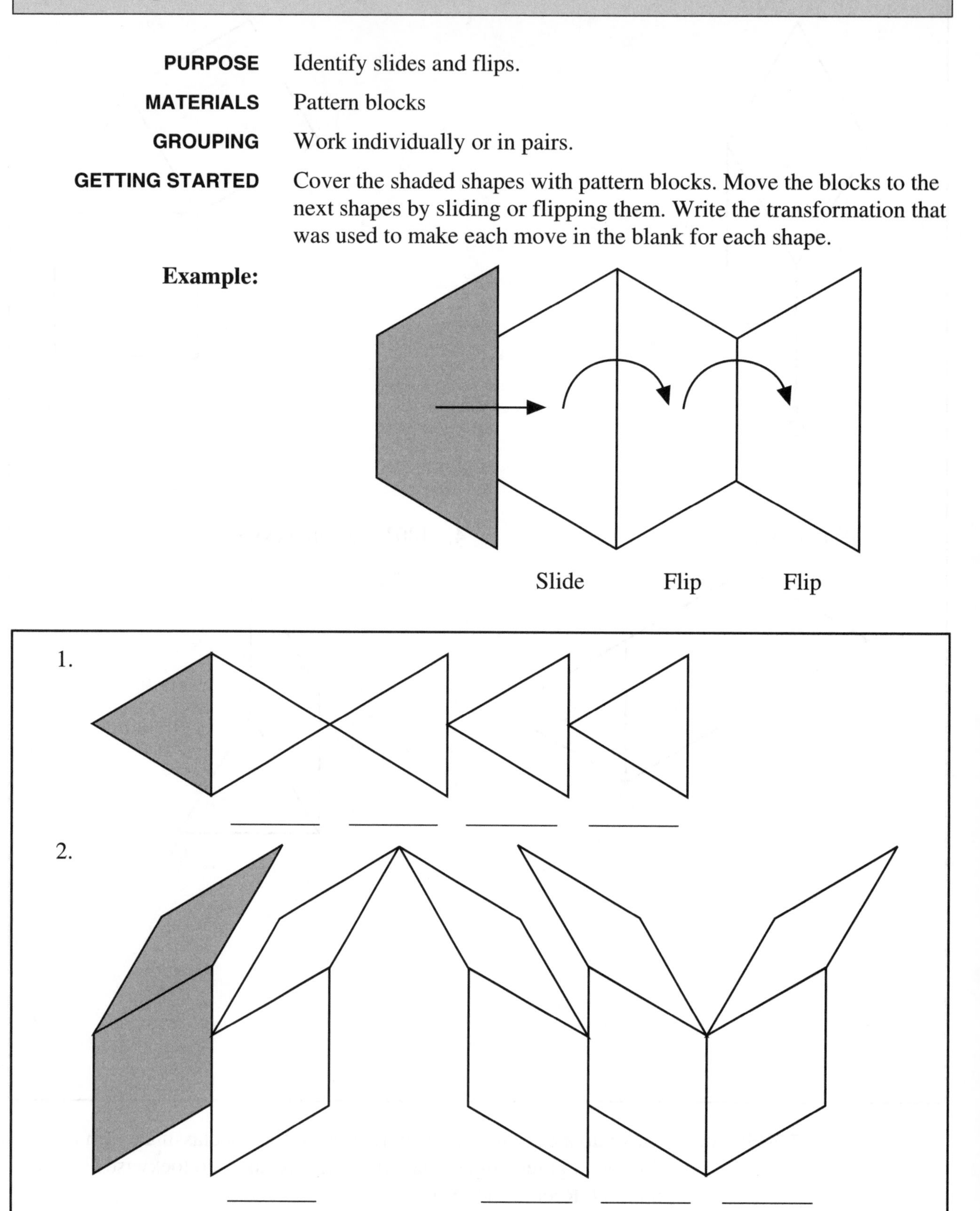

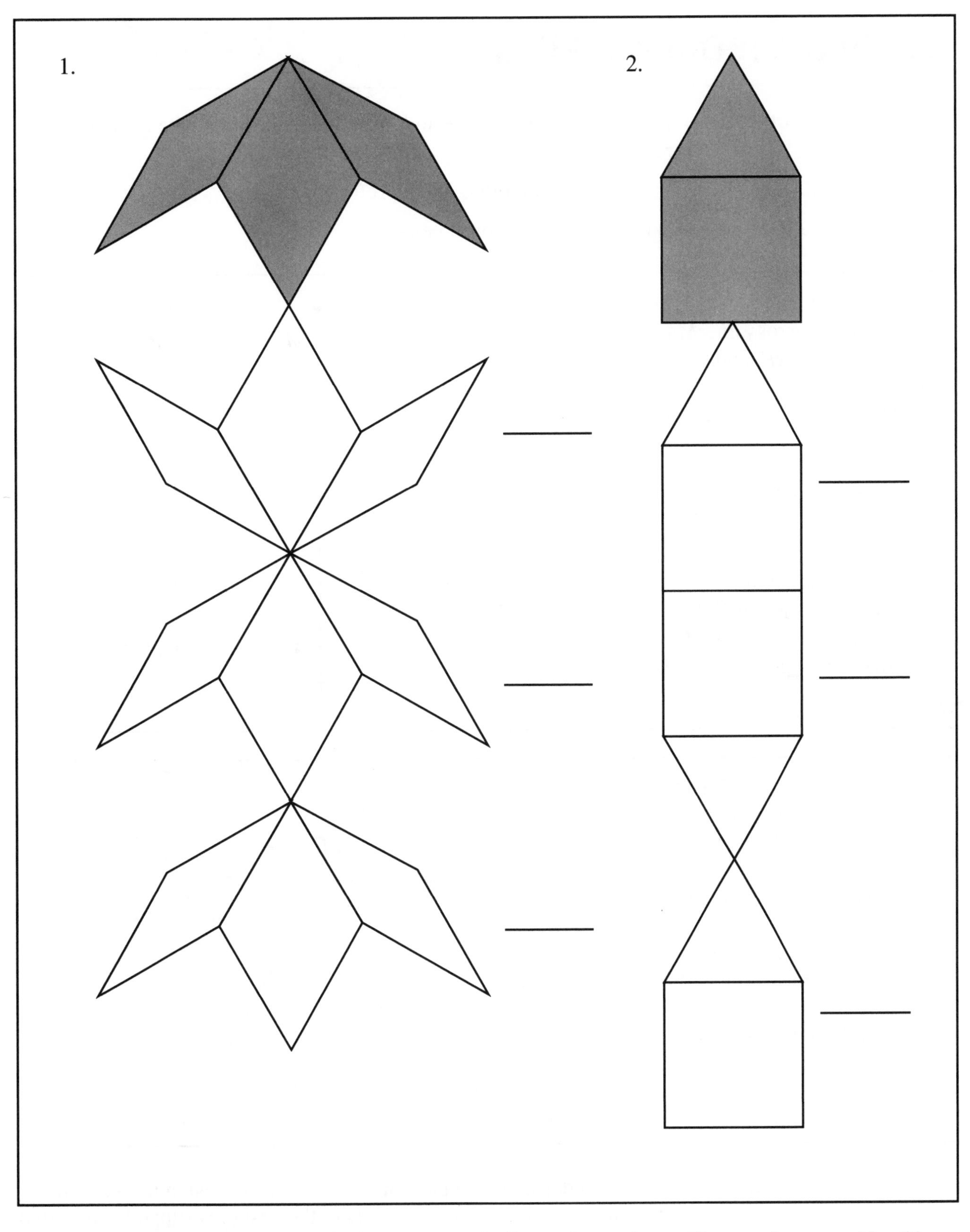

EXTENSIONS Make up patterns similar to those above. Have a classmate identify the transformations.

Activity 7: Double Flips

PURPOSE Identify transformations and investigate the composite of two reflections.

MATERIALS Pattern blocks, a ruler, and a protractor

GROUPING Work individually or in pairs.

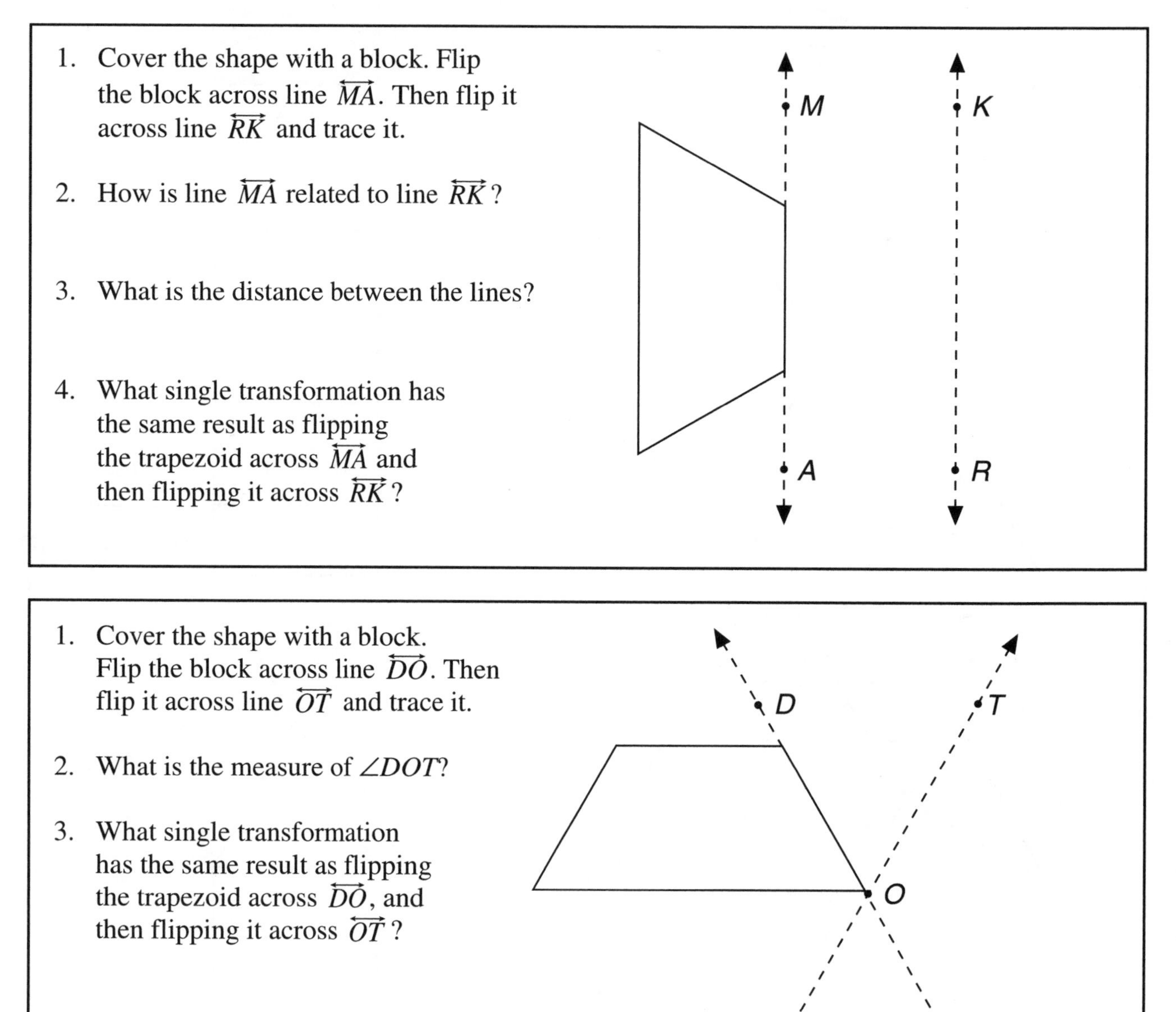

1. Cover the shape with a block. Flip the block across line $\overleftrightarrow{MA}$. Then flip it across line $\overleftrightarrow{RK}$ and trace it.
2. How is line $\overleftrightarrow{MA}$ related to line $\overleftrightarrow{RK}$?
3. What is the distance between the lines?
4. What single transformation has the same result as flipping the trapezoid across $\overleftrightarrow{MA}$ and then flipping it across $\overleftrightarrow{RK}$?

1. Cover the shape with a block. Flip the block across line $\overleftrightarrow{DO}$. Then flip it across line $\overleftrightarrow{OT}$ and trace it.
2. What is the measure of $\angle DOT$?
3. What single transformation has the same result as flipping the trapezoid across $\overleftrightarrow{DO}$, and then flipping it across $\overleftrightarrow{OT}$?

EXTENSIONS Make up a design using pattern blocks. Use the design to make two problems like those above, one where the design is flipped across two parallel lines and one where it is flipped across two intersecting lines. Have your partner perform the transformations, trace the results, and make the measurements as in the activities above.

Activity 8: Reflections

PURPOSE Explore reflections and their properties.

MATERIALS A compass, a centimeter ruler, a protractor, and a Mira

GROUPING Work individually or in pairs.

GETTING STARTED A Mira is a plastic drawing device that acts like a mirror. A Mira reflects objects, but since it is transparent, the image of an object reflected in it also appears behind the Mira.

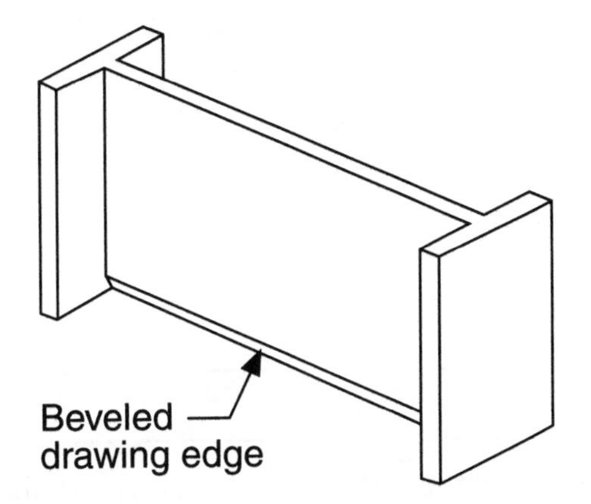

The drawing edge of a Mira is beveled. When using a Mira, place it with the beveled edge down. Look directly through the Mira from the side with the beveled edge to locate the image of the object behind the Mira.

Place your Mira so that the image of circle A fits on circle B. Hold the Mira steady with one hand and draw a line along the drawing edge.

B

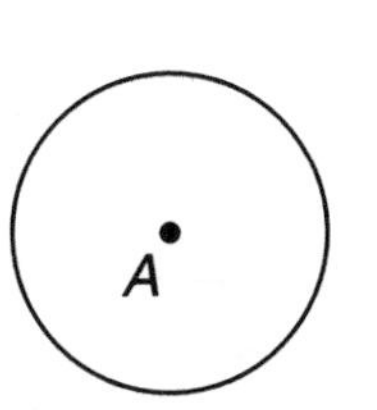

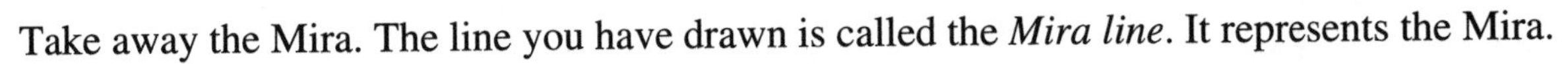

Take away the Mira. The line you have drawn is called the *Mira line*. It represents the Mira.

For each pair of figures below, use a Mira to fit the image of one of the figures onto the other. Then draw the Mira line.

a.

b.

c.

Use a Mira to draw the reflection of each figure through the given line.

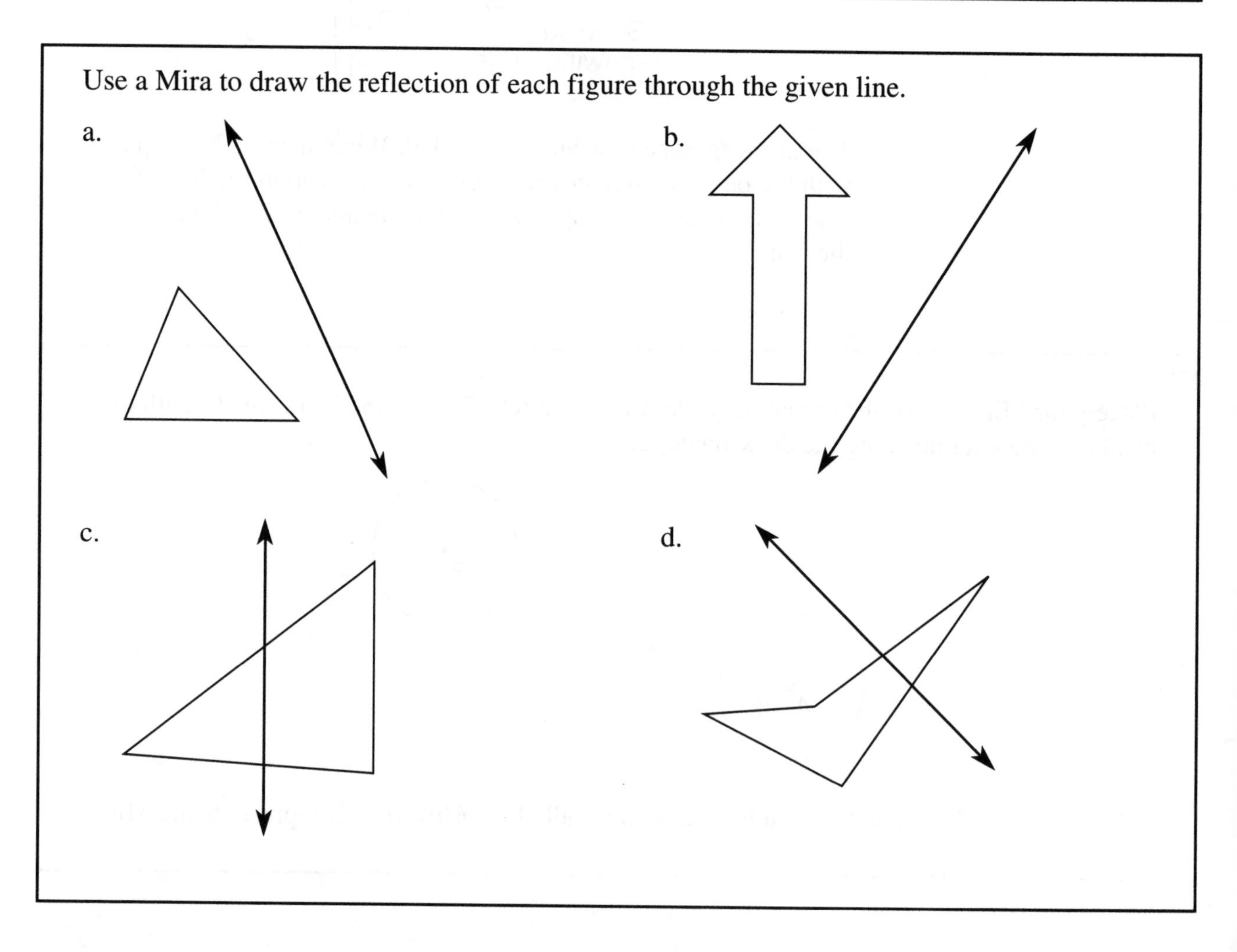

Use a Mira to mark the location of the reflection of each point through the line ℓ. Use prime notation to name each image point. For example, the image of point D would be named D'.

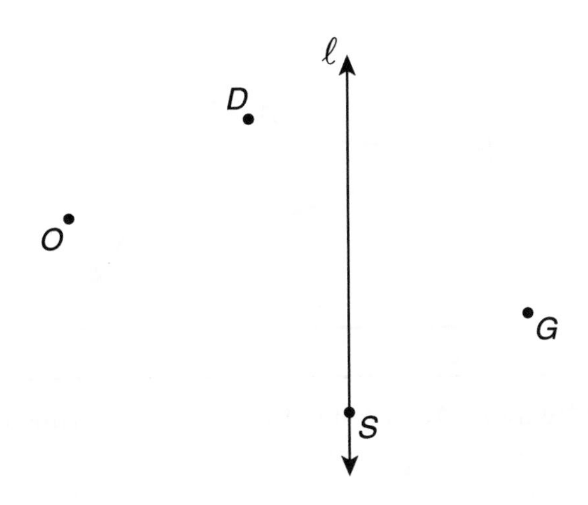

Draw line segments $\overline{DD'}$, $\overline{OO'}$, and $\overline{GG'}$. Label the points where line ℓ intersects these segments C, A, and T, respectively.

1. What is the relationship between the line ℓ and the segments $\overline{DD'}$, $\overline{OO'}$, and $\overline{GG'}$?
2. Where is point S located in relation to line ℓ?
3. What is the relationship between points S and S'?

Use a straightedge and compass to make the following constructions.

1. A line ℓ so that point P is the reflection of point A through ℓ.
2. A line ℓ so that pentagon R is the reflection of pentagon S through ℓ.

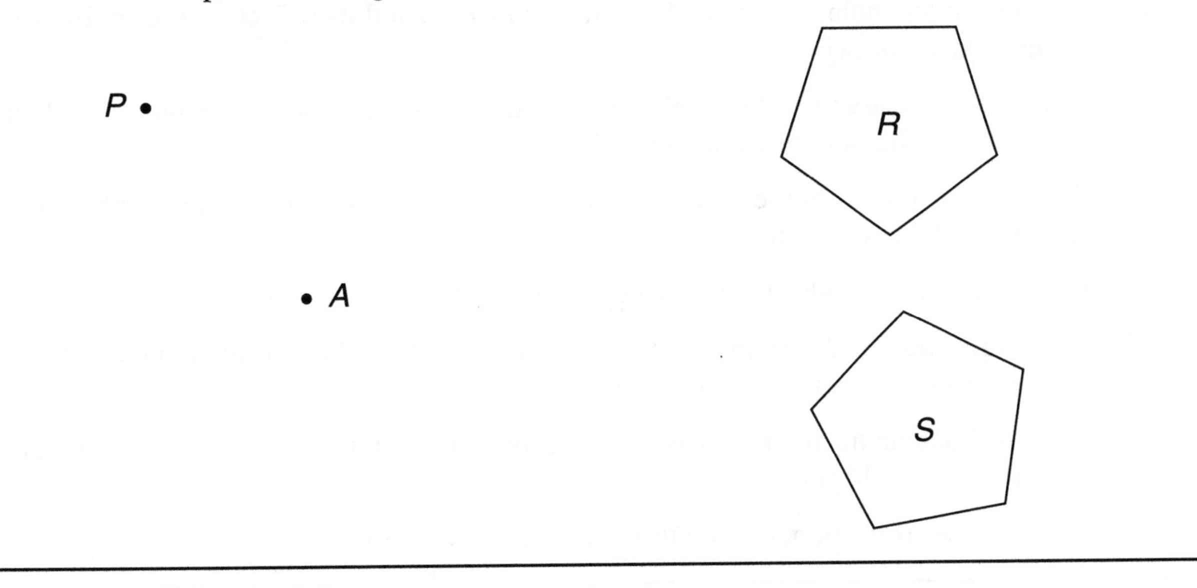

Use a straightedge and compass to construct the reflection of ΔABC through line ℓ.

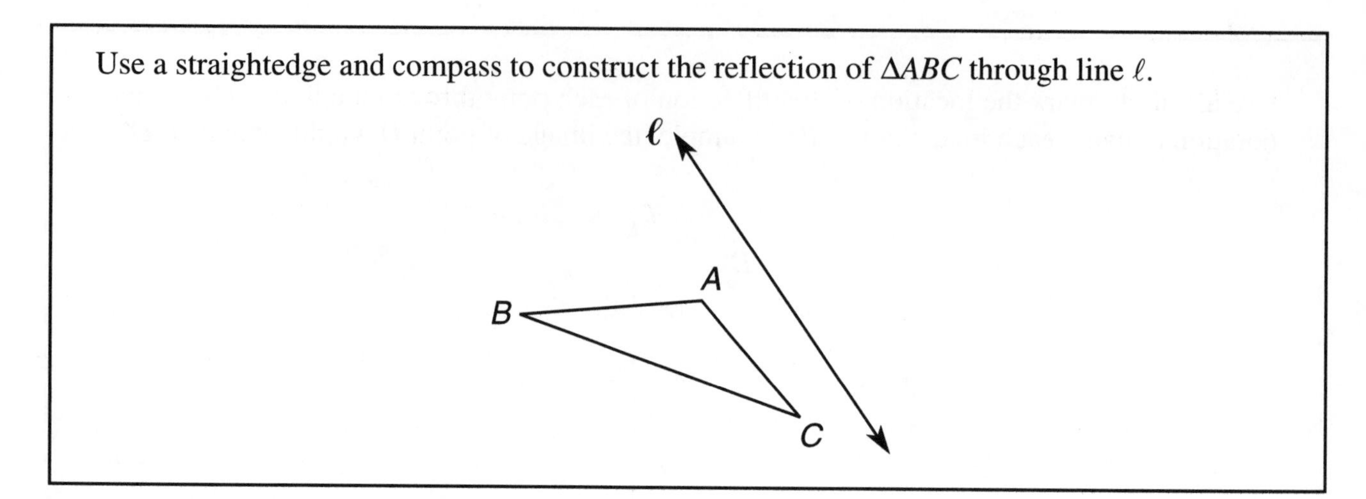

Use a Mira to reflect the figure *FLAG* through the line ℓ. Use prime notation to label the reflection.

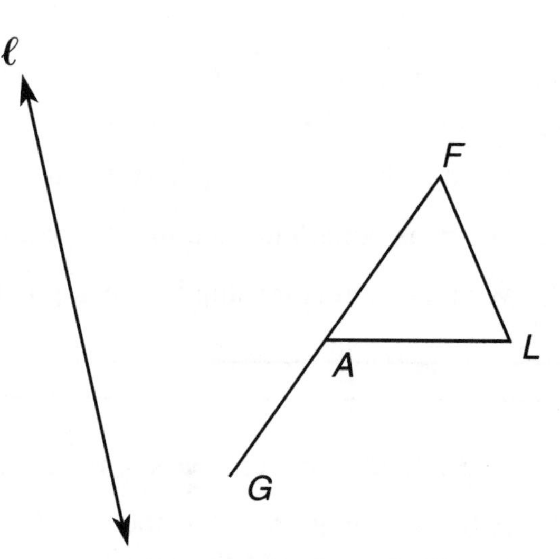

Measure the corresponding angles and segments in *FLAG* and its reflection image. Indicate the measures on the drawings.

1. Examine the measures of the angles. What can you conclude about the measure of an angle and the measure of its reflection?
2. Examine the lengths of the segments. What can you conclude about the length of a segment and the length of its reflection?
3. What can you conclude about a triangle and its reflection through a line?
4. a. Imagine tracing ΔFAL from F to A to L and back to F. What direction (clockwise or counterclockwise) would you move?

   b. Now imagine tracing the image $\Delta F'A'L'$ from F' to A' to L' and back to F'. What direction would you move?

   c. How does reflecting a figure through a line affect its orientation?

Activity 9: Translations

PURPOSE Explore translations and their properties.

MATERIALS A compass, a centimeter ruler, a piece of tracing paper, and a Mira

GROUPING Work individually or in pairs.

1. Reflect ΔMAT through line ℓ_1. Use prime notation to label the reflection.

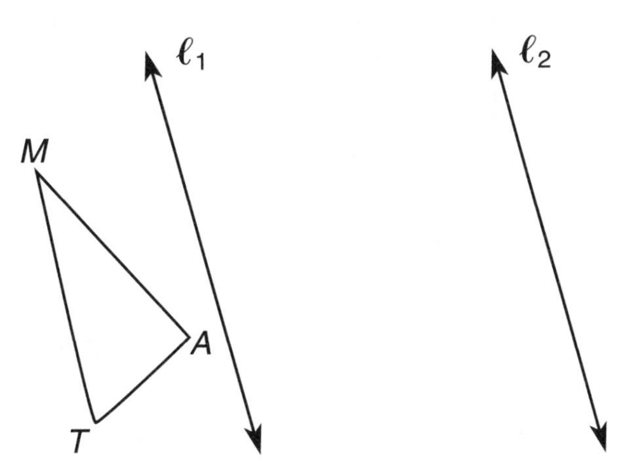

2. Reflect $\Delta M'A'T'$ through line ℓ_2. Let M'', A'', and T'' denote the images of M', A', and T', respectively.
3. Draw $\overline{MM''}$, $\overline{AA''}$, and $\overline{TT''}$.
4. Make a tracing of ΔMAT. Slide it onto $\Delta M''A''T''$ by moving its vertices along the three "tracks" you have just drawn. Is it necessary to flip or to turn the tracing to do this?
5. What two relationships do the "tracks" appear to have?

This transformation is called a *translation*. The exercises illustrate the following definition: **A *translation* is the composite of two reflections through parallel lines.**

1. Translate $\overline{ID}$ by reflecting it through line ℓ_3 and then reflecting the image through line ℓ_4. Let $\overline{ES}$ denote the final translation image of $\overline{ID}$ (E is the image of I and S is the image of D).

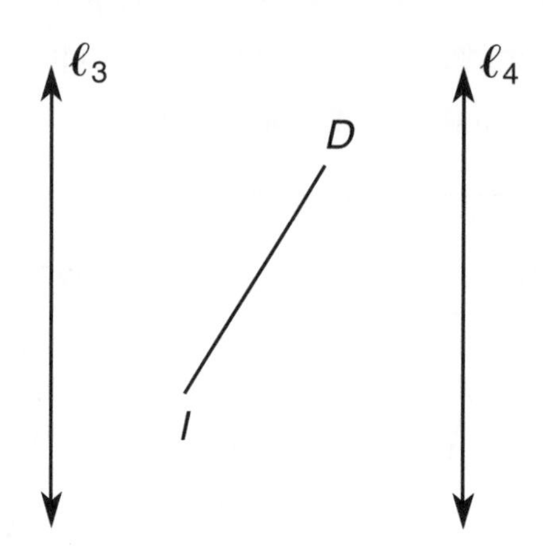

2. What is the distance between lines ℓ_3 and ℓ_4?

3. Measure IE and DS. How do these lengths compare with your answer in Exercise 2?

4. In what direction was $\overline{ID}$ translated?

5. a. Now translate $\overline{ID}$ by reflecting it through line ℓ_4 and then reflecting the image through line ℓ_3. Let $\overline{GL}$ denote the translation image of $\overline{ID}$ (G is the image of I and L is the image of D).

   b. Measure the lengths IG and DL. How do these lengths compare with the result in Exercise 2?

   c. In what direction was $\overline{ID}$ translated?

6. a. If a point is translated by reflecting it through two parallel lines that are x units apart, what is the distance between the point and its translation image?

   b. If a figure is translated by first reflecting it through line ℓ_5 and then reflecting its image through line ℓ_6, in what direction will the figure be translated?

   c. Use the translation of ΔMAT to check your conclusions in Parts (a) and (b).

Activity 10: Transforgeo

PURPOSE Determine the images that are defined by a reflection or a translation.

MATERIALS Two geoboards, six geo-bands, and dot paper

GROUPING Work in pairs.

GETTING STARTED *Transforgeo* is a game for two players. Each player needs a geoboard and three geo-bands.

GAME 1

- The players divide their geoboards in half by connecting the pegs in the middle row with a geo-band.
- Using one geo-band, each player constructs a figure on the top half of a geoboard.
- When both players have completed their figures, they trade geoboards.
- The first player to construct the reflection of the figure on the geoboard is the winner. (The reflection must be approved by the builder of the original figure.)

Example:

Player 1 builds this figure on the top half of the geoboard.

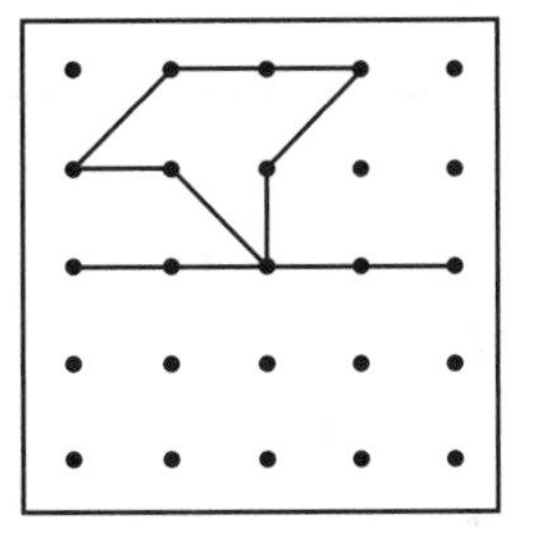

Player 2 must construct the reflection of the figure on the other half of the geoboard.

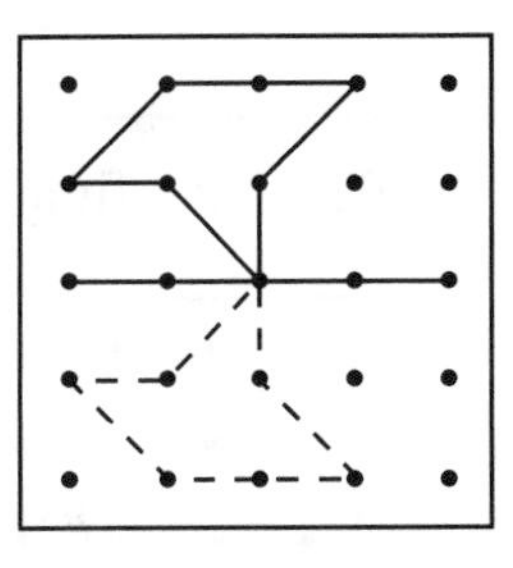

Play the game five times.

GAME 2

- The players divide their geoboards in half by connecting the pegs on a diagonal with a geo-band.
- Using one geo-band, each player constructs a figure on half of a geoboard.
- When both players have completed their figures, they exchange geoboards.
- The first player to construct the reflection of the figure on the geoboard is the winner. (The reflection must be approved by the builder of the original figure.)

Example:

Player 1 builds this figure on half of the geoboard.

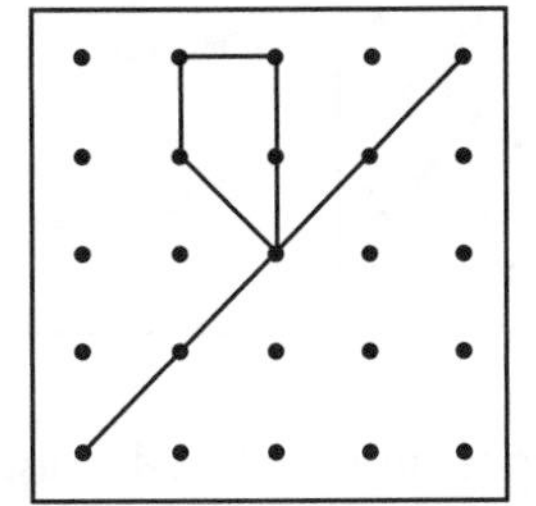

Player 2 must construct the reflection of the figure on the other half of the geoboard.

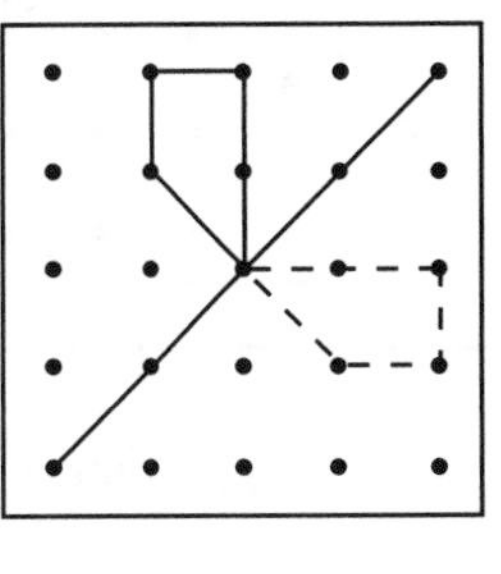

Play the game five times.

GAME 3

- Each player constructs a design on a geoboard using one geo-band.
- With the second geo-band, indicate the direction and distance the figure is to be translated. (The translation must not move the figure off the geoboard.)
- When both players have completed their figures, they exchange geoboards.
- The first player to construct the translation of the figure on the geoboard is the winner. (The translation of the figure must be approved by the builder of the original figure.)

Example:

Player 1 builds this figure on the geoboard. (Direction and distance of translation.)

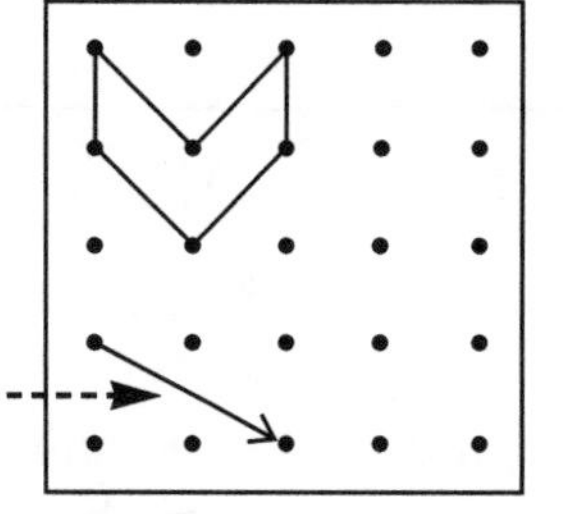

Player 2 must construct the translation of the figure on the geoboard.

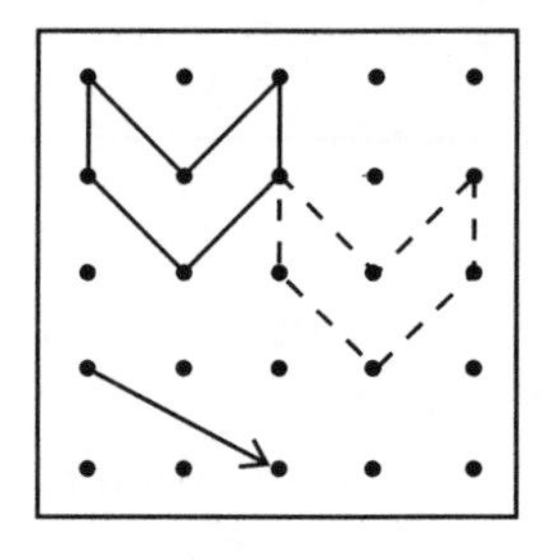

Play the game five times.

1. Reflect each figure through the given line.

a. b. c.

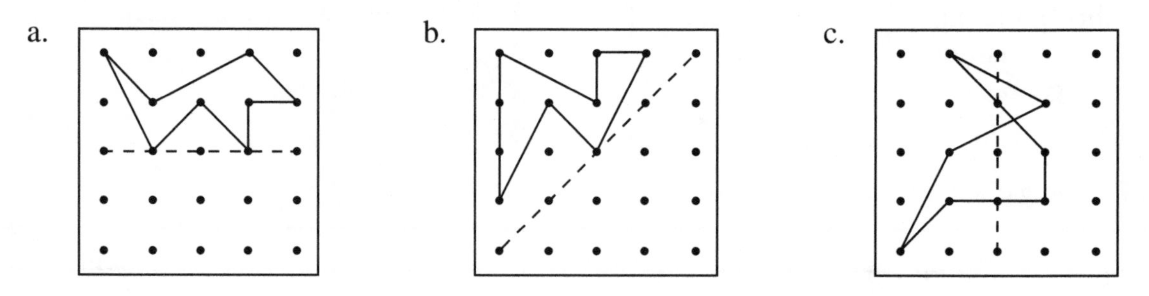

2. Translate each figure the distance and direction indicated by the arrow.

a.

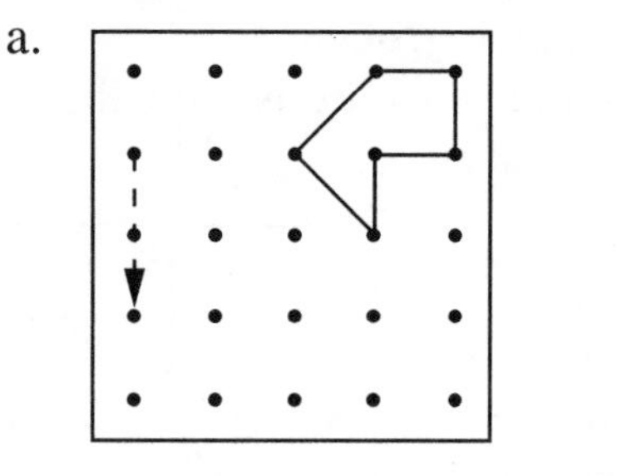

b.

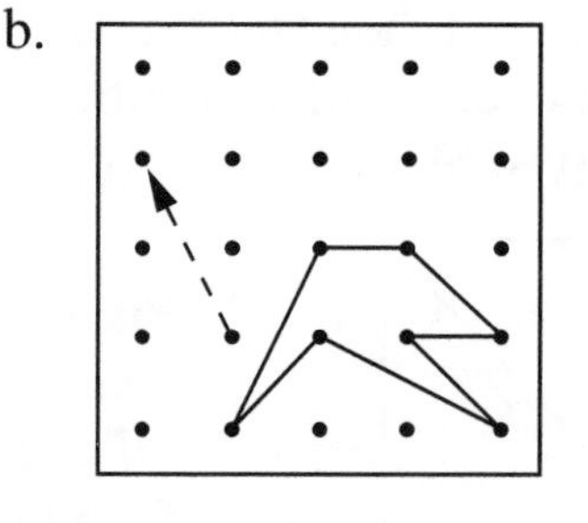

EXTENSIONS Play the games on dot paper using 10×10 squares to represent the geoboard.

Activity 11: Rotations

PURPOSE Explore rotations and their properties.

MATERIALS A compass, a centimeter ruler, a protractor, tracing paper, a straight pin or tack, and a Mira

GROUPING Work individually or in pairs.

GETTING STARTED The two drawings in Figure 1 are identical, but the duck is neither a reflection nor a translation of the rabbit. Use a Mira to verify that the duck is not a reflection of the rabbit and a tracing of the duck to verify that it is not a translation.

Place a piece of tracing paper over the rabbit. Pin it at point P. Trace the rabbit and then turn the paper about the pin until the rabbit coincides with the duck. This illustrates why the duck is called a *rotation image* of the rabbit.

Figure 1

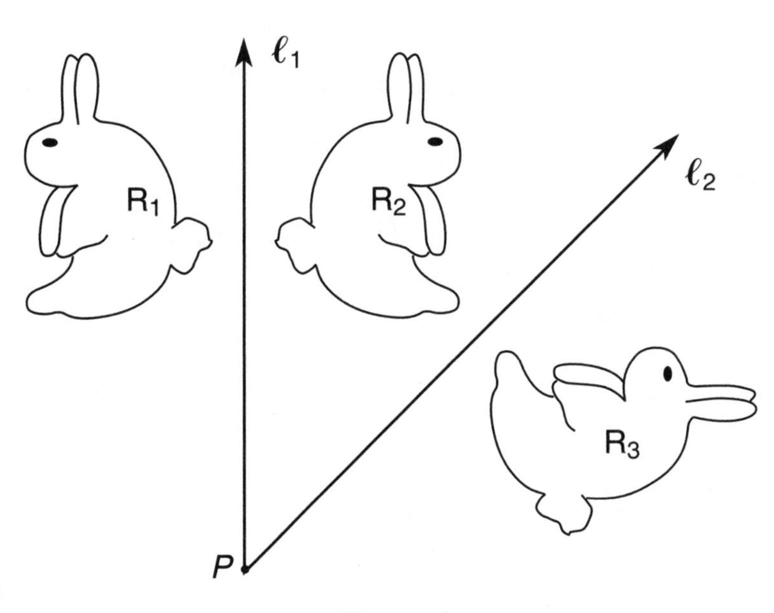

Figure 2

The drawings in Figure 2 show that if the rabbit, R_1 is reflected through line ℓ_1 and its image R_2 is reflected through line ℓ_2, then the result is the duck, R_3. Verify this with your Mira.

This example shows that the rotation that transforms the rabbit into the duck is the composite of two reflections through intersecting lines. This illustrates the following definitions:

A *rotation* is the composite of two reflections through intersecting lines.

The *center of rotation* is the point at which the two lines intersect.

1. What is the measure of the acute angle formed by lines ℓ_1 and ℓ_2?

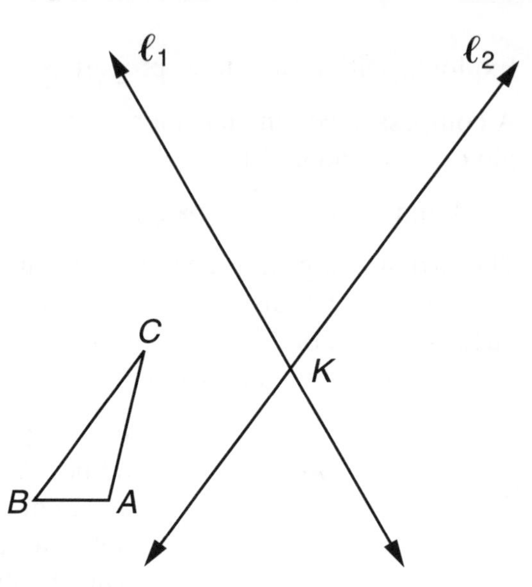

2. Rotate ΔABC by reflecting it through line ℓ_1 and then reflecting its image through line ℓ_2. Let X, Y, and Z be the rotation images of points A, B, and C, respectively.

   a. Draw and measure $\angle BKY$. (Note that the vertex of $\angle BKY$ is at the center of rotation and that one side of the angle contains a point B on the original figure. The other side contains the rotation image Y of point B. The measure of an angle formed this way is called the *magnitude of the rotation*.)

   b. How does the magnitude of the rotation compare with your answer to Exercise 1?

   c. In what direction, clockwise or counterclockwise, was ΔABC rotated?

3. Rotate ΔABC by reflecting it through line ℓ_2 and then reflecting is image through line ℓ_1. Let R, S, and T be the rotation images of points A, B, and C, respectively.

   a. What is the magnitude of the rotation? How does it compare with your answer to Exercise 1?

   b. In what direction was ΔABC rotated?

4. a. On the figure in Exercise 1, draw three circles with center K and radii $\overline{BK}$, $\overline{CK}$, and $\overline{AK}$, respectively. What do you notice about the circles?

   b. Make a tracing of ΔABC. Slide it onto ΔXYZ by moving its vertices along the circular "tracks" you have just drawn. Is it necessary to flip or to turn the tracing?

5. a. A point is rotated by reflecting it through two intersecting lines that form an acute angle with measure $x°$. What will be the magnitude of the rotation?

   b. If a figure is rotated by first reflecting it through line ℓ_3 and then reflecting its image through line ℓ_4, in what direction will the figure be rotated?

Activity 12: Glide Reflections

PURPOSE Explore glide reflections and their properties.

MATERIALS A compass, a centimeter ruler, a protractor, and a Mira

GROUPING Work individually or in pairs.

GETTING STARTED In Figure 1, footprint F_3 is a glide reflection of footprint F_1. As the name suggests, a glide reflection is a glide (or translation) followed by a reflection. However, the reflecting line must be parallel to the direction of the glide.

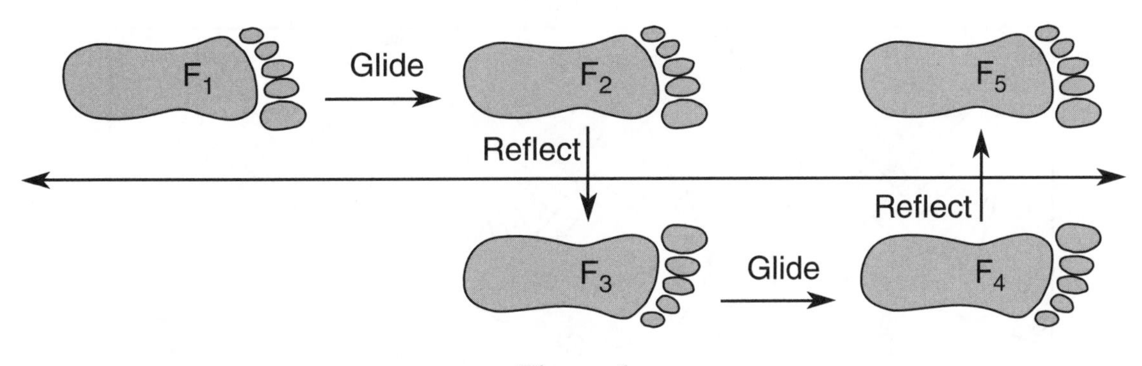

Figure 1

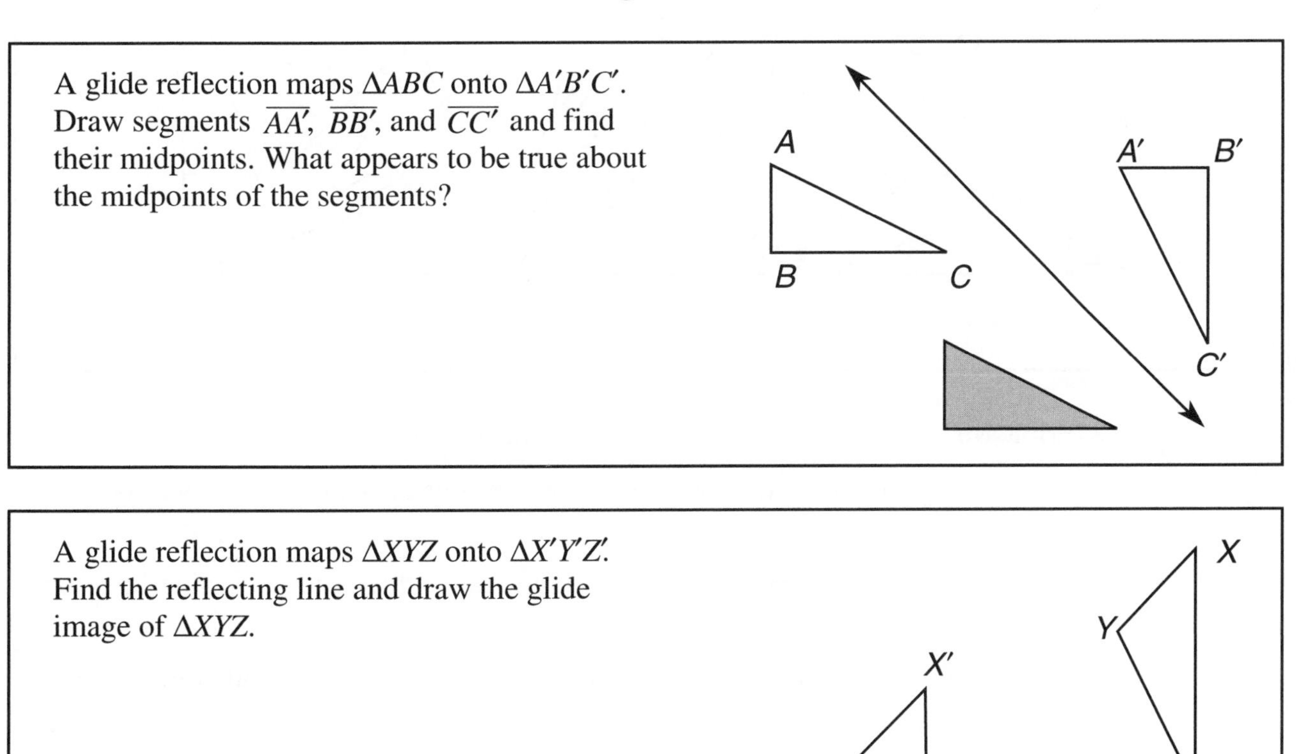

A glide reflection maps ΔABC onto $\Delta A'B'C'$. Draw segments $\overline{AA'}$, $\overline{BB'}$, and $\overline{CC'}$ and find their midpoints. What appears to be true about the midpoints of the segments?

A glide reflection maps ΔXYZ onto $\Delta X'Y'Z'$. Find the reflecting line and draw the glide image of ΔXYZ.

Study Figure 1. What single transformation is equivalent to the result of using a glide reflection twice?

Identify the transformation that will map the shaded hawk in Figure 2 onto each lettered hawk.

1. Hawk A ____________________
2. Hawk B ____________________
3. Hawk C ____________________
4. Hawk D ____________________
5. Hawk E ____________________
6. Hawk F ____________________

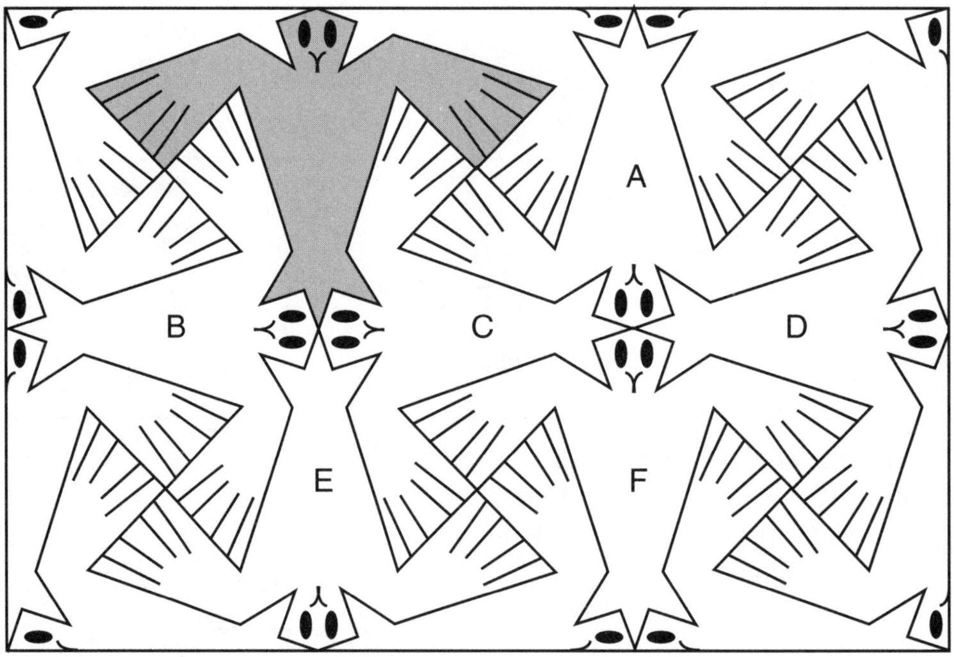

A
B
C
D
E
F

Figure 2

EXTENSIONS Complete the following statements:

1. The composite of two reflections through parallel lines is a ____________________.
2. The composite of two reflections through intersecting lines is a ____________________.
3. The composite of three reflections through parallel lines is a ____________________.
4. The composite of three reflections through concurrent lines is a ____________________.
5. The composite of three reflections through lines that are not all parallel and not all concurrent is a ____________________.

Activity 13: Coordinate Transformations

PURPOSE Investigate the effects of coordinate transformations on graphs.

MATERIALS Graph paper, tracing paper, a Mira, a ruler, and a protractor

GROUPING Work individually or in pairs.

1. Plot and label the points $A(1, 3)$, $B(2, {}^-3)$, and $C({}^-2, 1)$ on a pair of coordinate axes. Draw ΔABC.
2. Plot and label the points $A'(6, 3)$, $B'(7, {}^-3)$, and $C'(3, 1)$ on the same pair of coordinate axes. Draw $\Delta A'B'C'$.
3. How are the coordinates of A', B', and C' related to the coordinates of A, B, and C?
4. What transformation maps ΔABC onto $\Delta A'B'C'$?
5. a. The coordinates of points A'', B'', and C'' are formed from the coordinates of points A, B, and C by adding 3 to the respective y-coordinates. Without plotting points, predict the transformation that will map ΔABC onto $\Delta A''B''C''$.

   b. Check your prediction in Part (a) by finding the coordinates of points A'', B'', and C'' and plotting the points.
6. a. ΔXYZ is the translation image when ΔABC is translated 4 units to the left and 2 units down. Without performing the translation, how can you determine the coordinates of points X, Y, and Z from the coordinates of points A, B, and C?

   b. Check your answer to Part (a) by finding the coordinates of points X, Y, and Z and plotting the points.

---

1. Plot and label the points $P(1, 6)$, $Q(3, 5)$, and $R(4, 2)$ on a pair of coordinate axes. Draw ΔPQR.
2. a. The coordinates of points P'', Q'', and R'' are formed from the coordinates of P, Q, and R by multiplying the respective x-coordinates by ${}^-1$. Find the coordinates of P'', Q'', and R''.

   b. Plot and label the points P'', Q'', and R''. Draw $\Delta P''Q''R''$. What transformation maps ΔPQR onto $\Delta P''Q''R''$?
3. a. What would you do to the coordinates of points P, Q, and R to reflect ΔPQR through the x-axis?

   b. Check your answer to Part (a) by reflecting ΔPQR through the x-axis and finding the coordinates of the images of points P, Q, and R.

1. a. The vertices of a triangle are $S(2, 6)$, $T(4, 3)$, and $U(1, 4)$, and the points $S'(^-2, ^-6)$, $T'(^-4, ^-3)$, and $U'(^-1, ^-4)$ are their images under a transformation. Without plotting points, predict what transformation maps ΔSTU onto $\Delta S'T'U'$. **HINT:** This is the composite of two transformations.

   b. Check your prediction by plotting the points.

2. a. ΔSTU is mapped onto $\Delta S''T''U''$ by reflecting ΔSTU through the *y*-axis and then translating the image down 3 units (a glide reflection). Without performing the glide reflection, find the coordinates of points S'', T'', and U''.

   b. Check your answer by performing the glide reflection and finding the coordinates of S'', T'', and U''.

1. Plot and label the points $X(4, 1)$, $Y(4, 3)$, and $Z(7, 1)$ on a pair of coordinate axes. Draw ΔXYZ.
2. Plot and label the points $X'(1, 4)$, $Y'(3, 4)$, and $Z'(1, 7)$ on the same pair of coordinate axes. Draw $\Delta X'Y'Z'$.
3. How are the coordinates of X', Y', and Z' related to the coordinates of X, Y, and Z?
4. What transformation maps ΔXYZ onto $\Delta X'Y'Z'$?
5. Plot and label the points $X''(^-1, ^-4)$, $Y''(^-3, ^-4)$, and $Z''(^-1, ^-7)$ on the same pair of coordinate axes. Draw $\Delta X''Y''Z''$.
6. How are the coordinates of points X'', Y'', and Z'' related to the coordinates of X, Y, and Z?
7. What transformation maps ΔXYZ onto $\Delta X''Y''Z''$?

1. What single transformation has the same effect as reflecting a triangle through the line $y = x$ and then reflecting the image through the *y*-axis?
2. If the coordinates of the vertices of the triangle are (a, b), (c, d), and (e, f), what are the coordinates of the images of the vertices under this transformation?
3. Check your answers to Exercises 1 and 2 by drawing a triangle and using a Mira to perform the two reflections.

EXTENSIONS

1. Write a summary of the results of this investigation.
2. Investigate the effects on a triangle when the coordinates of its vertices are multiplied by positive constants.

Chapter Summary

Activities 1 and 2 introduced the concept of symmetry. There are three kinds of symmetry. An object has *line symmetry* when it can be divided into two halves in such a way that each half is the mirror image of the other.

An object has *rotational symmetry* when it can be made to coincide with itself by a rotation of less than 360° about a point. Objects with rotational symmetry of 180° are said to have *point symmetry*. An object has point symmetry when it appears the same whether viewed right-side-up or upside-down.

Line and rotational symmetry are very common in nature. The body structure of most animals, including humans, is symmetric about a line. Many flowers exhibit rotational symmetries, while some trees and the leaves of many plants also exhibit line symmetry.

In Activities 3–12, you studied four isometries—translations, reflections, rotations, and glide reflections—and discovered some of their properties. The first three transformations were introduced in Activities 3–6 through the use of pattern blocks and the less formal terminology of slides, flips, and turns. Because these terms are simple and very descriptive, they are the ones most often used with elementary students. The later activities employed formal terminology and used a variety of tools—tracing paper, Miras, geoboards, rulers, compasses, and protractors—to perform the transformations and to study their properties.

Activity 8 extended the idea of a flip to the more general concept of a reflection or mirror image. A fundamental property of reflections was developed in the activity. The reflecting line is the perpendicular bisector of the segment connecting a point and its reflected image. This property makes it possible to reflect objects using ruler and compass constructions or coordinate methods.

Activity 7 introduced the idea that slides and turns result from performing two successive reflections. This idea was explored in greater detail in Activities 9 and 11. In Activity 9, you discovered that a *translation* is the composite of two reflections through parallel lines, the distance the object is translated is twice the distance between the lines, and the translation is in the direction from the first reflecting line to the second. In Activity 11 you discovered that a *rotation* is the composite of two reflections through intersecting lines, the magnitude of rotation is twice the measure of the acute angle formed by the reflecting lines, and the rotation is in the direction from the first reflecting line to the second.

You also discovered that in addition to preserving the distance between points, all four transformations also preserved the measure of angles, collinearity of points, and congruence. The orientation of objects was preserved by translations and rotations but reversed by reflections and glide reflections.

Adding or subtracting a constant and multiplying or dividing by a constant are just two of the many transformations used in algebra. In Activity 13, you used coordinate methods to study the relationship between some algebraic transformations and geometric transformations.

ATTRIBUTE PEOPLE PIECES

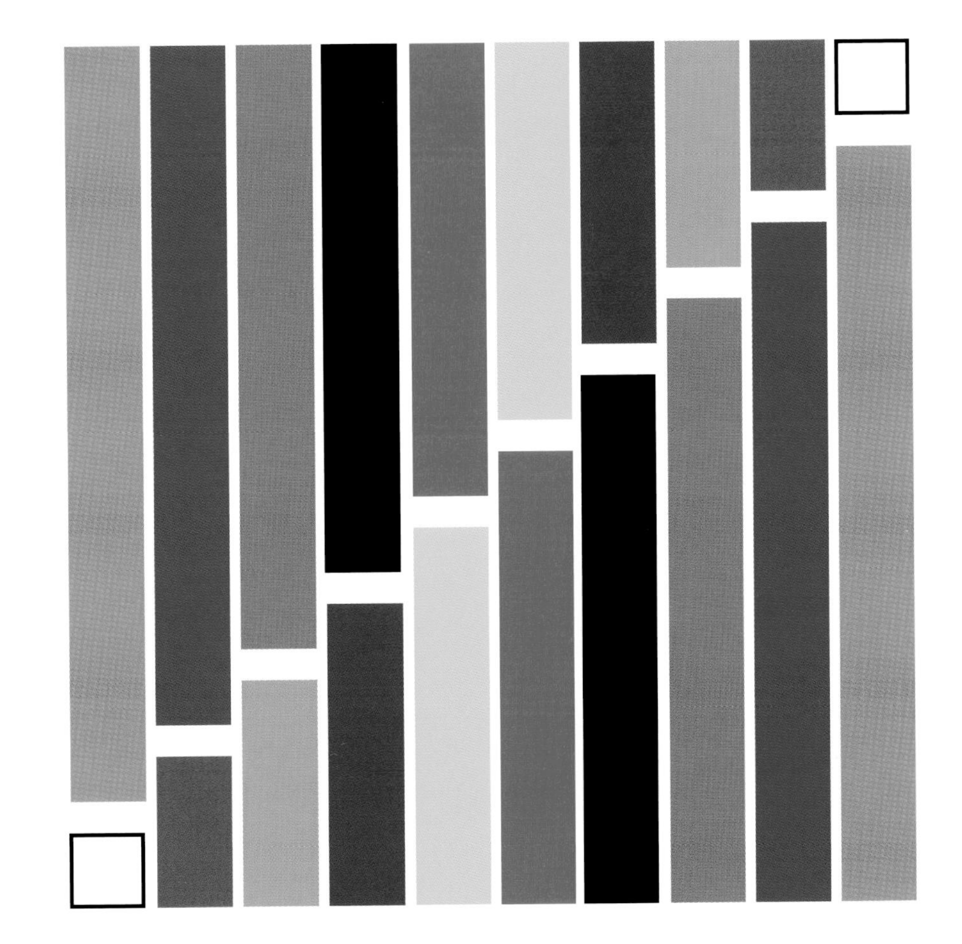

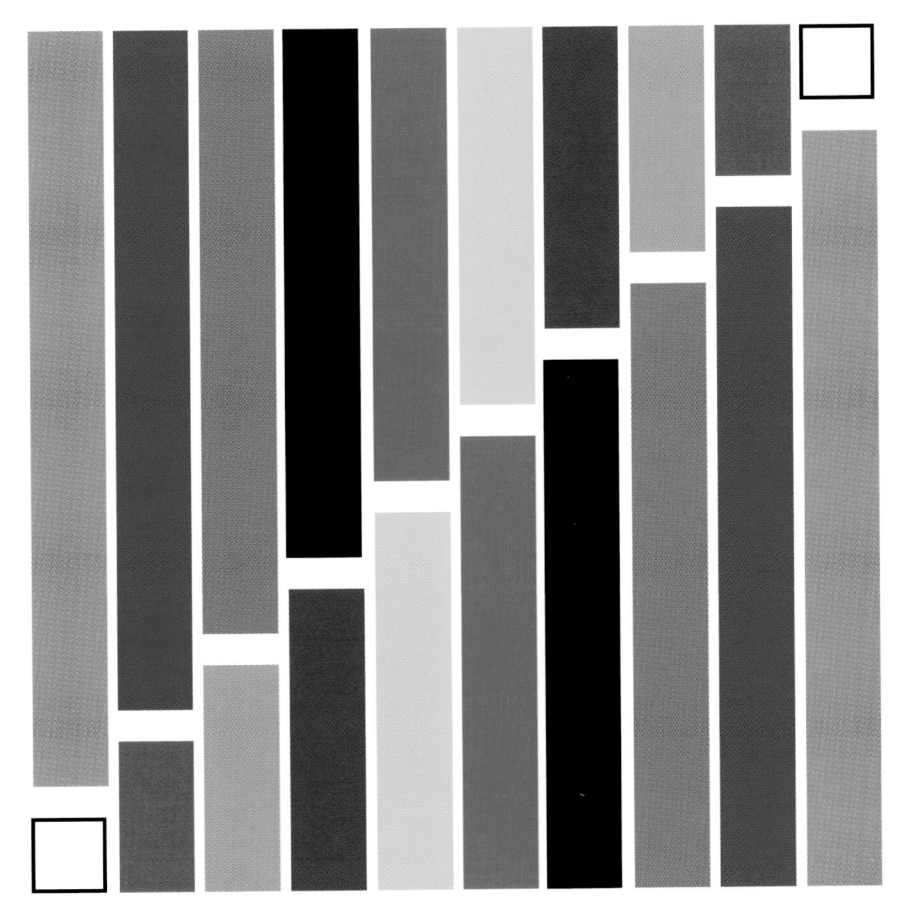

ATTRIBUTE PIECES

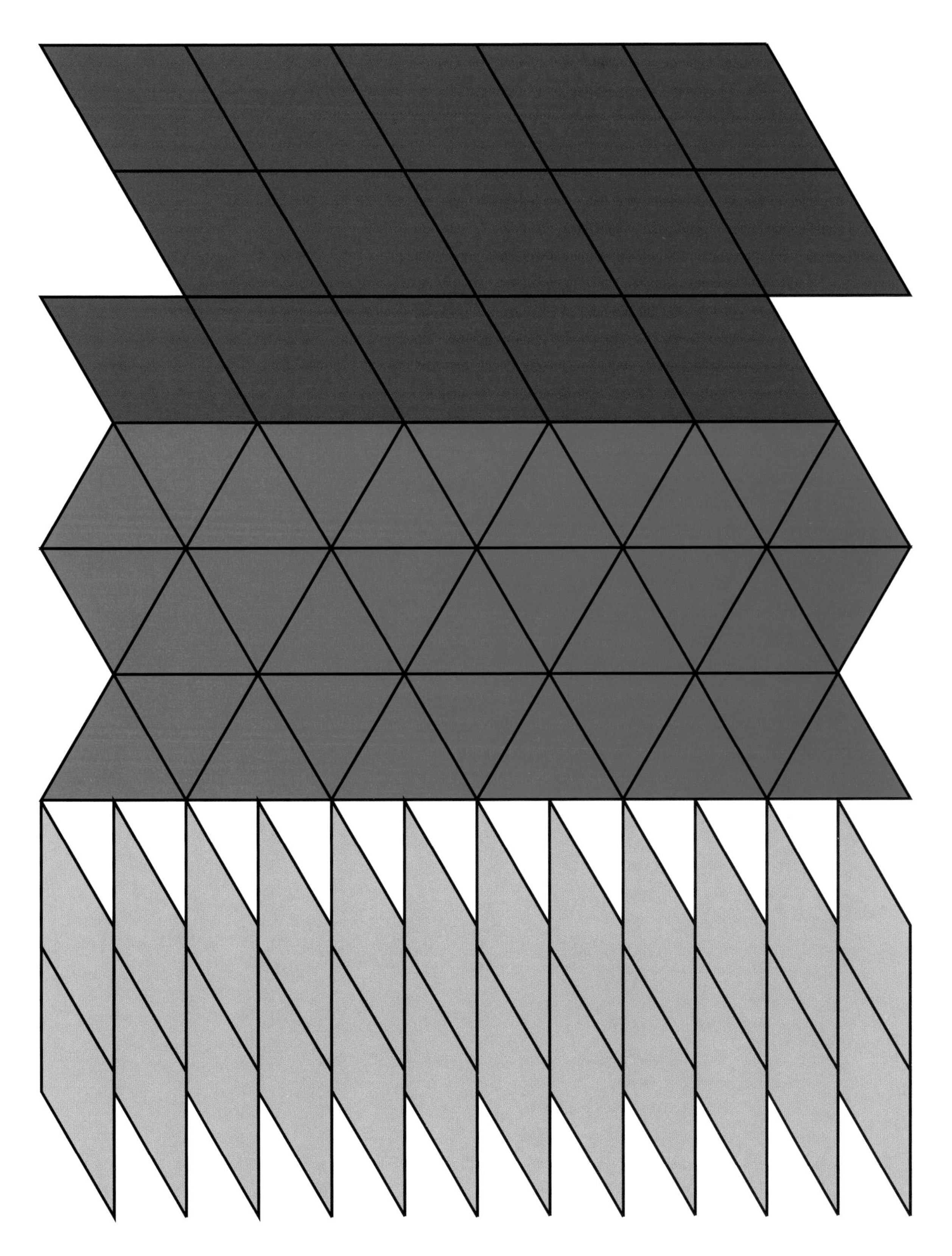

BASE TEN BLOCKS

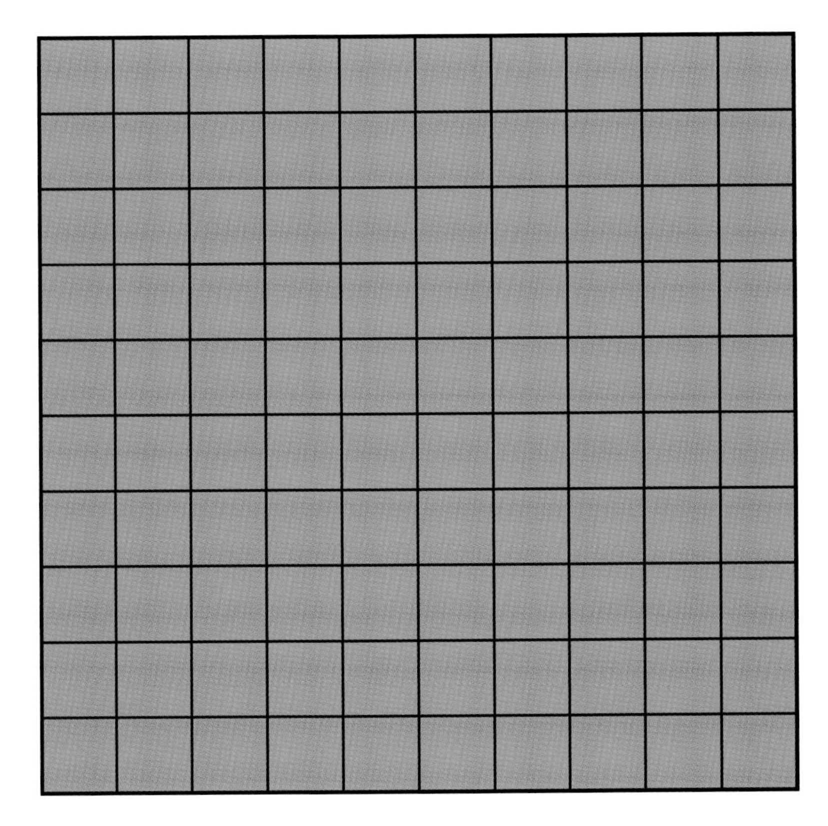

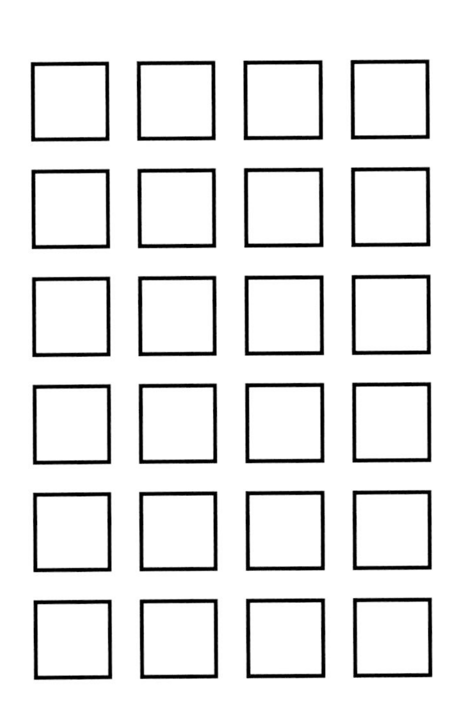

BASE TEN BLOCKS

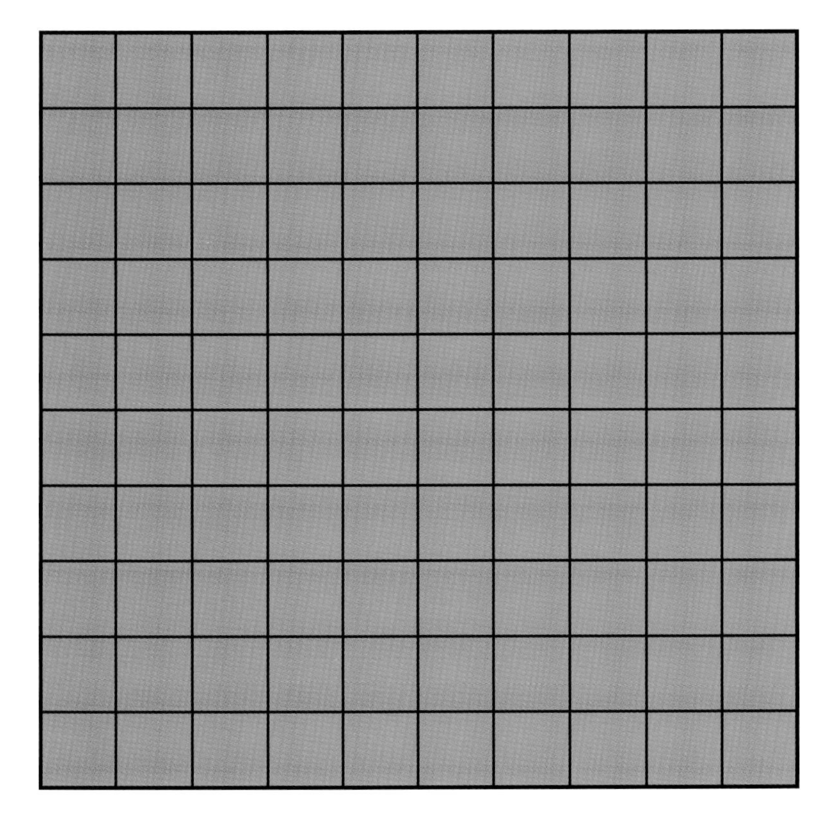

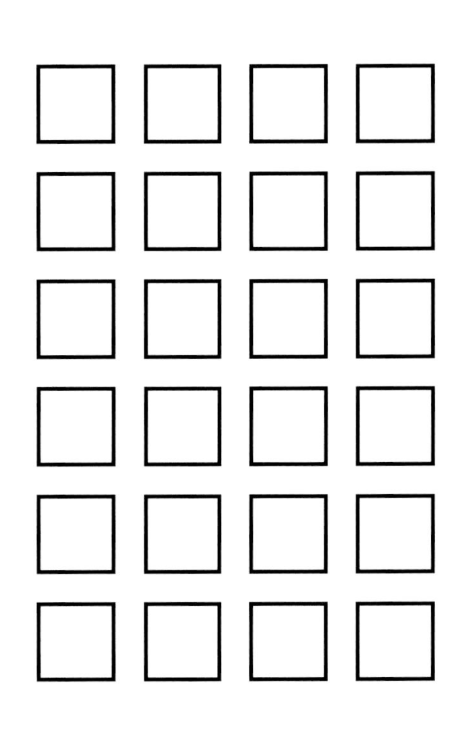

2×2 MATRIX

3×3 MATRIX

4×4 MATRIX

DIFFERENCE PUZZLE 1

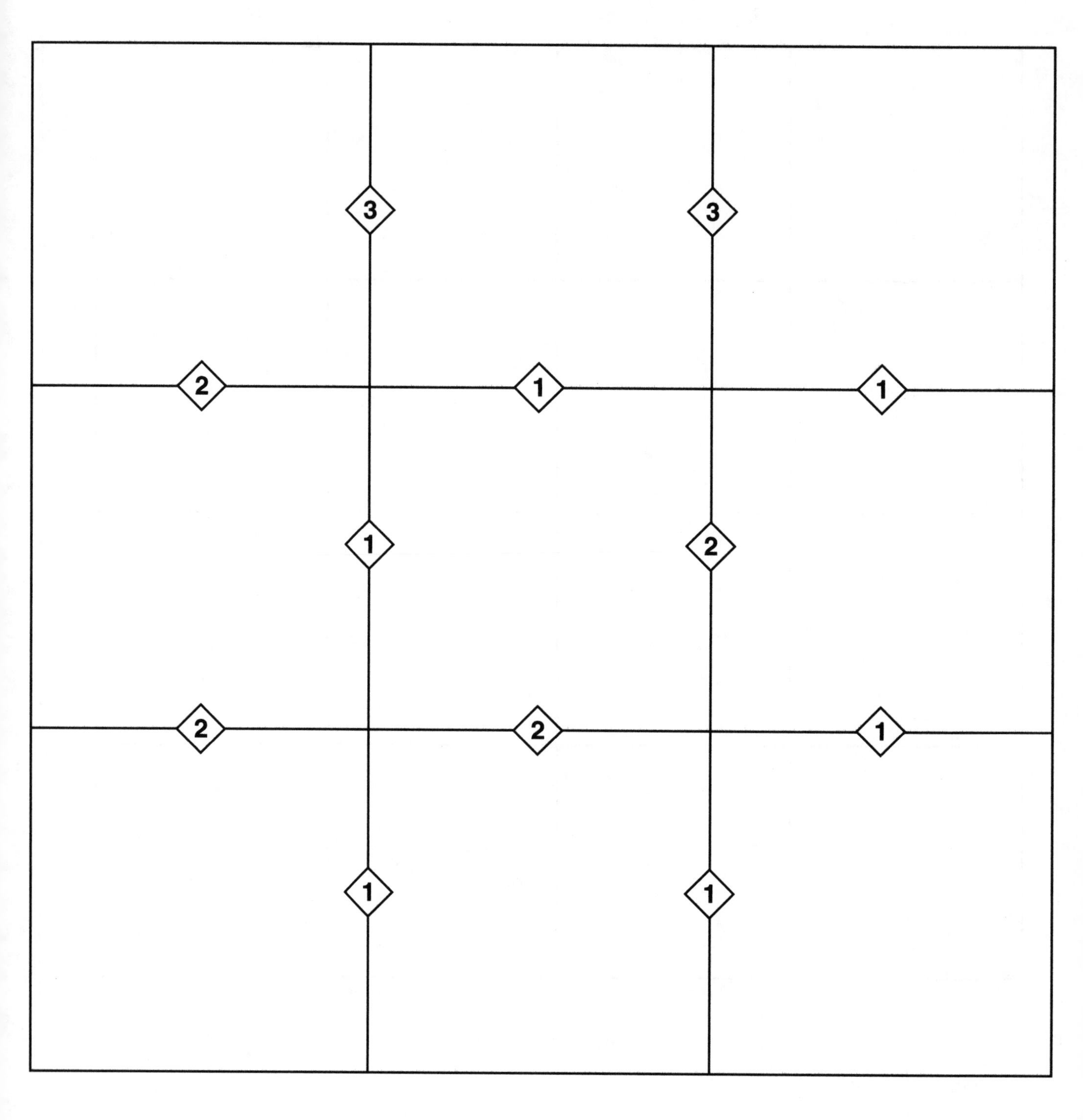

DIFFERENCE PUZZLE 2

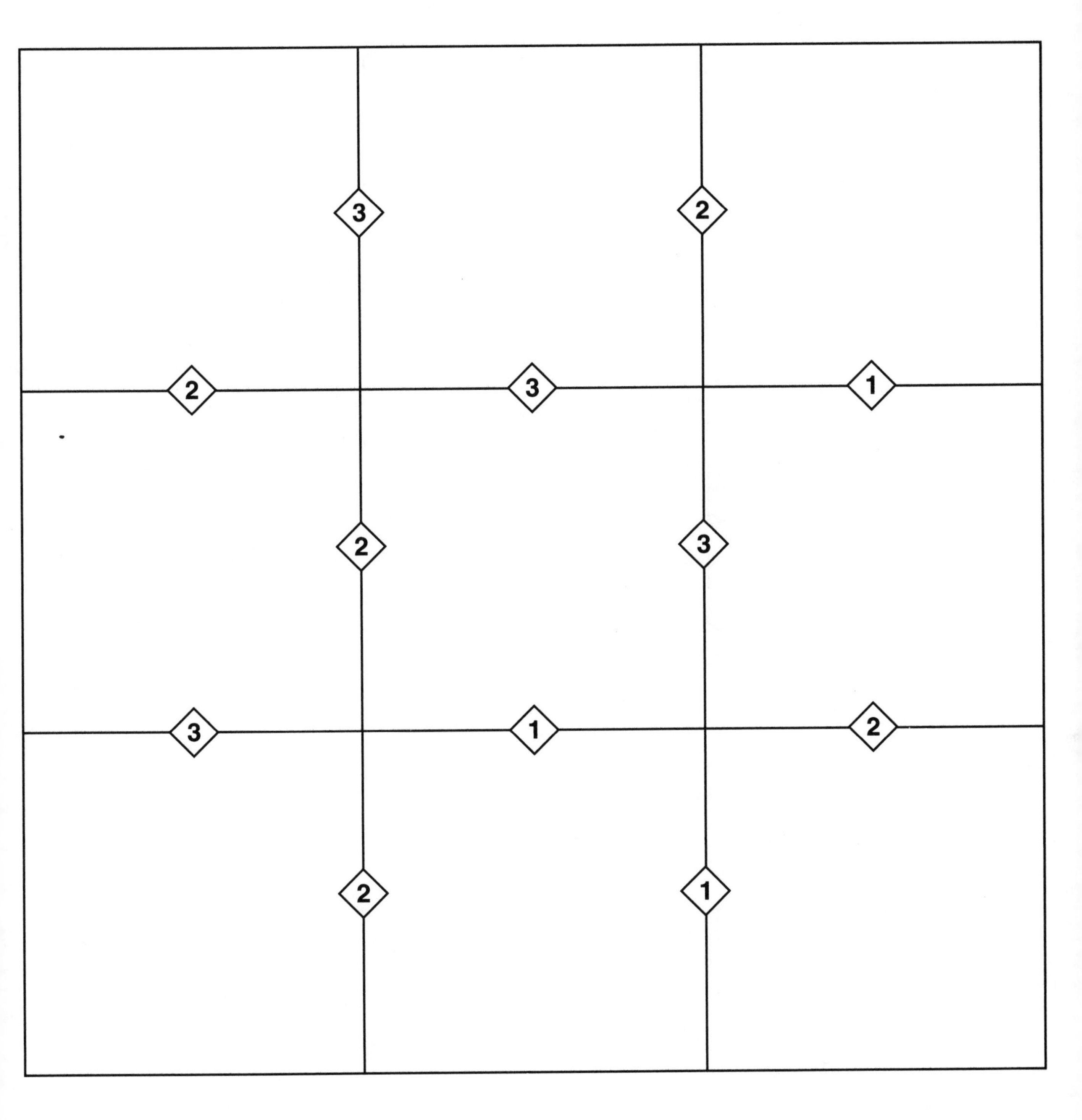

DIFFERENCE PUZZLE MASTER

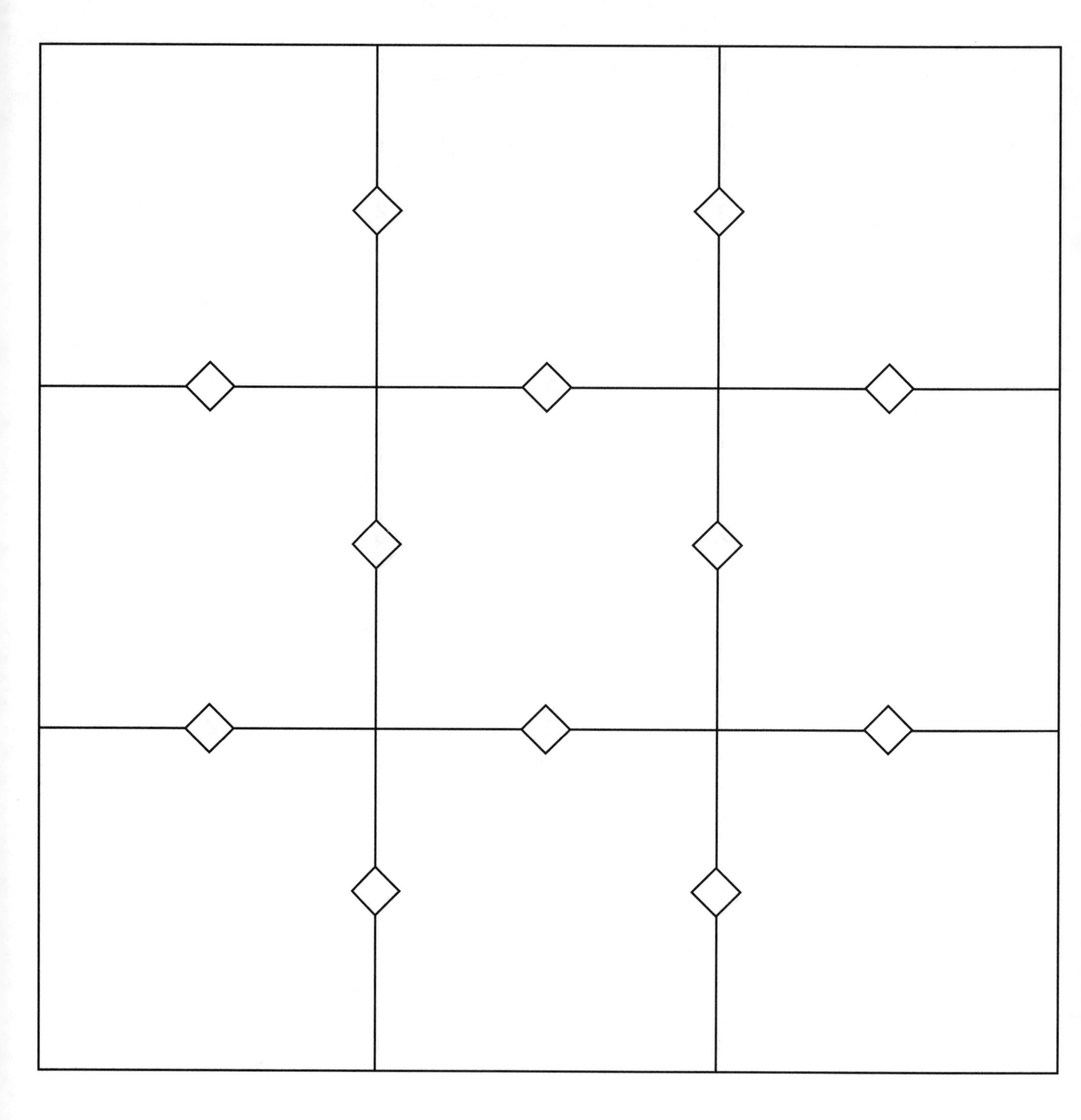

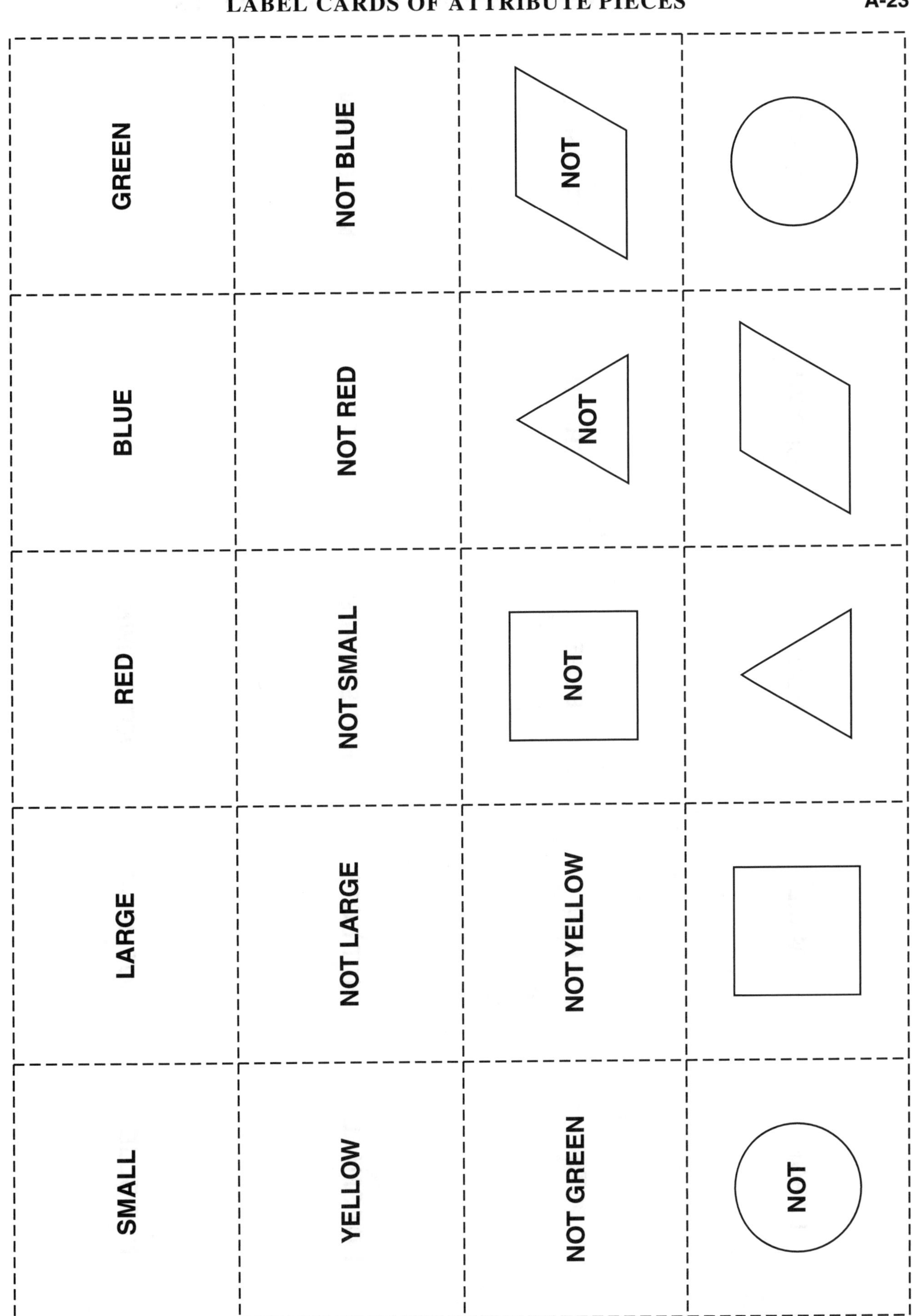
GREEN
NOT BLUE
NOT
BLUE
NOT RED
NOT
RED
NOT SMALL
NOT
LARGE
NOT LARGE
NOT YELLOW
SMALL
YELLOW
NOT GREEN
NOT

LABEL CARDS OF ATTRIBUTE PEOPLE PIECES

STRIPES	NOT LARGE	NOT GREEN	
CLOWN	GREEN	NOT RED	
HOBO	BLUE	NOT STRIPES	NOT CLOWN
SMALL	RED	NOT DOTS	NOT HOBO
LARGE	DOTS	NOT SMALL	NOT BLUE

What's My Function?	What's My Function?
Multiply input by 6. $y = 6x$ $x \rightarrow 6x$ $f(x) = 6x$	Add 7 to the input. $y = x + 7$ $x \rightarrow x + 7$ $f(x) = x + 7$
What's My Function? Multiply input by 4 and add 3. $y = 4x + 3$ $x \rightarrow 4x + 3$ $f(x) = 4x + 3$	**What's My Function?** Multiply input by 7 and subtract 5. $y = 7x - 5$ $x \rightarrow 7x - 5$ $f(x) = 7x - 5$
What's My Function? Divide input by 5. $y = x \div 5$ $x \rightarrow x \div 5$ $f(x) = \mathrm{x} \div 5$	**What's My Function?** Subtract 2 from the input. $y = x - 2$ $x \rightarrow x - 2$ $f(x) = x - 2$
What's My Function? Divide input by 2 and subtract 3. $y = x \div 2 - 3$ $x \rightarrow x \div 2 - 3$ $f(x) = x \div 2 - 3$	**What's My Function?** Add 3 to the input and divide by 2. $y = (x + 3) \div 2$ or $y = 0.5x + 1.5$ $x \rightarrow (x + 3) \div 2$ or $x \rightarrow 0.5x + 1.5$ $f(x) = (x + 3) \div 2$ or $f(x) = 0.5x + 1.5$
What's My Function? Multiply the input by itself. (Square the input.) $y = x \cdot \mathrm{x}$ or $y = x^2$ $x \rightarrow x \cdot x$ or $x \rightarrow x^2$ $f(x) = x \cdot x$ or $f(x) = x^2$	**What's My Function?** Multiply the input by itself and subtract 2. $y = x \cdot x - 2$ or $y = x^2 - 2$ $x \rightarrow x \cdot x - 2$ or $x \rightarrow x^2 - 2$ $f(x) = x \cdot x - 2$ or $f(x) = x^2 - 2$
What's My Function? Multiply the input by 1 more than the input. $y = x(x + 1)$ or $y = x^2 + x$ $x \rightarrow x(x + 1)$ or $x \rightarrow x^2 + x$ $f(x) = x(x + 1)$ or $f(x) = x^2 + x$	**What's My Function?** Multiply the input by 3 and add 4. $y = 3x + 4$ $x \rightarrow 3x + 4$ $f(x) = 3x + 4$

BEANSTICKS

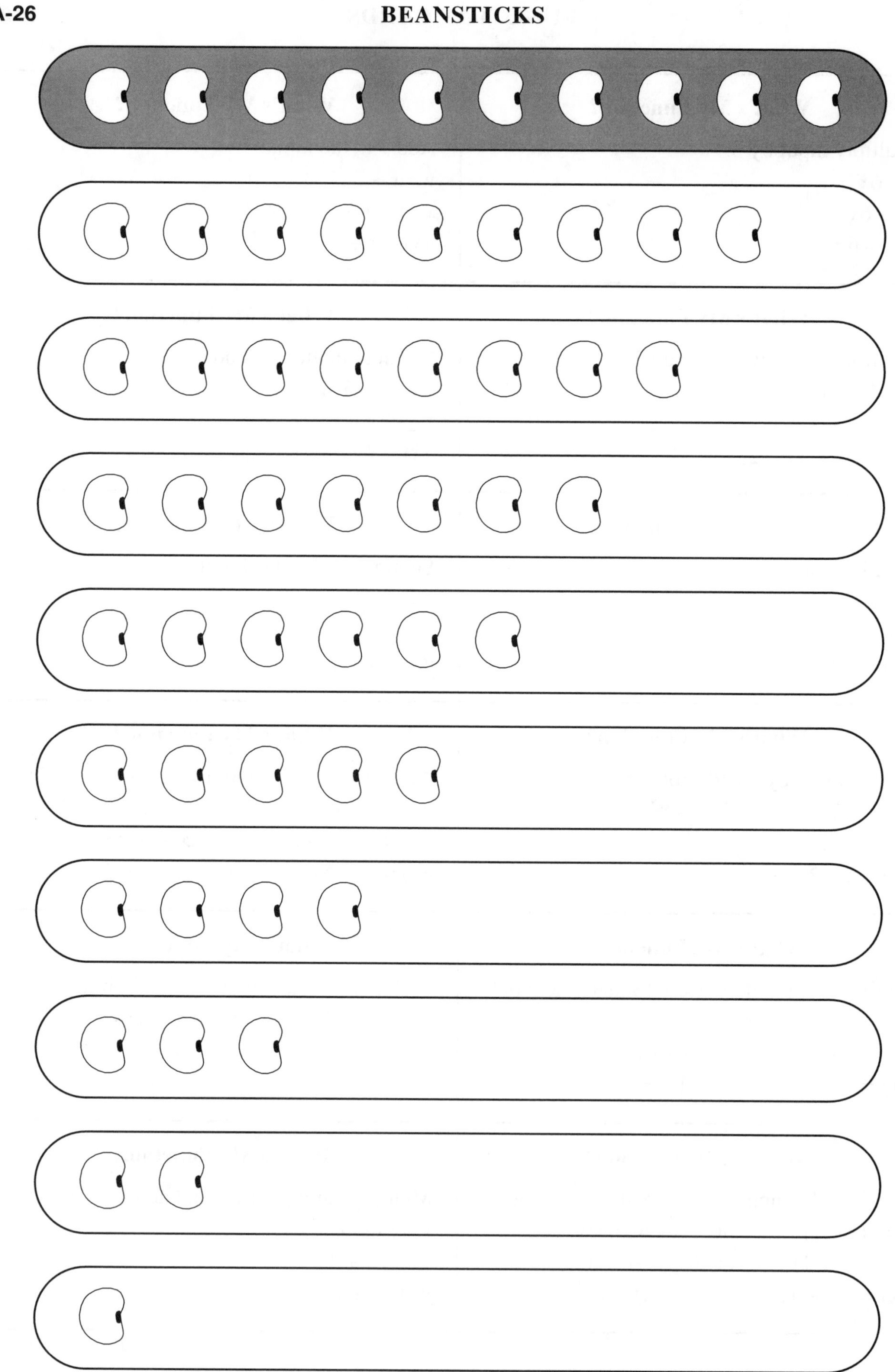

BEAN STICKS

MATERIALS Tongue depressors, Elmer's Glue, spray paint, and dried pinto beans

Spray paint one side of at least 10 sticks to code the tens stick. Place large dots of glue on each stick and line up beans horizontally on each glue dot. (Use the samples on page A-26 as models.) Allow the glue to dry thoroughly and store the beansticks in resealable plastic bags. Keep additional loose beans to use in place value and addition/subtraction activities.

Glue 10 tens sticks side by side on a piece of cardboard or on two blank tongue depressors to form a hundreds "raft." The raft corresponds to a flat in base ten materials.

Example:

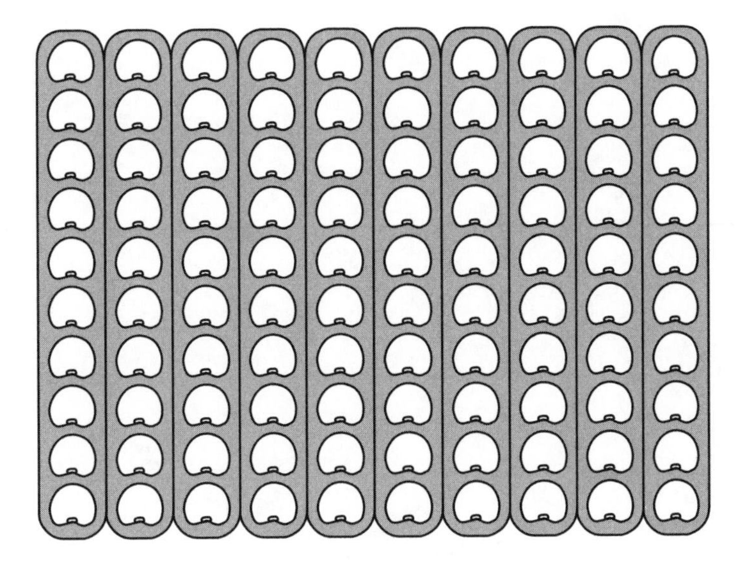

PLACE-VALUE DICE

MATERIALS Blank wooden or foam-rubber cubes and permanent markers

NOTE: Foam-rubber upholstery padding comes in 2 in. thickness. Mark off 2 in. squares and slice the foam with an electric carving knife.

Mark the cubes with random numerals of differing place values. When marking the foam, go over the marks several times so that the ink will seep into the foam.

Example:

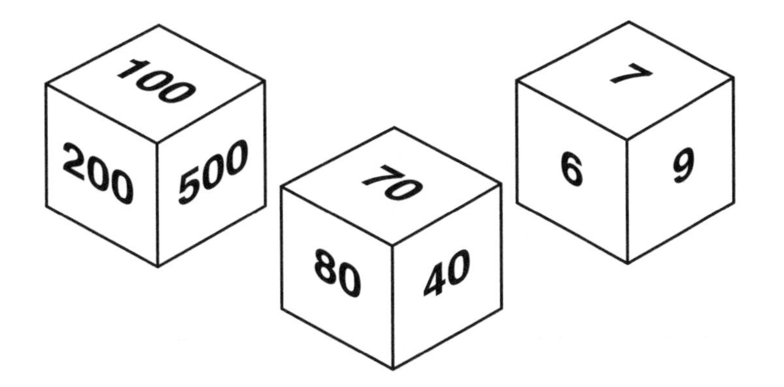

PLACE-VALUE OPERATIONS BOARD

FRACTION STRIPS

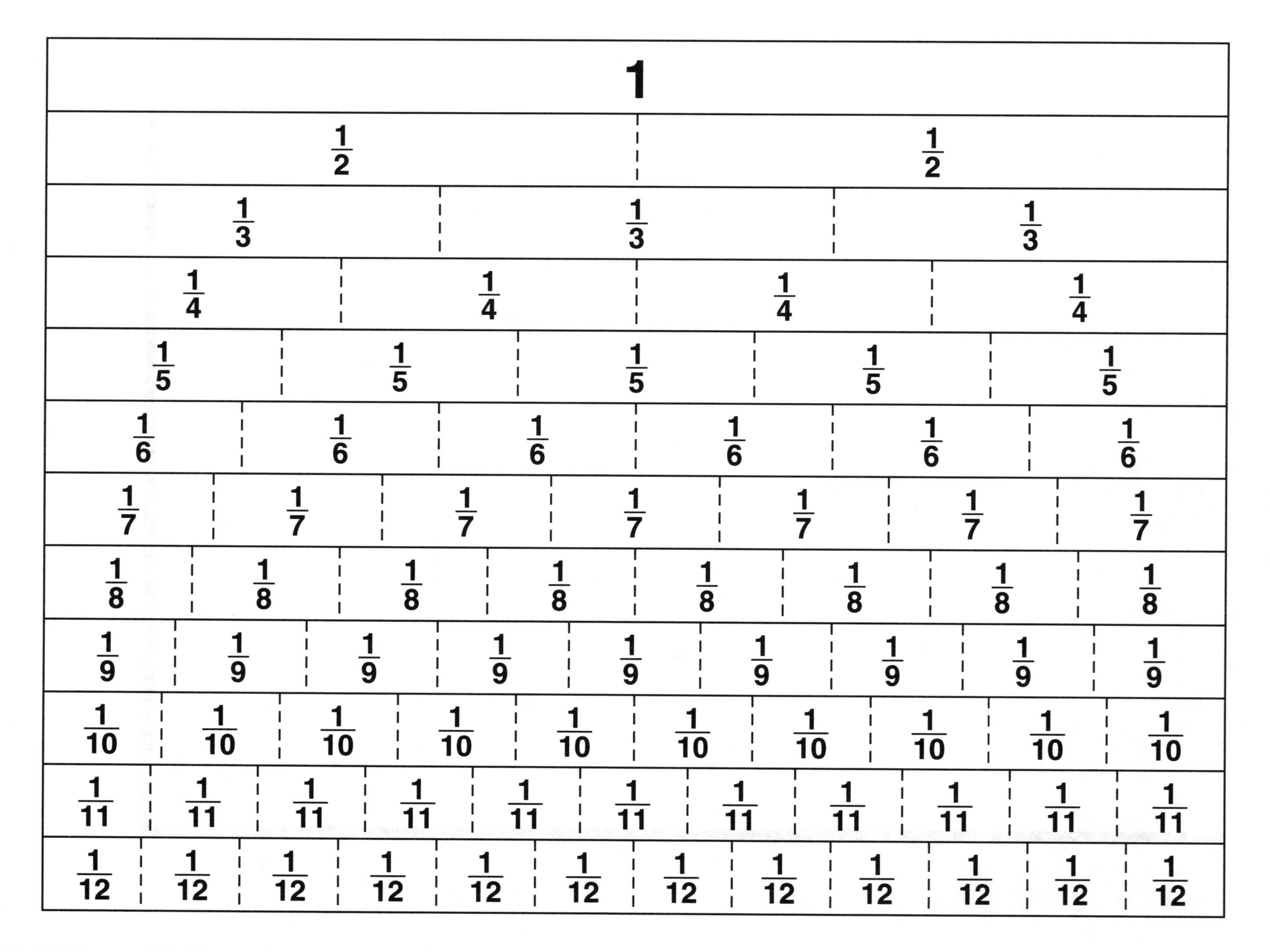

FRACTION CARDS – A

$\frac{5}{6}$	$\frac{4}{6}$	$\frac{2}{6}$	$\frac{1}{6}$
$\frac{1}{8}$	$\frac{2}{8}$	$\frac{3}{8}$	$\frac{5}{8}$
$\frac{6}{8}$	$\frac{7}{8}$	$\frac{1}{10}$	$\frac{2}{10}$
$\frac{4}{10}$	$\frac{3}{10}$	$\frac{6}{10}$	$\frac{7}{10}$

FRACTION CARDS – B

$\frac{8}{10}$	$\frac{9}{10}$	$\frac{1}{25}$	$\frac{2}{25}$
$\frac{3}{25}$	$\frac{24}{25}$	$\frac{23}{25}$	$\frac{22}{25}$
$\frac{21}{25}$	$\frac{20}{25}$	$\frac{4}{25}$	$\frac{5}{25}$
$\frac{11}{25}$	$\frac{12}{25}$	$\frac{13}{25}$	$\frac{14}{25}$

Close to 1	
Close to $\frac{1}{2}$	
Close to 0	

FRACTION ARRAYS

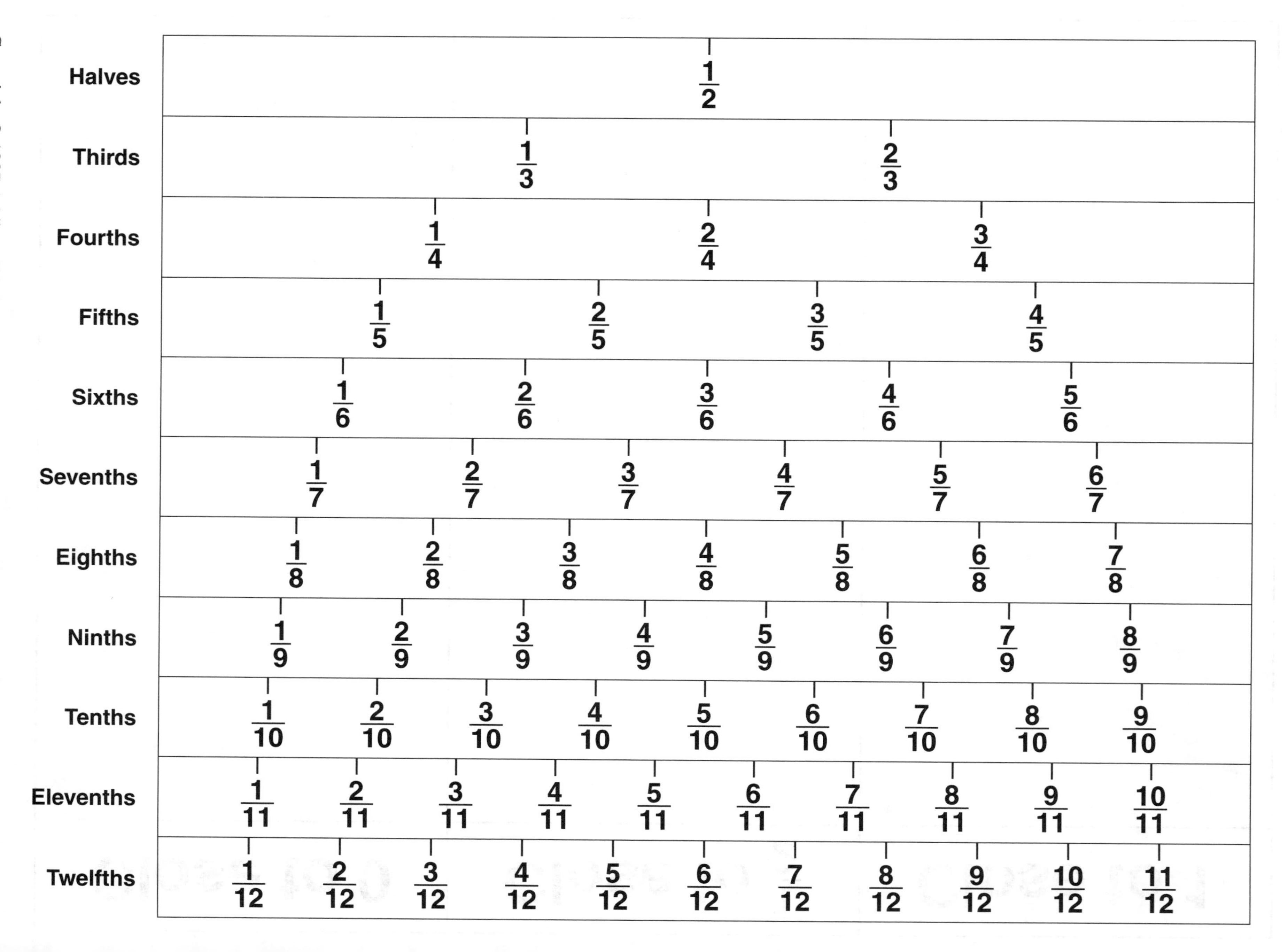

Player A
Player B

100 GRIDS

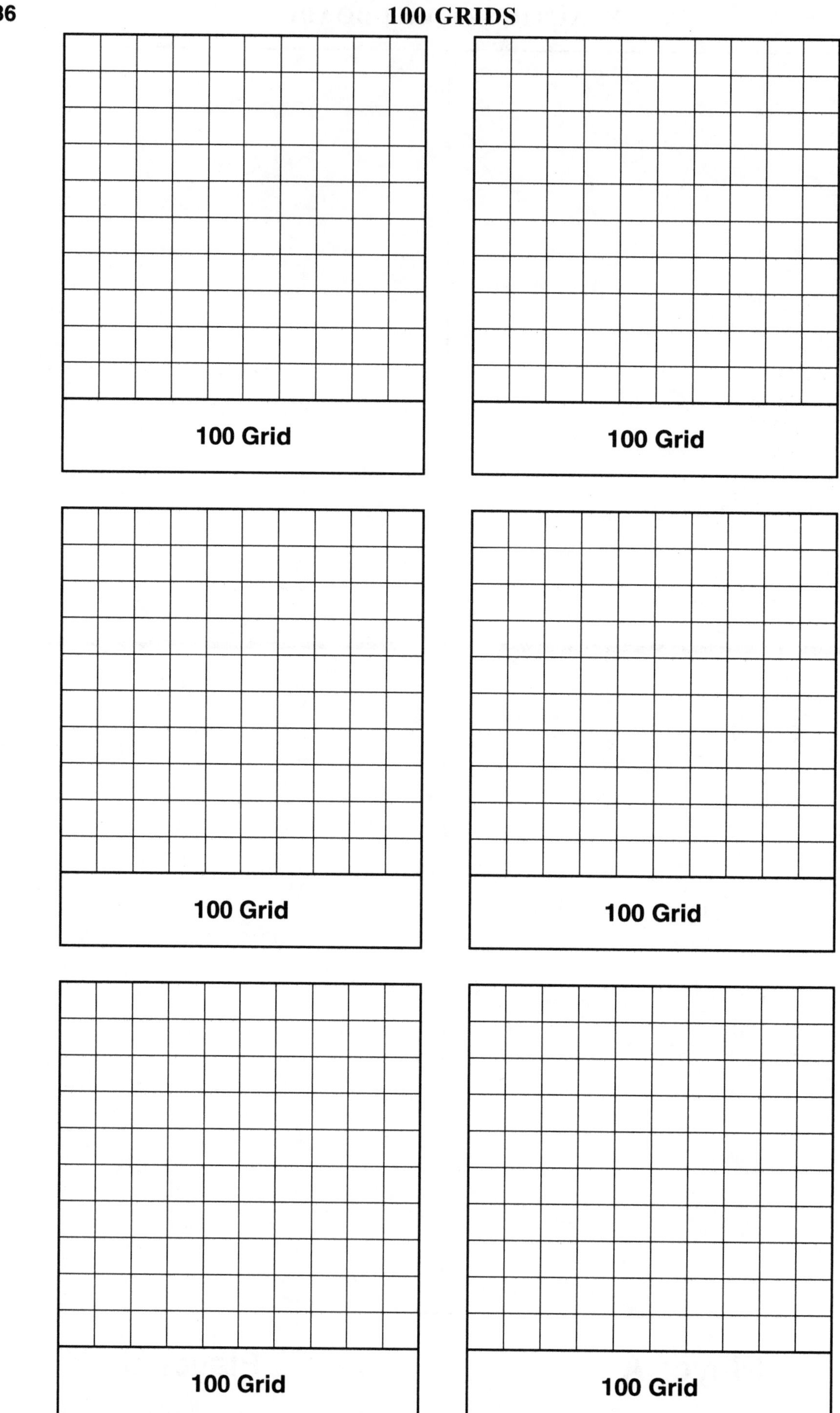

m&m's®

CIRCULAR GEOBOARDS

CIRCULAR GEOBOARDS

GEOBOARDS DOT PAPER

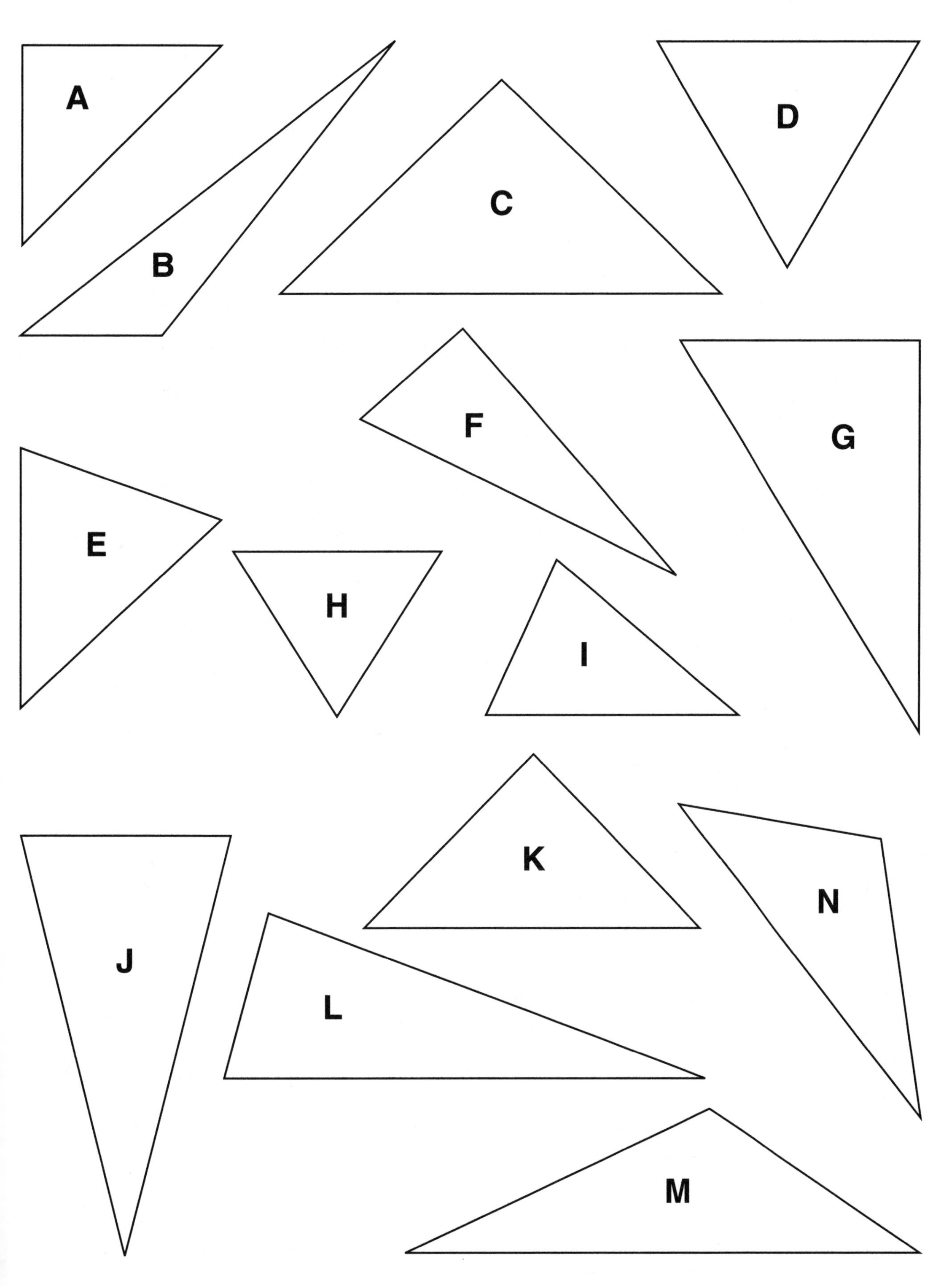
A
B
C
D
E
F
G
H
I
J
K
L
M
N

QUADRILATERALS

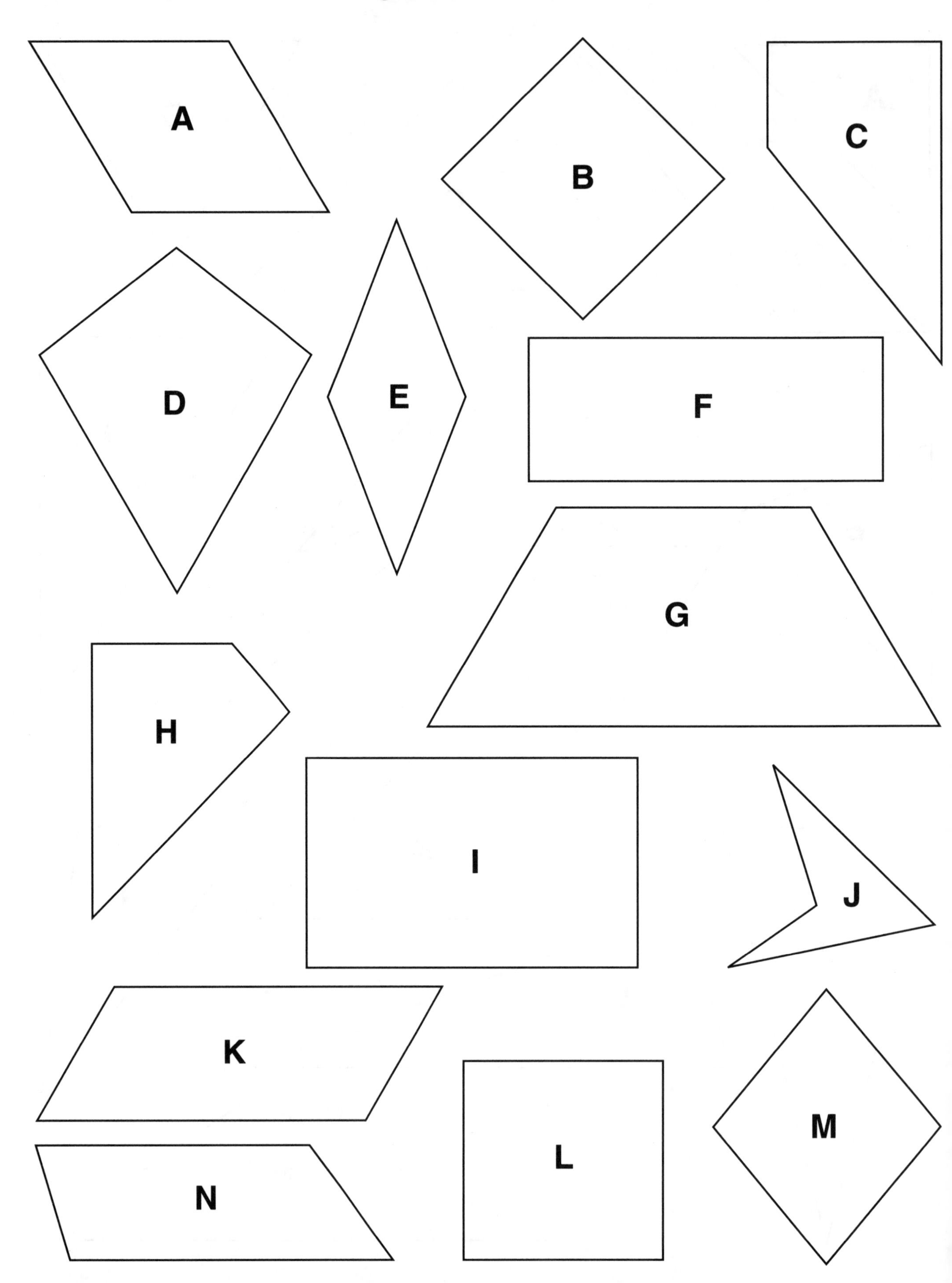